World Sustainability Series

Series editor

Walter Leal Filho, Hamburg, Germany

More information about this series at http://www.springer.com/series/13384

Walter Leal Filho
Editor

Sustainable Development Research at Universities in the United Kingdom

Approaches, Methods and Projects

Springer

Editor
Walter Leal Filho
FTZ-ALS
HAW Hamburg
Hamburg
Germany

and

School of Science and Environment
Manchester Metropolitan University
Manchester
UK

ISSN 2199-7373 ISSN 2199-7381 (electronic)
World Sustainability Series
ISBN 978-3-319-47882-1 ISBN 978-3-319-47883-8 (eBook)
DOI 10.1007/978-3-319-47883-8

Library of Congress Control Number: 2016954552

Printed on acid-free paper

This Springer imprint is published by Springer Nature
The registered company is Springer International Publishing AG
The registered company address is: Gewerbestrasse 11, 6330 Cham, Switzerland

Preface

Many universities in the United Kingdom perform high-quality research on matters related to sustainable development. Yet, there are relatively few events where a multidisciplinary overview of research efforts and projects has taken place, and where researchers from across the spectrum of the natural and social sciences have had the opportunity to come together to discuss research methods, the results of empirical research or exchange ideas about ongoing and future research initiatives focusing on sustainable development.

It is against this background that the "Symposium on Sustainable Development Research at Universities in the United Kingdom" was organised by Manchester Metropolitan University, in cooperation with a number of institutions of higher education active in this field, from across the country. It involved researchers in the field of sustainable development in the widest sense, from business and economics, to arts and fashion, administration, environment, languages and media studies.

The symposium focused on "Sustainable Development Research in the United Kingdom" and was expected to contribute to the further development of this fast-growing field.

This book, titled "Sustainable Development Research at Universities in the United Kingdom", has three main aims. First, it intends to provide an opportunity to document and promote the variety of works in this field in the UK today, including matters related to curriculum innovation, empirical work, activities, case studies and practical projects. Second, the book intends to offer a platform for the exchange of information, ideas and experiences acquired in the execution of research projects, especially successful initiatives and good practice. Finally, it offers a sound basis for readers to inform themselves about the various methodological approaches and projects taking place in the UK today, offering a better understanding of sustainable development across society and across economic sectors.

I thank the authors for their willingness to share their knowledge, know-how and experiences, as well as the many peer reviewers, which have helped us to ensure the quality of the manuscripts. Thanks are also due to Svenja Scheday for the support in handling the manuscripts.

Enjoy your reading!

Manchester, UK
Winter 2016/2017

Walter Leal Filho

Contents

An Inspired Education, the University of Wales Trinity Saint David

Carolyn S. Hayles

Abstract

In this paper the results of a centrally mandated, systematic and institution-wide policy to embed sustainability across all aspects of a university's life is described. Using a descriptive case study approach, this paper provides evidence that it is both feasible and beneficial to embed sustainability within an institution at a time of organisational change using a structured top-down approach. The Institute of Sustainable Practice, Innovation and Resource Effectiveness (INSPIRE) at the University of Wales Trinity Saint David (UWTSD) was established in January 2012. INSPIRE is a virtual institute which provides a focus for sustainable development activities across UWTSD. INSPIRE's role is to work across academic and support structures to deliver on the University's strategic priorities and embed sustainable development through its learning, teaching, curricula, campus, community and culture. In this paper key initiatives delivered through the INSPIRE model are presented using both primary and secondary data collection methods. By sharing UWTSD's whole-institution approach and in particular the work of INSPIRE to deliver on the University's strategic plan for embedding sustainability, it is hoped that other institutions will feel empowered to make changes to embed sustainability strategically.

Keywords

Community engagement · Embedding · Regional · Sustainable development · University

C.S. Hayles (✉)
Academic Lead, Institute of Sustainable Practice, Innovation and Resource Effectiveness (INSPIRE), University of Wales Trinity Saint David, Cardiff, UK
e-mail: carolyn.hayles@uwtsd.ac.uk

W. Leal Filho (ed.), *Sustainable Development Research at Universities in the United Kingdom*, World Sustainability Series,
DOI 10.1007/978-3-319-47883-8_1

1 Introduction

The University of Wales Trinity Saint David (UWTSD) is a new university with an historic past. UWTSD was formed on 18 November 2010 through the merger of the University of Wales Lampeter and Trinity University College Carmarthen, under Lampeter's Royal Charter of 1828. On the 1 August 2013, Swansea Metropolitan University became part of UWTSD. The University's Royal Charter is the oldest in Wales and England after the universities of Oxford and Cambridge. In 2011 HRH the Prince of Wales became its Royal Patron.

The UWTSD Group includes Coleg Sir Gâr and Coleg Ceredigion as part of a dual sector (HE and FE) group structure comprising Further Education (FE) Colleges and the University. The UWTSD Group has over 25,000 learners across 17 campuses in rural and city locations. UWTSD aims to deliver clear, tangible benefits for learners, employers, industry and communities by offering a vocational approach from entry level to post-doctoral research. The Group will be further strengthened with the merger of University of Wales into UWTSD, now scheduled for 2017 (UWTSD 2015).

The University's main campuses are situated in various locations in and around Swansea's city centre as well as in the rural towns of Lampeter and Carmarthen in South West Wales. The Wales International Academy of Voice is located in Cardiff and in addition the University has a Business School in London for international students. UWTSD has a clear national profile, with many of its staff and students speaking the Welsh language and there are opportunities for students to undertake their studies through the medium of Welsh. Indeed the University's strong presence in South West Wales alongside its dual sector delivery makes it an important voice in the region. Many students are locals, living and working in the region. They also intend to live and work in the region on completion of their studies. Indeed their contribution to the local region, its economy, environment and culture are readily identifiable (UWTSD 2015).

As a newly formed institution that has placed itself at the heart of a region, the university has an integral role to play in delivering a sustainable future for South West Wales. In this paper the results of a centrally mandated, systematic and institution-wide policy to embed sustainability across all aspects of a university's life is discussed, from campus initiatives to pedagogy developments as well as student-led initiatives. This is timely with the recent introduction of the Well-being of Future Generations (Wales) Act 2015, the first of its kind in the World, where the well-being of future generations will be considered at the heart of government decision making.

Seven development goals have been identified by the Welsh Government: a prosperous Wales; a resilient Wales; a healthier Wales; a more equal Wales; a Wales of cohesive communities; a Wales of vibrant culture and thriving Welsh language; and a globally responsible Wales. Going forward there will be a requirement for public bodies to make sure that when making decisions they take into account the impact they could have on people living their lives in Wales in the future. It will expect them to work together better, to involve people reflecting the diversity of communities, to look to the long term as well as focusing on now and to take action to try and stop problems getting worse, or even stop them happening in the first place (Welsh Government 2015). The university is therefore well placed to not only meet the requirements of this act but also to support others to do so through its learning, teaching and research activities (Hayles 2015).

2 Embedding Sustainability

2.1 A Systematic Institution-Wide Sustainability Policy

Transformation, adaptability and flexibility are familiar words to organisations working in times of merger. Indeed as a newly formed institution experiencing widespread organisational change, the University has had to make a number of key decisions in shaping the new university (Hayles 2015). This has brought about many opportunities, one of which has been to place sustainability at the core of its strategic planning; embedding sustainability within its core operations and culture. The core values of the University as outlined in its Strategic Plan: 2013–2017 are as follows:

1. *Collaboration*: Through the establishment of a range of strategic relationships at regional, national and international level. Such networks will have the potential to inspire learners, staff and partners to create exciting new learning futures.
2. *Inclusivity*: Through putting learners first and championing lifelong learning without barriers; and supporting students from all backgrounds and at all stages of their education.
3. *Employability and creativity*: By harnessing the entrepreneurial, research, creative and enterprising skills of its learners, the university will offer educational programmes that allow students to have the best opportunities to gain employment and develop their transferable skills.
4. *Sustainable development*: Through a system-based approach to delivering meaningful and relevant educational pathways, promote learning and social responsibility that supports 'development that meets the needs of the present without compromising the ability of future generations to meet their own needs' (World Commission on Environment and Development 1987).
5. *Wales and its distinctiveness*: Through celebrating the distinctive linguistic and cultural assets and heritage of Wales.
6. *The concept of global citizenship*: Through the development of further multi-national activities and opportunities for learners, staff and partners.
7. *Research and its impact on policy*: By ensuring that its research activity and outcomes influence the evidence base of policies developed in Wales and beyond (UWTSD 2013).

Therefore the University's strategic plan makes the commitment of UWTSD clear, namely to embed sustainability as a core principle across all aspects of the University.

2.2 The Institute of Sustainable Practice, Innovation and Resource Effectiveness (INSPIRE)

The Institute of Sustainable Practice, Innovation and Resource Effectiveness (INSPIRE) is a virtual institute which provides a focus for sustainable development activities across the University and the wider UWTSD group including Coleg Sir Gar and Coleg Ceredigion. INSPIRE was established by Jane Davidson, a former Minister for Environment and Sustainability in Wales, in January 2012. In 2013 INSPIRE became a strategic sustainability directorate and a Sustainability Committee, which serves the whole university, was established. INSPIRE's role is to work across academic and support structures to deliver on the University's strategic priorities and embed sustainable development through its learning, teaching, curricula, campus, community and culture.

Through INSPIRE the university aims to:

- Develop curriculum-related delivery to ensure that students are provided with the knowledge, skills and attitudes that will equip them for their future contribution to the economy, community and environment;
- Develop a research and innovation capacity focused on the core strengths of the University;
- Develop its campuses to the highest standards of environmental performance; and
- Contribute to local communities by giving particular regard to issues of sustainable rural and urban communities and the development of South West Wales as a low carbon region.

In this paper a number of initiatives delivered through the INSPIRE model will be described using a case study approach.

The descriptive case study, as a research approach, allows for the systematic identification of a process that has taken place using both primary and secondary data collection approaches. It looks at who, what, where and when. The aim is not to test a theory or hypothesis, but to record and share an in-depth knowledge of and insight into the process that has taken place (Fellows and Lui 2003; Naoum 2013).

By sharing UWTSD's whole-institution approach and in particular the work of INSPIRE to deliver on the University's strategic plan for embedding sustainability, it is hoped that other colleagues and institutions will feel empowered to make changes to embed sustainability strategically. INSPIRE at UWTSD actively welcomes the development of collaborations and partnership with other institutions that promote and support sustainability embedding initiatives and willingly shares its embedding experiences, including the challenges it has faced.

3 Descriptive Case Study

The descriptive case study is presented in five discrete sections, exploring the work that has been achieved to embed sustainability across the whole institution. These are as follows:

a. *Faculty Level* engagement including the development of faculty sustainability plans;
b. *Framework Approach* to the design of new and existing teaching programmes;
c. *University-wide Academic Engagement* through staff surveys and sustainability ambassadors;
d. *University-wide Academic Activities* including symposium and research groups; and
e. *Student Initiatives* including the Green Impact programme and internships.

In each instance the 'who, what, where and when' of each activity has been reported.

3.1 Faculty Level

UWTSD recognises 'Education for Sustainable Development and Global Citizenship' (ESDGC) holistically, whilst deliberately and consciously acknowledging the need for a balance between society, economy and environment to contribute towards individual and community wellbeing, a reduction in environmental impacts and consequently a more resilient future (Hayles 2015).

The University's intention is to ensure that it embeds a framework for ESDGC in a way that delivers; where the emphasis is not merely on the content of the modules and programmes of study offered, but on the entire learning and teaching experience. Indeed, the University's agreed Learning, Teaching and Enhancement (LTE) Strategy includes 'sustainability conscious learning' to enable students to have a clear understanding of the impacts of their future actions on the physical, social and economic environments (INSPIRE 2015).

This commitment to strategically introduce ESDGC through its teaching and learning is recorded in Faculty Sustainability Plans, which were introduced in 2013 and provide a mechanism for annual reporting on faculty level ESDGC through the University Sustainability Committee. Each University faculty is required the produce a plan, to a common template, outlining how they are working to embed ESDGC within subject disciplines as well as identifying cross curricular opportunities. The plans also reflect environmental commitments including resource conservation and the practice of video conferencing, to reduce travel between campuses.

Each faculty plan is required to provide a summary of the key ethos and pedagogical approach of the faculty and how the faculty as a whole intends to take the sustainability commitment forward, including arrangements for plan delivery and reporting structure; e.g. area/activity, sustainability element and faculty wide and interdisciplinary approaches and concepts.

Specifically the faculty makes commitments in relation to:

- Sustainability: working within environmental limits;
- Sustainability: how the faculty teaches;
- Sustainability: what the faculty teaches;
- Sustainability and research and development activity;
- Sustainability, the Faculty and the wider community; and
- Sustainability: competitive advantage.

Faculty Plans demonstrate the link between faculties and the University's strategic agenda on sustainability. The documents are available on the INSPIRE web page and can be used publicly to demonstrate the University's practical application of its commitment to sustainability (Davidson 2014).

3.2 INSPIRE's Framework Approach

Academics are encouraged to work within existing frameworks when developing new programmes and updating existing programmes. As part of the validation and revalidation of teaching programmes, course directors are required to provide a sustainability statement, demonstrating their pedagogical approach and how sustainability has been embedded in the programme, its modules, their learning outcomes and assessment. As part of the validation process, all programme validation documentation is reviewed by the academic lead for INSPIRE to ensure consistency with the University's sustainability commitments. This is a very valuable process and is proving to be a useful intervention point with which to shape curriculum in relation to ESDGC. Going forward validation will also need to consider a programme's contribution to meeting the Well-Being of Future Generations (Wales) Act development goals.

Academics have been encouraged to use UNESCO's five pedagogic principles to support the development of ESDGC curricula. These are as follows:

(1) *Futures thinking*: engages people in imagining preferred visions for the future. It involves the exploration of assumptions and of meaningful understandings and interpretations of sustainable development. This process of envisioning futures leads people to take ownership and responsibility for more sustainable futures.
(2) *Critical and creative thinking*: enables people to explore new ways of thinking and acting, make informed decisions and create alternatives to present choices. It involves reflecting on how people interrelate with one other, understanding cultural differences and creating alternative ways to live together.

(3) *Participation and participatory learning*: The engagement of people is needed to build sustainable futures collectively. Engaging diverse stakeholders and communities is essential, as they value and include differing knowledge systems and perspectives. The process of participation is also important for creating ownership and empowerment.
(4) *Thinking systemically*: is essential to sustainable development, as piecemeal approaches have proved not to work - instead resolving one issue while creating other problems. Sustainable development requires approaches, which go beyond analysis in terms of 'problem-solving' and/or 'cause-effect'.
(5) *Partnerships*: a motivating force towards change. They empower people and groups to take action, to take part in decision-making processes and to build capacity for sustainable development. Intercultural and multi-sectoral partnerships in particular are often highlighted as critical in Education for Sustainability approaches (UNESCO 2002).

In addition the university recognises the Higher Education Academy (HEA)'s Future Fit Framework (Stirling 2012); and the HEA's Quality Assurance Agency for Higher Education (QAA) guidance, 'Education for Sustainable Development Guidance' (2014), which was formally adopted by the University in March 2015 as the framework for ESDGC curriculum design, delivery and review within the University. Educators are encouraged to use it as a framework, within their own disciplinary context, rather than as a prescription of a curriculum or pedagogic approach. INSPIRE runs workshops on the 'Education for Sustainable Development Guidance' and provides academic staff with support and an implementation toolkit.

In the first instance, the university made a commitment that 15 % of the total student experience for each undergraduate would include ESDGC (Davidson 2014). In order to monitor and evaluate progress towards this target, a review of all degree programmes was undertaken in 2014 to ascertain where ESDGC was part of the learning and teaching experience. In this initial review, only core and compulsory modules on each programme were reviewed. A methodology developed by Bristol University (Willmore 2015) was utilised and details of this can be found on the INSPIRE website.

The results of the curriculum audit exceeded expectation at the time, showing for example that 93 % of modules in Teacher Education contained teaching and learning on ESDGC. Other faculties with high scores included Business (77 %), Architecture, Computing and Engineering (67 %), Art and Design (58 %), and Social Sciences (51 %). The lower scores were recorded as 39 % by Performance and 29 % by Humanities (Davidson 2014). This approach produced a quantitative assessment of the progress the University has made to embed sustainability within its curriculum (Hayles 2015).

3.3 University-Wide Academic Engagement

At the beginning of the embedding process a sustainability skills survey was conducted to develop an evidence base of existing expertise, experience and interest in sustainability across the institution. The results from the survey that have been published indicate a significant potential within the institution to take the sustainability agenda forwards. 78 % expressed an interest in ESDGC whist 49 % documented experiences of working on ESDGC already (Davidson 2014). This process was also used to identify sustainability link contacts.

Sustainability link contacts are observers on the University Sustainability Committee, who act as a liaison for their faculty and meet with INSPIRE to discuss issues arising and identify the support and training needed at an individual, group and faculty level. Each faculty has one or more sustainability link contacts.

3.4 University-Wide Academic Activities

In 2015 an annual ESDGC symposium was launched. The symposium, which is open to all academics within the university, showcases best practice from across the university's campuses, faculties and disciplines. This is seen as an opportunity for academic staff to exchange experiences and ideas, find commonalities and form alliances with staff from elsewhere in the University. All academics are welcome to attend the event and learn from the experiences of others. The first INSPIRE conference took place in June 2015. Educators shared experiences of cultural and social sustainability within the classroom; looked at mental health and safety issues and the well-being of students; curriculum content from engaging students with rubbish, farming and the wider natural environment; student IT projects which delivered on all aspects of sustainability through the support of charity; and early learner, infant and senior school teacher training.

In 2016 the symposium took place in March and formed part of the Wales NEXUS conference, the University's annual teaching and learning conference. Educators shared new research on the 'cost' of email wastage and the impact email as a 24/7 access point to academics can have on their physical and mental wellbeing. They spoke about the importance of creativity in developing sustainability skills, showcasing student work using only recycled materials; stressed the importance of experiential learning, with anthropology students delivering their own lectures and making items to support their understanding of materiality; whilst there was feedback from a two-day workshop run with students from the Arts and Design Faculty on engaging with sustainability and shared the student's self-motivation and enthusiasm to support and promote sustainably across the institution. Other academics talked about mainstreaming sustainability in work-based learning, and the promotion of the Eco-code, a document that provides a series of reminders about the sustainability policies and the practical steps that are expected. The last two speakers looked at early learning, discussing how children and young people can learn from educational activities that involve animals and animal welfare; and the School of Early Childhood's approach to supporting the development of students' own awareness of sustainability issues and how they can transfer this interest to the young children they work with in the future.

All the presentations have been made available on the UWTSD and INSPIRE's web pages and YouTube. Written papers are submitted to the Wales Journal of Learning and Teaching in Higher Education. Presenters have also been encouraged to publish their ESDGC work outside of the University.

A University-wide interdisciplinary ESDGC research group was also formed in 2015 to facilitate the development of ESDGC research collaborations and bids for research funding and scholarly activity.

3.5 Student Initiatives

3.5.1 National Union of Students (NUS) 'Green Impact' Programme

A number of student initiatives have also been established within UWTSD. In 2013 the Students' Union and the University jointly signed up to the NUS 'Green Impact' programme, an environmental accreditation programme, which brings staff and students together within their wider communities to enable and showcase positive changes in environmental practice and to make simple, tangible and potentially powerful changes in behaviour and policy documented via an online workbook (Hayles 2015).

3.5.2 INSPIRE Student Internships

To incentivise students to participate in sustainability embedding activities, internship opportunities were set up in 2013. These internships, created in partnership with the Students' Union, are open to any student prepared to make 'One Planet Living' commitments. Nine annual internships positions were created, three students on each of the University's principal campuses, namely in Lampeter, Carmarthen and Swansea. They were tasked with working on Fair-trade, Green Impact and Sustainability Exchange programmes:

- *Fair-trade interns*: promote Fair-trade and develop the University's commitment to Fair-trade activity, through organising and hosting events and boosting student involvement.
- *Green Impact interns*: have more of an operational and organisational focus, playing an active role in making the University more sustainable through the delivery of the 'green impact' programme; and
- *Sustainability Exchange interns*: promote and organise opportunities for staff and students to share ideas, news and views on sustainability issues within the University and the wider community.

In 2015 a tenth internship was added to the portfolio. This student works on developing the on-line presence of INSPIRE including the interns blog.

The interns are expected to work both independently and in teams, meeting regularly with INSPIRE staff, a nominated Students' Union representative and each other. All interns are encouraged to be dynamic, creative and to forge links with the student body and relevant external groups and organisations (Hayles 2015).

3.5.3 Student Employability Award

The University introduced a Student Employability Award in 2013, again in partnership with the Students' Union. The link between the sustainability and employability agenda is made explicit in the student attribute requirements, namely:

- Active Citizenship: able to appreciate the importance of environmental, social and political contexts to their studies;
- Creative Problem Solving: able to think creatively, holistically, and systemically and make critical judgements on issues;
- Teamwork: able to work collaboratively and work in interdisciplinary teams;
- Learning and Personal Development: able to develop a high level of self-reflection at a personal and professional level; and
- Communication: able to understand, critically evaluate, adopt thoughtfully and communicate sustainability values (Davidson 2014).

4 A Sustainable Campus

Campus sustainability, and in particular environmental sustainability, is usually the first thing university's tackle, as good energy and environmental management systems (EMS) contribute towards the efficient and effective use of resources. It is undeniable that addressing environmental challenges is key to building a more resilient and sustainable future. The university needs to be an exemplar of best practice as students expect to see in 'practice', what the academics 'preach' (Hayles 2015).

The University has recognised that environmental issues are fundamental to the future health and wellbeing of all those involved within the institution, the wider community and society as a whole, and accepted its responsibility to demonstrate leadership in sustainability and build resilience through innovation and enhancement of sustainable solutions to environmental concerns. To this end, the University produced an institutional action plan, 'Towards Living within our Environmental Limits' (UWTSD 2012), which details the University's immediate priorities and longer-term plans for addressing environmental challenges. Areas covered included: the MS; the roles and responsibilities of the members of the institution; student and staff engagement; carbon management; procurement; buildings and information technology; waste management; travel; and transport.

A review undertaken in 2014 recognised that UWTSD is at the beginning of its journey relative to many other UK universities as a result of the recent mergers and the challenges of dispersed campuses. However, the review commended the work of INSPIRE in relation to outreach and the work done to make staff and students more aware of environmental issues. Indeed, a number of initiatives have been put in place. For example, the University Council adopted the 'University's Carbon Management Plan' in July 2014 (UWTSD 2014). The policy considers both the need for environmental management systems and behaviour change programmes. Many meetings involving staff from different campuses now take place via the video conferencing suits, Lync or Skype, which has reduced the amount of travel between campuses significantly (Hayles 2015).

5 Conclusions and Way Forward

This paper uses a case study approach to describe a centrally mandated, systematic and institution-wide approach to embedding sustainability across all aspects of a university's life. The paper focusses on the work of INSPIRE, the virtual institute tasked with delivering the University's strategic plan for embedding sustainability.

Presented as a descriptive case study, the work of INSPIRE has not been critically analysed, but reported as an example of what has been achieved in one institution. Whilst the presentation of a single case study as a research output can be subject to a number of criticisms, the most common of which being those of methodological rigour, researcher subjectivity and external validity; as presented, the value of this case study is that it allows for information and knowledge transfer and ultimately provides evidence that it is both feasible and beneficial to embed sustainability within an institution at a time of organisational change.

By using INSPIRE as its delivery mechanism, UWTSD's aim is to *inspire* individuals (students, graduates and practitioners), all of whom can make a difference in society. Through work-based learning, research and knowledge transfer networks, the University aims to play a pivotal role in the promotion of the Well-Being of Future Generations (Wales) Act.

Indeed, the introduction of the Well-being of Future Generations (Wales) Act has brought new challenges to the institution. A number of activities have already taken place, including a series of workshops on what the Act means to the University and its academics were run by INSPIRE in 2015 and early 2016. Adherence to the principles of the Act will strengthen UWTSD's sustainability commitment to its staff, students and the wider community, in particular through a focus on issues relating to sustainable rural and urban communities in South West Wales and its development as a low carbon region.

References

Davidson, J. (2014). *Sustainability Strategy 2014*. Available online at: www.uwtsd.ac.uk/…/strategies…/Sustainability–Strategy–Approved-by-Council-10–7-14.pdf Accessed February 01, 2016.

Fellows, R., & Lui, A. (2003). *Research methods for construction*. Oxford: Blackwell.

Hayles, C. S. (2015). A case study of capacity building: Embedding sustainable development principles and practices at the university of Wales, Trinity Saint David. In *Proceedings of the 5th International Conference on Building Resilience*, Newcastle, New South Wales: Springer

INSPIRE (2015). About INSPIRE. Available online at: http://www.uwtsd.ac.uk/inspire/about-inspire/ Accessed February 01, 2016.

Naoum, S. G. (2013). *Dissertation research and writing for construction students*. Abingdon: Routledge.

Sterling, S (2012). *The Future Fit Framework – An introductory guide to teaching and learning for sustainability in HE,* The Higher Education Academy. Available online at: https://www.heacademy.ac.uk/node/3573 Accessed February 01, 2016.

UNESCO (2002). *Teaching and Learning for a Sustainable Future*. Available online at: www.unesco.org/education/tlsf Accessed February 01, 2016.

UWTSD (2012). *Living within environmental limits*. Available online at http://www.uwtsd.ac.uk/inspire/inspire-activity/living-within-our-environmental-limits/ Accessed February 01, 2016.

UWTSD (2013). *University of Wales: Trinity Saint David Strategic Plan: 2013 to 2017*. Available online at: www.uwtsd.ac.uk/media/uwtsd-website/content-assets/documents/inspire/Strategic-Plan—FINAL-AGREED-version-7-Nov-13.pdf Accessed February 01, 2016.

UWTSD (2014). Carbon Management Plan. Available online at: http://www.uwtsd.ac.uk/about/strategies-and-policies/ Accessed February 01, 2016.

UWTSD (2015). *About University of Wales: Trinity Saint David*. Available online at: http://www.uwtsd.ac.uk/about/ Accessed February 01, 2016.

Welsh Government (2015). *Future Generations Bill*. Available online at: http://gov.wales/topics/people-and-communities/people/future-generations-bill/?lang=en Accessed February 01, 2016.

Willmore, C. J., Tweddell, H. M., & Tierney, A. E. P. (2015). Measuring education for sustainable development: Experiences from the University of Bristol. *International Journal of Sustainability in Higher Education, 16*(4), 507–522.

World Commission on Environment and Development. (1987). *Our common future*. New York: Oxford University Press.

An Integrative Approach to Sustainable Development Within a University: A Step-Change to Extend Progress on Multiple Fronts

Chris Shiel and Neil Smith

Abstract

This paper adopts a case-study approach describing how one institution has sought to maintain an integrative approach to sustainable development, in an institutional context that has served to fragment holistic ways of working. The paper sets out the institutional context before outlining the interventions, designed to achieve a step-change and to take engagement with sustainable development to a further level. It is suggested that achieving awards such as 'EcoCampus Platinum' are important to demonstrate environmental credentials however, securing the support of a university's senior educational committee is vital, if all students are to experience education for sustainable development (ESD). Working across the institution, particularly in partnership with academic groups and the Students Union is a further way to increase engagement and momentum. The paper argues for the importance of integrative approaches but suggests that maintaining integration poses challenges; initial successes should not be taken for-granted; maintaining momentum across all fronts requires substantial effort from academics and environmental managers. An evaluation will be provided of the strategies adopted to achieve both an award and the support of a broader group of academics engaging with ESD. A summary of the lessons learned from the experience will be of value to others.

Keywords

Integrative approaches to sustainability · Eco-campus · Leading change

C. Shiel (✉)
Faculty of Science & Technology, Bournemouth University, Poole, UK
e-mail: cshiel@bournemouth.ac.uk

N. Smith
Sustainability Manager Estates, Bournemouth University, Poole, UK
e-mail: Nsmith@bournemouth.ac.uk

W. Leal Filho (ed.), *Sustainable Development Research at Universities in the United Kingdom*, World Sustainability Series,
DOI 10.1007/978-3-319-47883-8_2

1 Introduction

The need for integrative approaches to sustainability within higher education has been argued previously (Leal Filho et al. 2015; Sterling et al. 2013). Universities need to contribute to sustainable development through research, by ensuring that sustainable development is considered within the entire curriculum and in the extra-curricular sphere, and through working in the community to both educate and build capacity. Synergies will be created by working holistically. At the same time, institutions need to manage their estates in ways that exemplify best environmental practice and strive to achieve a culture such that sustainability is embedded in the fabric of every university activity—something that remains a challenge to achieve (Sterling et al. 2013). However, across the world, and particularly in the UK, it is quite evident that while many universities have exemplified 'campus-greening' and focused on environmental management, there are fewer examples of integrative approaches (Leal Filho et al. 2015). There is still, much further to go (Brennan et al. 2015; Amaral et al. 2015) if higher education is to make a full contribution to sustainable development.

It is in the context of a desire to work holistically and 'go further', that this paper has been developed. The case study considered represents the learning from a single university setting where historically, engagement with sustainable development has been ambitious and extended from the outset to encompass all aspects of university life. A single site case study obviously has limitations but as Sharpe (2002) suggests such learning is important to inform processes of systemic transformation across the higher education sector; therein, lies the value of this paper. Further, it sets out a number of actions that were taken to advance sustainable development across institutional domains exemplifying a collaborative endeavour between an academic and a practitioner, seeking to align an integrative agenda. The reflection on experience will be of relevance to those seeking to develop integrative approaches and cross-boundary relationships.

2 Integrative Approaches to Sustainability

A holistic and transformational approach to sustainable development within a university requires systemic change and new ways of working (Sterling et al. 2013; Wals and Corcoran 2006). Champions of change need to challenge silo mentalities and to develop processes which encourage synergies across university functions, striving to re-align systems and goals towards the common endeavour of sustainability (Shiel and Williams 2014). The aim is to move beyond one dimensional approaches, such as campus-greening (which is important but not enough on its own) and curriculum initiatives, where "integrating sustainability" merely results in the development of a single module as an "add-on", or "package of knowledge" (Haigh 2005) to approaches that synchronise the efforts applied in any one dimension to other dimensions. The ambition is to drive whole-institutional change, systemic transformation, and to encourage others to engage in a radical re-thinking of the purpose of education.

In a sector that is "notoriously resistant to change" (Wals and Blewitt 2010, p57) achieving such a step-change is not an easy task. The evidence continues to suggest (at least within the UK) that while a few institutions exemplify such ways of working and are exploring institutional change (as Walls and Blewitt note, "third-wave sustainability"), there are far fewer examples of what might truly be described as 'the sustainable university' (Sterling et al. 2013). Most universities find it easier to focus on campus greening and environmental management (Shiel et al. 2015) as singular initiatives; curriculum change is sometimes opportunistic rather than part of a strategic and integrative endeavour, rarely linked to campus greening. Addressing sustainability across campus, curriculum and community (Jones et al. 2010) means pushing boundaries and overcoming challenges.

Finding new ways to align campus, education, and community is essential; combining academic and practitioner knowledge is important for sustainability research (White 2013) but will also be valuable for enhancing the learning experience of students and the institution. While there is no single way to achieve an integrative approach, if the aim is to develop a culture where sustainability is owned by all stakeholders and permeates the institution, the efforts of professional services/administrative staff and academics need to be aligned; maintaining collaborative relationships across boundaries is an essential element of working holistically (Shiel and Williams 2014).

A brief account of the institutional context follows before collaborative actions taken to address a step change are described.

3 The Context

Bournemouth University (BU) is a medium-sized UK university, inaugurated in 1992, with around 17,000 students, 650 academic staff and 800 professional and support staff. Environmental issues became a focus of attention at the end of the nineties with a concern for saving resources, particularly utilities. Engagement with the broader concept of sustainable development was not a significantly strategic issue until 2005, when a strategy was developed for the whole institution; from 2006 this strategy embraced both global citizenship and sustainability (Shiel 2007). The strategy outlined the importance of a holistic approach and emphasised integrative ways of working on over-lapping agendas (Shiel et al. 2005). Since then, a variety of initiatives have been pursued to enable the institution to progress towards being a sustainable university (in the sense used by Sterling et al. 2013). The success of the approach, which is not dissimilar to the "4C" model at Plymouth University (Jones et al. 2010, p7) has resulted in a number of institutional awards, a consistent placing in the top ten, of the People and Planet University League, and substantial journey of change. The current strategic vision for the university now makes clear commitment to sustainable development, with the aim of "inspiring our students, graduates and staff to enrich the world", and the assurance that: "we will ensure our environmental credentials are held in high esteem" (BU 2018). Further, the 2012–2018 Strategic Plan refers explicitly to "a holistic approach to SD" (p30), the need to "ensure that graduates develop a global perspective and understand the need for sustainable development by seeking to embed sustainable development across the curriculum" (p19) and the need to "ensure BU operates an affordable, sustainable and secure estate" (p53).

An appraisal would suggest that the institution has done more than many universities, and moved much further than a campus-greening approach since 2005. However, maintaining momentum has not always been easy, as an evaluation of the challenges revealed in 2013 (Shiel and Williams 2014); those leading the agenda have to continually critique their approaches and instigate new initiatives if progress is to be maintained. Since 2014, a new appointment to the role of Environmental Manager has contributed to refocusing efforts: the job title was changed to Sustainability Manager; the Environment Strategy Committee became the Sustainability Strategy Committee as a consequence, with a smaller membership but with greater academic representation and a more strategic focus. At the same time, commitment to education for sustainable development (ESD) has been made more specific in policy documents with a goal of achieving more critical engagement. While there are undoubtedly several courses that exemplify sustainable development, for example, the MSc Green Economy (Newton et al. 2014), sustainability is less considered in some provision. In short, ESD needed a further push to extend engagement; further work was needed to raise the profile of the academic agenda and to seek alignment with campus greening efforts.

4 Interventions to Take Sustainable Development to a New Level—Greener Campus and 'ESD +'

In order to gain further traction and develop integration further, three particular courses of action were pursued to contribute to change.

- Reinvigorating the education agenda
- Achieving the highest credential to exemplify best practice in the environmental management of the Estates (EcoCampus Platinum and ISO 14001)
- Developing the culture and building capacity by working in the extra-curricular sphere—initiating Green Impact teams across the university.

The three actions will be commented upon in turn. They each contribute towards two further objectives:

- Exemplifying holistic ways of working by creating synergy between the academic endeavour and the professional services responsibility
- Communicating across the university the sustainability agenda.

5 Reinvigorating the Education Agenda

Although BU was one of the first institutions to implement institutional curriculum guidelines to ensure that all course teams consider how to incorporate ESD when developing new provision or at the re-validation of existing provision (see Bourn and Shiel 2009; 672), it has not necessarily resulted in full coverage across all programmes. The Sustainability Strategy group concluded that a further push was needed to engage all Faculties; the best way to achieve that goal was to raise ESD at the Education and Student Engagement Committee (ESEC) and to stimulate an academic debate. ESEC is chaired by the Deputy Pro Vice Chancellor (Education) and has members from across the institution, with all Faculties and those in Professional Services with educational responsibility. Students elected to the Students Union also participate. Achieving approval to schedule a debate item in what is always a very full committee agenda, was an objective that was not immediately achieved. Several approaches had to be made to the senior team, however, once agreed, the debate was scheduled as a substantive item with time allowed for a short-presentation followed by a formal debate to discuss further actions.

The authors prepared a presentation that highlighted the drivers for an integrative approach, the report from the National Union of Students (NUS 2015), an analysis of the current situation, opportunities for doing things differently i.e. going beyond the current position, and the potential for moving from addressing ESD, to an 'ESD+' approach. The latter would promote academic and practitioner collaboration, greater collaboration between campus and curriculum, and greater participation of students.

The debate was successful in raising awareness, securing engagement and developing actions. Formal actions agreed in the minutes note:

- The provision of sustainability staff development, through the "PG Cert in Education" module and/or provide lunchtime training sessions in order to introduce the change in staff culture which would be passed on to students;
- The Green Task Force providing workshops for staff and students to attend which in turn would have a good impact within the University;
- Strengthening guidance for programme development;
- Raising awareness of sustainability and how to promote the legacy messages on the hoardings which currently border the new "Fusion Building 1" (a new build where sustainability messages have been writ large);
- Consideration of brave and bold statements and initiatives for sustainability e.g. a bottle free campus (suggested by students as an action they would like made compulsory).

Further, the Chair hoped that the sustainability message could be driven forward effectively across the University community, and members were requested to disseminate this essence of the discussion across the institution.

It is too soon to comment on whether the approach will result in further innovation however, all Faculties are now required to respond to ESEC actions and report back. The presentation also served to introduce the Sustainability Manager to the Faculties, to reinforce success to date and projects in development, and to sow the seeds for extending collaborative learning opportunities for sustainability projects (within the curriculum and in the extra-curricular sphere).

The approval of inclusion of a sustainability focus on the PG Cert in Education was an important step, as all new staff participate in the programme. Further invitations to deliver staff development have also resulted, plus an invite to write a blog for the Centre of Excellence in Education. Workshops for staff and students are being developed and the guidance for curriculum development will also benefit from being strengthened. As communication is key for success (Djordjevic and Cotton 2011) presenting at ESEC was a message in itself; the sanction to develop sustainability messages (effective messages are currently being used to screen building developments on campus) so that sustainability efforts are more visible in the future, was also an important outcome.

Students who were on the committee were active participants in the discussion. They reinforced for academic colleagues that students want to learn more and engage with change. They confirmed support for the outcomes of the National Union of Students Survey (NUS 2015) but also suggested that sometimes it would be better if top-management made decisions that are enforceable, i.e. "a ban" on unsustainable products/actions as a way forward.

6 EcoCampus

In parallel to efforts to enhance ESD an important goal was to validate the university's practice in relation to environmental management through "EcoCampus" accreditation. This would further reinforce that while the university advocates greener behaviour for students and staff, it is also striving to manage its business in ways that are sustainable.

EcoCampus was designed by the sector to help universities implement environmental management systems (EMS). An EMS is a risk management tool to minimise the impact on the environment whilst also promoting positive impacts, such as Education for Sustainable Development (ESD).

EcoCampus splits the international standard for EMSs (ISO 14001) into four bite sized pieces: Bronze (Planning); Silver (Implementation); Gold (Operating) and Platinum (Checking and Correcting), where the Platinum award is the equivalent of ISO 14001.

BU started implementing its EMS following the EcoCampus model in autumn 2008 and re-secured "Gold" in July 2014. Developments to take the university to the next level slowed in 2014, but further actions for progress were carried through in 2015. EcoCampus Platinum and ISO 14001 certification were awarded at the end of the year, following an external audit. This was an important achievement as BU is now one of only 15 Universities to achieve the result of dual certification.

Adopting a more integrative approach means that BU's Sustainability Policy and EMS scope includes embedding sustainability in the curriculum. This is reflected in the aspects and objectives and targets' registers. These are now key elements of the EMS, where ESD sits alongside the more standard tasks of minimising the harmful impacts the University has on the environment, such as energy and water use.

BU's EMS now provides a structured approach, supported by senior management, to continual improvement with its ESD programme. The three year external audit cycle for retaining certification will also provide checks to ensure BU continues to innovate in its curriculum offer (further reinforcing integration).

7 Staff and Student Awareness and Engagement

The third intervention related to a number of actions to build capacity and encourage more sustainable behaviour. BU recognises that whilst implementing technological solutions will help reduce its environmental impact, it also needs its staff and students to do their bit by adopting more sustainable habits, such as switching off equipment and recycling. There are great opportunities to link the development of such behaviours through the curriculum, extra-curriculum and campus management.

BU has a calendar of events planned throughout the academic year to raise awareness of, and engage with, staff and students about sustainability. This year BU has signed up to the NUS's staff engagement programme, Green Impact to encourage and reward positive sustainability behaviour. Staff teams implement sustainability initiatives following criteria in a workbook which has been tailored to the institution and is split into bronze, silver and gold award levels.

BU students will be trained as auditors to check the evidence provided by staff to show how the teams have met the criteria. Staff involvement and their achievements are recognised and rewarded.

Many Universities have used this model to engage with their staff and it is hoped BU staff will deliver change at a local level and have some fun at the same time.

BU has also signed up to the NUS run Student Switch-off inter-halls energy competition. Students signed up at Freshers' Fair and take part in mini competitions to show off their energy saving behaviour using social media. The hall that saves the most amount of energy together with good evidence of student engagement through the year wins the competition and will be rewarded at the end of the year.

Again this model is based on encouraging individuals to adopt more sustainable habits whilst having some fun and winning prizes.

8 Discussion and Lessons Learned

The approach at BU, as might be expected in an institution that has sought a holistic approach from the outset, has exemplified each of the patterns that Barth and Rieckmann (2013) suggests are distinct ways that institutions engage with sustainability: top down institutional approaches, bottom up, and sustainability as the environmental management of estates. The approach followed hitherto, has also acknowledged the importance of those in middle management roles (Brinkhurst et al. 2011) and strategies to 'middle-out' initiatives. Such approaches are challenging to maintain (Shiel and Williams 2015) so those leading change need to continually find ways to ensure that momentum is not lost. The actions outlined in this paper have sought to make a step-change on multiple fronts. Top-management support has been visibly reinforced; further bottom-up initiatives will result from students; Faculty staff will engage further with the agenda across BU (middling out). Further, the interventions deployed during 2015 have enhanced communication about the university's commitment to sustainable development.

The importance of engaging senior management in taking forward sustainable development is critical in the early stages (Kemp et al. 2012) but it is also worth noting that commitment has to be reignited from time to time as leadership and strategy changes (Shiel and Williams 2014), or other agendas overshadow the focus on sustainability. In this case study the very act of seeking to get an item on the strategic committee for education, served to engage the Deputy Vice Chancellor Education. It also secured the attention of the Deputy Deans Education in the Faculties to revitalise the agenda. Achieving EcoCampus Platinum has also secured further acknowledgment and commitment from the senior team.

It is also important to remember, when working with Students' Unions that leadership of the Union also changes; commitment thus, also needs reinforcing. The presentation to ESEC together with participation in Green Impact with students auditing staff endeavours has served to create further synergies with the student body. This will facilitate further campaigns.

Staff development is vital for capacity building (Desha and Hargroves 2012) and for transforming the curriculum (Cebrián et al. 2012). Staff development has been a key feature of BU's efforts but the opportunity to contribute further and through a formal programme will extend reach.

In relation to EcoCampus, celebrating progress at each stage of the model with implementing the EMS was vital in maintaining momentum with the scheme. However, there needed to be greater involvement and engagement with staff on the implementation of the EMS to ensure staff are aware of their responsibilities for managing their activities. It is not the responsibility of one person or one team to do this and without that wider participation the full benefits of the EMS will not be realised.

Embedding an EMS into the culture of the organisation will take time. One of the paper's authors recognised it took five years after implementing an EMS in the National Oceanographic Centre, Southampton to see a change in the culture of that organisation.

With regard to Green Impact and building staff and students' capacity, the resources needed to support staff in signing up to the scheme and then for maintaining their participation should not be under-estimated. The Sustainability Team managed to recruit a further member of staff to support the scheme, just before launch, and so communication was not as well planned as it could have been. However, now the new team member is working well with the NUS to encourage more teams to sign up and to support those currently recruited.

The scheme has been taken up by teams in Professional Services but less so by academics. Having the time to complete the workbook has been cited as a reason for this lack of engagement. It has also been suggested that an engagement scheme that is more in tune with academics work may yield a more positive result.

While students have generally been enthusiastic about engaging with initiatives the "Student Switch-off" campaign has not been entirely positive. There were problems with the heating controls in student rooms in one of the participating halls, leading to the overheating of rooms. As a result students have questioned the value of them trying to save energy through the scheme when they see such wastage.

On the whole, the interventions taken to achieve a step-change have contributed to moving things forward and expanding engagement. They have undoubtedly extended communication about sustainable development and will in time, contribute to further organisational learning. They have exemplified partnership working across organisational boundaries, something that is at the heart of an integrative approach. Further developing the initiatives and reflecting on experience for this paper, has served to develop further the relationship between the academic lead for the agenda and the Sustainability Manager, a relationship which has been highlighted previously as essential to an integrative approach (Shiel and Williams 2014) and which continues to be important for progress. It will lead to further research and will result in co-creation of research projects with students.

9 Sustainable Development Research

While much could have been said in this paper about the substantial research at BU that is discipline based and addresses directly the various components of sustainability, for example, coastal conservation, sustainable design, engineering solutions, etc., the authors have omitted such inclusion in this paper. That is not to say that such research is not essential and valuable but to highlight that to achieve sustainable development, what is also needed is research that focuses on leadership and change management and particularly research to build capacity and holistic ways of working. As White (2013; 171) notes "Sustainability research is about much more than merely knowledge domains": researchers themselves can participate in the change process; sustainability research needs to extend across disciplines and structural boundaries.

As befits an 'integrative approach', this paper has focused on the research and actions needed to take forward sustainable development across institutional domaines. The authors have contributed to the change process. Actions cannot be taken forward without researching change; actions themselves (with reflective processes embedded) lead to further change; evaluation reveals what does and does not work, leads to further research, and informs better approaches for collective action. The ongoing approach at BU seeks "to bring together action and reflection, theory and practice, in participation with others, in the pursuit of practical solutions to issues of pressing concern to people, and more generally the flourishing of individual persons and their communities" (Reason and Bradbury 2001, p1). A critical inquiry methodology, participative action research and cooperative inquiry have supported the development of the approach (Shiel 2013) to mobilise change (Shiel and Williams 2014). Such research is not only valuable in that it contributes to an emerging dialogue on how we build capacity for sustainable development but has the potential to support the discipline based research. Discipline based research provides the scientific data and new technological solutions, but this may not be enough to achieve a sustainable future. 'Sustainability research' (in the sense used by White 2013) needs a combination of approaches and efforts. This paper has outlined a combined effort to contribute positively towards sustainability within a higher education setting.

10 Conclusion

This paper has demonstrated how one institution has sought to re-energise its efforts to exemplify a sustainable university. It has argued that integrative approaches to sustainable development are important however implementing an integrative approach to sustainability requires substantial efforts and on-going actions if momentum is to be maintained. Actions need to be addressed across multiple fronts and serve to visibly reinforce holistic ways of working by combining the efforts of academics and practitioners in professional services functions. Initial successes should not be taken for-granted. It is too easy to sit back once a sustainability policy has been endorsed and think that sufficient actions will flow as a consequence. Maintaining traction requires continual evaluation of progress and the development of new initiatives that encourage the entire academic community to participate. It is important to continually reinforce the message that the agenda is not just one person, or one team's responsibility.

Three initiatives have been described: one to take ESD to a further level, one to exemplify excellence in environmental management but which also combines ESD; a third to build capacity which in turn will impact on environmental behaviours such as increasing recycling and reducing energy use. In totality, the interventions have extended communication and debate about sustainability issues.

The importance of working through committees that lead the educational agenda has been reinforced, as has the need to continually re-engage leaders. Every initiative needs the backing of those at the top; more initiatives are undoubtedly necessary to build capacity amongst staff and students. It is easier to strive for external certification of the environmental management of the campus, albeit that that requires considerable efforts, than it is to secure hearts and minds of all stakeholders across an institution. Gaining institutional certification lies within the control of a smaller team but still requires many stakeholders to take responsibility or continual improvements will fail. Any scheme or plan for sustainable development has to be embedded into the culture of the organisation to deliver real change; culture change is achieved more easily when it is supported by leadership from the academic and professional services areas working in partnership. Addressing sustainable development within higher education involves working in areas which yield the greatest traction but also seeking synergy by working in partnership across multiple fronts. It is also critical to engage with students to encourage a bottom-up approach to stimulate change if sustainability is to be addressed fully within the curriculum, extra-curricular and across the campus.

References

Amaral, L. P., Martins, N., Gouveia, J. B., Leal Filho, W., & Sima, M. (2015). Quest for a sustainable university: A review. *International Journal of Sustainability in Higher Education, 16*(2), 001. doi:10.1108/IJSHE-1102-2013-0017

Barth, M., & Rieckmann, M. (2013). *A Review on Research in Higher Education for Sustainable Development.* Paper presented at the 7th World Environmental Education Congress, Marrakech, Morocco, 9–14th June 2013. Available at: http://www.weec2013.org/adminweec/UserFiles/ABSTRACT/365_presentation.pdf Accessed on September 11, 2013.

Brinkhurst, M., Rose, P., Maurice, G., & Ackerman, J. D. (2011). Achieving campus sustainability: top-down, bottom-up, or neither? *International Journal of Sustainability in Higher Education, 12*(4), 338–354.

Brennan, L., Binney, W., Hall, J., & Hall, M. (2015). Whose job is that? *Saving the Biosphere Starts at Work, Journal of Nonprofit & Public Sector Marketing, 27*(3), 307–330. doi:10.1080/10495142.2015.1053348

BU. (2018). *Bournemouth University Strategic Plan.* Available at: http://strategicplan.bournemouth.ac.uk/ Accessed on February 2014.

Bourn, D., & Shiel, C. (2009). Global perspectives: Aligning agendas. *Environmental Education Research Journal, 15*(6), 661–676.

Cebrián, G., Grace, M., & Humphris, D. (2012). developing people and transforming the curriculum: action research as a method to foster professional and curriculum development in education for sustainable development in higher education. In W. Leal Filho (Ed.). *Sustainable Development at Universities: New Horizons*, (pp. 273–283) Berlin: Peter Lang Scientific Publishers.

Desha, C.J., & Hargroves, K. (2012). Fostering rapid transitions to education for sustainable development through a whole-system approach to curriculum and organizational change. In W. Leal Filho (Ed.), *Sustainable Development at Universities New Horizons,* (pp. 29–46). Berlin: Peter Lang Publishing Group.

Djordjevic, A., & Cotton, D. R. E. (2011). Communicating the sustainability message in higher education institutions. *International Journal of Sustainability in Higher Education, 12*(4), 381–394.

Haigh, M. (2005) Greening the University curriculum: Appraising an international movement. *Journal Of Geography in Higher Education, 29*(1), 31–48.

Jones, P., Selby, D., & Sterling, S. (Eds.). (2010). *Sustainability education: Perspectives and practice across higher education*. London: Earthscan.

Kemp, S., Scoffham, S., Rands, P., Robertson, A., Robinson, Z., Speight, S., et al. (2012). a national programme to support education for sustainable development in the United Kingdom: University Experiences of the HEA Green Academy Programme. In W. L. Filho (Ed.), *Sustainable development at universities: New Horizons* (pp. 363–372). Frankfurt: Peter Lang Scientific Publishers.

Leal Filho, W., Shiel, C., & Paço, A. D. (2015). Integrative approaches to environmental sustainability at universities: An overview of challenges and priorities. *Journal of Integrative Environmental Sciences, 12*, 1–14.

National Union of Students (NUS). (2015). *Student attitudes towards, and skills for, sustainable development*. Available at: https://www.heacademy.ac.uk/resource/student-attitudes-towards-and-skills-sustainable-development-2015#sthash.wuBFFwLk.dpuf Accessed on January 2016.

Newton, A., Cantarello, E., Shiel, C., & Hodder, K. (2014). Lessons learned from developing a new distance-learning masters course in the green economy. *Sustainability, 6*(4), 2118–2132.

Reason, P., & Bradbury, H. (Eds.). (2001). *Handbook of action research: Participative inquiry and practice*. London: Sage.

Sharp, L. (2002). Green campuses: The road from little victories to systemic transformation. *International Journal of Sustainability in Higher Education, 3*(2), 128–145. doi:10.1108/14676370210422357

Shiel, C. (2007). Developing and embedding global perspectives across the university. In S. Marshall (Ed.), *Strategic leadership of change in higher education* (pp. 158–173). London and New York: Routledge.

Shiel, C. (2013) Developing global perspectives: Global citizenship and sustainable development within higher education. PhD thesis. Bournemouth University, School of Applied Sciences.

Shiel, C., Leal Filho, W., & do Paço, A. (2015). Putting sustainable development in practice: Campus greening as a tool for institutional sustainability efforts. In *Sustainability in Higher Education*, Kiddlington, (pp. 1–19). UK: Shandos Publishing.

Shiel, C., & Williams, A. (2014). Working together, driven apart: Reflecting on a joint endeavour to address sustainable development within a university. In W. Leal Filho, et al. (Eds.), *Integrative approaches to sustainable development at university level,* (pp. 425–447). Cham, Switzerland: Springer.

Shiel, C., Williams, A., & Mann, S. (2005). Global perspectives and sustainable development in the curriculum: enhanced employability, more thoughtful society? In B. Newland K. Norman & M. Johnson (Eds.). *Enhancing Graduate Employability, the Roles of Learning, Teaching, Research and Knowledge Transfer: Proceedings of the Bournemouth University Learning and Teaching Conference* (pp. 63). Poole: Bournemouth University.

Sterling, S., Maxey, L., & Luna, H. (2013). *The Sustainable University: Progress and prospects*. London and New York: Earthscan/Routledge.

Wals, A. E. J., & Blewitt, J. (2010). Third-wave Sustainability in Higher Education: Some (Inter)national Trends and Developments. In P. Jones, D. Selby, & S. Sterling (Eds.), *Sustainability: Perspectives and Practice Across Higher Education* (pp. 55–74). London: Earthscan.

Wals, A. E. J., & Corcoran, P. B. (2006). Sustainability as an outcome of transformative learning. In Holmberg, J. & Samuelsson (Eds.). *Drivers and Barriers for Implementing Sustainable development in Higher Education*. ESD in Action Technical Paper Number 3. Paris: UNESCO.

White, R. M. (2013). Sustainability research: a novel mode of knowledge generation to explore alternative ways for people and planet. In: Sterling S. Maxey L. & Luna H. (Eds.). *The Sustainable University: Progress and prospects*. London & New York: Routledge/Earthscan pp. 168–191.

Author Biographies

Dr. Chris Shiel is a Professor in Sustainability & Globalisation in the Faculty of Applied Sciences, Bournemouth University. She is the former Director of the Centre for Global Perspectives, a Leadership Foundation Fellow and a Principal Fellow of the Higher Education Academy. She has led ESD within the UK for almost two decades.

Dr. Neil Smith is the Sustainability Manager at Bournemouth University having been the Environment Manager at Southampton University for over nine years. He chairs the EAUC, Southern Central Environment Managers Group.

Benchmarking Sustainability Research: A Methodology for Reviewing Sustainable Development Research in Universities

Victoria Hands and Richard Anderson

Abstract

The need for high quality research to impact sustainability policy and action has been identified in international frameworks for environmental sustainability in higher education from 1977 to 1990 (Wright in Int J Sustain Higher Educ 3:203–220, 2002), and repeated by the UK Research Councils (Research Council UK in Research Council UK Submission to House of Commons Environmental Audit Committee Inquiry into Sustainable Development Goals (SDGs), 2014). With the emergence of indicators on measuring research for sustainable development in universities (AULSF in Sustainability Assessment Questionnaire, 2009), the research study which forms the focus of this paper, is a first attempt to establish a practical methodology to provide such data. The aim of this study was to investigate the extent to which sustainable development research was already being carried out across a large university, and whether it was possible to devise a relatively quick and reliable methodology to identify the scope and areas of research being undertaken, which would provide the university with a baseline of existing sustainable development research. The object was to capture and report the existing contributions to sustainable development research and to make an initial assessment of its current impact and contribution towards research excellence at the university. The work of 465 staff was analysed using content and thematic analysis to identify those relating to sustainable development, broadly defined as 'economic, social, environmental, community, wellbeing, global and future equity'. The analysis identified both researchers interested in sustainable development research issues, and those currently

V. Hands (✉) · R. Anderson
Sustainability Hub, Kingston University, River House, 53-57 High Street, Kingston upon Thames, Surrey KT1 1LQ, UK
e-mail: v.hands@kingston.ac.uk

R. Anderson
e-mail: richard.anderson@kingston.ac.uk

W. Leal Filho (ed.), *Sustainable Development Research at Universities in the United Kingdom*, World Sustainability Series,
DOI 10.1007/978-3-319-47883-8_3

researching sustainable development. The research also identified the degree to which published research showed evidence of a set of key external viability factors identified as: 'sustainability content', 'research impact', and 'knowledge transfer viability'. The methodology is intended to be replicable at other times and in other universities. It promises to raise the profile of sustainable development research internally, enabling further and meaningful engagement with the researchers it identifies, and encouraging cross-faculty working, potentially providing a rationale for researchers to engage in sustainable development as an exciting discipline in it's own right, contributing solutions to contemporary issues.

Keywords
Sustainability · Sustainable development · Research · Benchmarking · Higher education · Universities

1 Introduction

A review of the literature has identified a growing body of research on sustainability implementation in universities spanning operations, community, education and research, for instance; Leal Fihlo and Davim (2015) and Sterling et al. (2013). Whilst the literature on Education for Sustainable Development (ESD) encompasses both curriculum content and pedagogy for transformative learning, such as Winter et al. (2015), that on research, points to the transdisciplinary nature of sustainability research (Lang et al. 2012) and the acknowledged challenges this involves. White (2013) indicated that the growth in sustainability research is evidenced by the appearance of more dedicated journals, specific research funding and calls for 'impact' assessment by funders. Kordestani et al. (2015) suggest that the evolution of sustainability research can be traced and assessed via the '4Ps': "principles; policy; practice and; performance", based on their content analysis of 1,502 peer-reviewed articles on sustainability in business and management over a 20-year period. The call to engage with sustainable development in research is traced by Wright (2002), who reviewed international frameworks for environmental sustainability in higher education noting "…the encouragement of academic research related to sustainability…" This ranged from: *The Tblisi Declaration* (UNEP 1977) which included the need for environmental education for "…scientists and technicians whose specialized research and work will lay the foundations of knowledge on which education, training, and efficient management of the environment should be based." (Clause 8, UNEP 1977); to *The Talloires Declaration* (AULSF 1990) which called for a "culture of sustainability" and "interdisciplinary research" to "move toward global sustainability"; and the *Kyoto Declaration* (UNFCCC 1997) Clause 4 of which Wright stated "…implores universities to undertake research and action in sustainable development." (Wright 2002). However, for some authors, research based on knowledge generation has been regarded as "...in sharp contrast to our real needs" (Orr 2004). Indeed, the calls for high quality research continue: the UK Government's 'Stern Review on the Economics of Climate Change' focused on the growing need for high quality research relating to sustainability noting that:

> In preparing to manage the severe risks of climate change, the world needs the very best researchers to work on the crucial challenges. (Stern 2007).

At the global level, the critical and continuing need for Education for Sustainable Development (ESD) and high quality research to impact policy has most recently been confirmed with global agreement on the 17 intergovernmental Sustainable Development Goals and 169 associated targets published in *Transforming Our World: The 2030 Agenda for Sustainable Development* (United Nations 2015). The commitment in Clause 4 focuses on education:

> By 2030, ensure that all learners acquire the knowledge and skills needed to promote sustainable development, including, among others, through education for sustainable development and sustainable lifestyles, human rights, gender equality, promotion of a culture of peace and non-violence, global citizenship and appreciation of cultural diversity and of culture's contribution to sustainable development. (Clause 4, United Nations 2015).

2 The UK Policy Context

In the UK, this need has also been recognised across research support, funding and assessment agencies. For example, the Higher Education Funding Council for England (HEFCE) published a *Sustainable Development Framework-Policy Guide* observing that "Universities and colleges are well positioned to make a key contribution to the challenges and opportunities via teaching and research" (HEFCE 2014a). While their "Sustainable development in higher education—2008 update to strategic statement and action plan" stated:

> Within the next 10 years, the higher education sector in this country will be recognised as a major contributor to society's efforts to achieve sustainability – through the skills and knowledge that its graduates learn and put into practice, its research and exchange of knowledge through business, community and public policy engagement, and through its own strategies and operations. (HEFCE 2009).

The Universities That Count (UTC) project, a partnership between the Environmental Association for Universities and Colleges (EAUC), an umbrella group for university and college sustainability staff and students, Business in the Community and Corporate Social Responsibility (CSR) Consultancy, to create a benchmark for the university sector stated:

> … it is increasingly recognised that education for sustainable development and research into a sustainable future are the most significant contributions that universities can make to the problems of sustainability. (UTC 2010, Sect.1).

The Research Councils UK (RCUK), which represents the seven major UK research funding councils, recently stated in their submission to the House of Commons Environmental Audit Committee Inquiry into Sustainable Development Goals, that the:

> Research Councils will use the Sustainable Development Goals to inform research questions in existing and future joint activities to help ensure the evidence, tools and solutions are available to those implementing the new goals. (Research Council UK 2014).

In addition, the Research Excellence Framework (REF), the key agency in assessing the quality of research in UK higher education institutions, has stated in the introduction of the recent *Sector impact assessment of research* that:

> The REF is used to identify research of the highest quality and benefit to the environment, society and the economy, broadly defined. The introduction of impact assessment into the REF will, therefore, explicitly reward research that has sustainability benefits. (REF 2014).

The REF attempts to assess the impact of research outside of academia, that is the extent to which research has "an effect on, change or benefit to the economy, society, culture, public policy or services, health, the environment or quality of life, beyond academia". (HEFCE 2014b).

There is an increased focus in UK funding policy on: proven impact; innovation; and universities and business collaboration; ogether with the trend for funding bodies to concentrate support on research centres of 'proven excellence'. An example of this trend is 'The Urban Living Partnership' formed by the RCUK and Innovate UK to "harness the broad spectrum of UK research and innovation expertise to help cities realise their aspiration of healthy, prosperous and sustainable living." (EPSRC 2016). The Partnership is the first time all seven UK Research Councils and Innovate UK have come together to address the complex challenges and opportunities of urban living, initially by creating pilot city projects, led by a consortium of researchers, local authorities, service providers and businesses. However, despite these positive indicators, sustainable development research still faces significant challenges.

3 Measuring Sustainability Research

A commonly quoted challenge for sustainable development is its definition which can lead to a perceived lack of relevance (Dawe et al. 2005). Given the many different definitions of sustainability and sustainable development, for the purposes of this research, the definition is based on the Kingston University sustainability goals which focus on creating an institutional culture which: "collectively works to continually improve our environmental, social, ethical, global and long-term impacts as responsible global citizens." Kingston University (2013) and the United Nations 17 intergovernmental Sustainable Development Goals (United Nations 2015), in an attempt to reflect policy relevance at the global scale. However, in practical terms, indicators for measuring progress on sustainable development and research are only just emerging and require further research (Bullock and Wilder 2016). Most pertinent to this study is the proposal of 13 indicators for measuring sustainability in research by Lozano (2006), which included an indicator on "percentage of faculty doing research in sustainability issues" (RE3) and related information such as the names of faculty members and areas under study. The Association of University Leaders for a Sustainable Future (AULSF 2009) issued a '*Sustainability Assessment Questionnaire*' to signatories of *The Taillories Declaration*. This included 5 questions on sustainability research requesting estimations of "the amount of faculty research or scholarship being done in various disciplines in the area of sustainability" (question 6a, AULSF 2009) and a question on "development opportunities to enhance understanding, teaching and research in sustainability" (question 14, AULSF 2009). Example areas of sustainability research included "…renewable energy, sustainable building design, ecological economics, indigenous wisdom and technologies, population and development…". Response ranges went from 0 (don't know) 1 (none) 2 (a little) 3 (quite a bit) 4 (a great deal).

The desk-based study which forms the focus of this paper, is a first attempt by the authors, informed by the literature, to establish a practical methodology to engage with these emerging indicators for sustainable development research. The aim of this desk-based study was to investigate the extent to which sustainable development research was already being carried out across a large university, and whether it was possible to devise a relatively quick and reliable methodology to identify the scope and areas of research being undertaken, which would provide the university with a baseline of existing sustainable development research. The object was to capture and report the existing contributions to sustainable development research and to make an initial assessment of its current impact and contribution towards research excellence at the university. This as an outcome would enable further and meaningful engagement with the researchers identified. It was also the intention that the methodology could be replicable by other universities.

4 The Research Context at Kingston University

Along similar lines to many universities in the UK, the Kingston University Research Strategy 2015–2020 aims to develop and enhance a collaborative, outward-looking research culture across the university's academic disciplines and centres of research excellence. It recognises the role of high quality research as critical to the university's

> …distinctiveness, success and sustainability in an increasingly competitive Higher Education environment. (Kingston University 2015a).

Note that the term 'sustainability' here is used to indicate long-term viability, exemplifying the contested nature of the term and the need for awareness raising on the scope of definitions for sustainable development, discussed and clarified in relation to the empirical research later. Among the key aims of the university's Research Strategy 2015–2020 there is an emphasis on research which makes a difference in the real world, exemplified by the commitment to:

> …enhance the reputation and impact of our research nationally and internationally and maximise the synergies between our research and enterprise activities…. (Kingston University 2015a).

Sustainability has been a key focus of activity at Kingston University since 2002, championed by academics, making incremental improvements to operational areas of work and benefiting from a Centre of Excellence for Teaching and Learning (CETL) grant from 2006 to 2009 which saw the launch of several sustainability related courses, and attracting research grants and funding, as noted by Taylor (2013).

In common with the majority of universities in the UK, sustainability forms an important focus of the university's overall vision and strategy 'Led by Learning' (Kingston University 2012), with three of the objectives being directly associated with its delivery:

> 2.6 - We will demonstrate the economic, social and cultural impact of our research and how it benefits individuals, the community and the environment.
>
> 3.2 - We will act ethically to minimise our impact on the environment; we will include issues relating to sustainability and ethics in the curriculum.
>
> 3.7 - We will operate efficiently, optimise our resources (including diversifying our income sources) and ensure the future viability of the university.

However, a key difference for Kingston University, is that whilst many universities have made commitments to embedding sustainability in the operational aspects of their work, it is a relatively recent phenomenon to specify so clearly the embedding of sustainability and ethics in the core business of the university, that is in the research (as per clause 2.6) and education (as per clause 3.2) activities. Whilst there is an existing body of sustainability indicators to measure progress on operational issues, indicators to measure progress on sustainability in education and research are only just emerging and this study responds to a research need to benchmark existing data to measure sustainable development research.

5 Background to Kingston University, London

Kingston University, London was originally founded as Kingston Technical Institute in 1899, and later established as a Polytechnic in 1970, gaining university status in 1992 (Gibson 2001). It can be regarded as typical of a large multi-faculty metropolitan university in the UK, with a student population of (16,092 undergraduate/foundation and 3,826 postgraduate students), and 2,040 staff

(Kingston University 2015b). The university has 27 schools across five faculties: Art, Design & Architecture; Arts and Social Sciences; Business and Law; Science, Engineering and Computing; and Health, Social Care and Education which is a joint Faculty with St George's London. (Kingston University 2016a).

To give an indication of Kingston University in the global and UK context, the *Times Higher Education World University Rankings 2016* ranked Kingston University in the top 200 most outward-looking institutions (Bothwell 2016). The *Complete University Guide University League Table 2016* ranked Kingston University 104 of 126 UK universities listed (CUG 2016), and while, although Kingston University chose to opt out due to concerns with methodology, the *2015 People and Planet Green League*, an independent assessment of the environmental and ethical performance of every UK university, ranked Kingston University 108 of the 151 UK universities listed[1] (People and Planet 2015).

6 Research Question and Methods

The research question is to assess the extent to which sustainable development research is already being carried out across the university, based on publicly available web-based information. The research method was therefore chosen to enable the analysis and interpretation of a large quantity of written text material (described below) and to quickly provide a benchmark that could be replicated in future years and by other institutions. A combination of Content Analysis (Krippendorff 2013) and Thematic Analysis (Patton 2002) was employed to investigate the themes, patterns and concepts emerging from the data. The focus of the analysis was not on specific research projects, but rather on the areas of research work and interest, and the discourse of the specific researchers, in order to identify those academic staff who were currently researching potentially high-profile issues related to sustainable development. The aim was to identify key research activity, researchers, and to assess research and researcher profile and impact in relation to sustainability content, research impact and knowledge transfer viability. Together these were referred to as 'external viability factors' and are described below. The analysis of sustainability content was defined through the use of key words associated with sustainable development at two levels: the international level and the institutional level, also described below.

The web-based content provided in the researcher profiles included: research interest(s); teaching; memberships; research outputs; recent publications; and external activities. Each researcher profile was rated, firstly against the three external key viability factors as either "1" for having or "0" for not having evidence of covering each factor; and then for evidence of 'Sustainability' key words as either "5" for having strong evidence of covering each of these themes to "0" for

[1]Kingston University did not participate in the Green League in 2014 and the results were based on publicly available information on the website.

having no evidence. In addition, information was also collected on a number of factors including the use of key words and their context, the names of Networks and Academic and Professional Memberships, significant aspects of the stated Study Areas, Commentaries and Relevant Publications. This additional information was collected to also assist with support and future dialogue with the researchers, which forms a follow up research study.

To test and validate the replicability, the method of identification and evaluation was tested on a small random sample of academics known for their participation in sustainable development research from both the Sustainability Knowledge Alliance (SKA), which is an independent network of researchers offering expertise and evidence-based advice on sustainability issues. SKA (2016), and the Centre for Climate Change Economics and Policy (CCCEP), which is a grouping of "some of the world's leading researchers on climate change economics and policy, from many different disciplines." (CCCEP 2016).

7 Data Sources

The study focused on the information published through the Kingston University website summarised in Table 1. The *Find a Researcher* directory included research subject areas, researchers, and their research interests. It listed the interests of research staff under Subject Areas (Kingston University 2016b), ranging from "Abandoned Spaces" and "Abolitionism" to "Youth Violence" and "Zoonosis", and covered the research interests of over 465 academic research staff, covering 4,136 "Subject Area" titles.

In addition, data was gathered from the university's eight Centres of Research Excellence and seven postgraduate courses relating to sustainable development (often informed by research) and shown in Table 2.

8 Data Analysis—External Viability Factors and Sustainability Key Words

Data analysis focused on the degree to which the data showed evidence of a set of key external viability factors derived from the Institutional Sustainability Policy and the aims of the university's Research Strategy 2015–2020. These three key viability factors are: Sustainability Content; Research Impact; and Knowledge Transfer Viability. The definitions were further clarified with the use of external guidance documents. The Sustainability Content was defined via the identification of key words from the latest internationally agreed agenda (Table 3) and the Kingston University Sustainability Policy (Table 4), summarised as: Economic; Social; Environmental; Global and Future Equity. Research Impact was defined as research that has "an effect on, change or benefit to the economy, society, culture, public policy or services,

Table 1 Sources of documents and texts

Description	Documents and text sources
Kingston University Strategies and Policies (Main university website)	• Kingston University's vision and strategy: Led by Learning • Kingston University Sustainability Policy, Institutional Sustainability Goals • Kingston University Research strategy 2015–2020
Faculty Websites (Websites of the five university faculties)	• Faculty Background, News and Events • Undergraduate courses • Postgraduate courses • Research • 8 Centres of Excellence
Find a Researcher (Website Directory)	• Researchers and Research Subject Areas
Post-Graduate Courses	• 7 Sustainable Development related (MA and MSc) Course Materials

Source Authors

Table 2 Faculties, Centres of Research Excellence and post-graduate courses relating to sustainable development

Faculty	Centres of Excellence (focus of the research)	Sustainable Development related Post-Graduate Courses (often research informed)
Art, Design & Architecture (FADA)	Contemporary Art Research Centre (CARC)	Sustainable Design (MA)
		Sustainable Building Design and Performance (MSc)
Arts and Social Sciences (FASS)	Modern Interiors Research Centre (MIRC)	
	Visual and Material Culture Research Centre (VMCRC)	
Business and Law (FBL)	Small Business Research Centre (SBRC)	Sustainable Environmental Development with Management Studies (MSc)
		Environmental Law and Sustainability Masters (LLM)
Health, Social Care and Education (FHSCE)	Centre for Health and Social Care Research (CHSCR), Centre for Research in Modern European Philosophy (CRMEP)	
Science, Engineering and Computing (SEC)	Digital Information Research Centre (DIRC)	Sustainability & Environmental Change Masters (MSc)
	Interdisciplinary Hub for the Study of Health and Age-related conditions (IhSHA)	Environmental & Earth Resource Management Masters (MSc)
	Centre for Engineering, Environment and Society Research (CEESR)	Design & Construction Management with Sustainability (MSc)

Source Authors

health, the environment or quality of life, beyond academia" (HEFCE 2014b). Knowledge Transfer encompassed a very broad range of activities to support mutually beneficial collaborations between universities, businesses and the public sector, often linking to the Enterprise agenda.

Table 3 Keywords derived from the UN Sustainable Development Goals

Goal 1—Poverty	**Goal 10—Inequality**
Poverty	Reduce inequality
Goal 2—Food	Inequality
Hunger	**Goal 11—Habitation**
Food security	Inclusive human settlements
Nutrition	Inclusive cities
Sustainable agriculture	Cities
Goal 3—Health	Human settlements
Healthy lives	**Goal 12—Consumption**
Well-being	Sustainable consumption
All ages—elderly	Consumption
Goal 4—Education	Production patterns
Equitable education	Production patterns
Inclusive education	**Goal 13—Climate**
Opportunities for all	Climate change
Goal 5—Women	**Goal 14—Marine-ecosystems**
Gender equality	Conserve oceans
Empower women	Sustainably oceans
Women	Oceans
Girls	Marine
Goal 6—Water	Seas
Water	**Goal 15—Ecosystems**
Sanitation	Terrestrial ecosystems
Goal 7—Energy	Ecosystems
Affordable energy	Manage forests
Reliable energy	Desertification
Sustainable energy	Land degradation
Energy	Land
Goal 8—Economy	Biodiversity
Sustainable economic growth	**Goal 16—Institutions**
Sustainable growth	Peaceful societies
Economic growth	Inclusive societies
Productive employment	Access to justice
Employment	Justice
Decent work	Inclusive institutions
Work	Accountable institutions

(continued)

Table 3 (continued)

Goal 9—Infrastructure	**Goal 17—Sustainability**
Resilient infrastructure	Global Partnership for Sustainable Development
Infrastructure	
Sustainable industrialization	
Industrialization	
Foster innovation	
Innovation	

Summarised from The United Nations, General Assembly, "*Transforming Our World: The 2030 Agenda for Sustainable Development*" 17 Sustainable Development Goals (United Nations 2015)

Table 4 Keywords derived from the Kingston University Sustainability Policy, Institutional Sustainability Goals

Sustainable development	Ethical
Sustainability	Ethics
Environment	Global
Environmental	Equity
Economic	Future
Social	Resources
Community	Long-term impact

Based on KU Sustainability Policy, Institutional Sustainability Goals (Kingston University 2013)

9 Data Analysis

The research method involved a series of analytical stages as follows:

Stage One: In the first stage of the research, subject area titles were analysed to identify those focusing on issues relating to sustainable development, using a key words search (Tables 3 and 4). From this stage of the analysis the total number of 4,136 Subject Areas and 465 researchers was reduced to 321 titles, with 159 researchers listed as involved in research in these areas. These 159 researchers represented a core group who could be expected to be interested in, and carrying out research work in the field of sustainable development.

Stage Two: The second stage involved the analysis of the full research profiles of each of the 159 research staff identified. The *Find a Researcher* area of the Kingston University website provides a "Researcher Profile" for each of the university academic research staff. The Researcher Profile provides the following information for each member of staff: Name and Contact Details; Biography, Educational and Professional Qualifications; Expertise; Research Interest(s); Teaching; Memberships; Research Outputs and Recent Publications (Articles, Book Section, Conferences or Workshop Items, Monographs); and

External Activities (Kingston University 2016a). The researcher profile of each research staff was evaluated to identify the degree to which the profile showed evidence of key external viability factors.

10 Analysis of Findings

The aim of this research was to identify from the full directory of 465 academic research staff, those who would be the most appropriate to build a profile of the current research on sustainable development being undertaken at the university and at a later stage of the research, to be interviewed.

From the 465 researchers working in the over 4,100 subject areas analysed it was possible to identify 159 researchers with research work relating to sustainable development issues. From closer content and thematic analysis, it was possible to group these researchers into three categories: those with 'high profile' sustainable development research (SDR); those with 'potential' to develop a high profile; and those whose work had links to sustainable development research (SDR). Table 5 shows the range of scores by university faculty.

It can be seen that the current researchers work on sustainable development with a 'high profile', and those with 'potential' to develop a high profile, tends to focus in the Faculties of Art, Design and Architecture with 20 researchers, and Science Engineering and Computing with 21 researchers, reflecting a concentration on research covering issues related to the physical environment. The Faculty of Business and Law has 25 'potential' and 'high profile' researchers, reflecting research covering economics, governance and ethics. From the analysis the research identified a cohort of 54 academic staff from across each of the five faculties, who are currently conducting 'high profile' research on sustainable development.

The table also shows the high number of research projects with links to sustainable development across the differing faculties, offering the possibility to develop cross-faculty and cross-disciplinary collaborative research opportunities. The analysis found few examples of work related to the areas of health and well-being within the researchers analysed. Table 6 shows a sample of the breadth of research subjects covered by researchers across the five faculties.

11 Limitations of the Research

The main limitations of the research methodology are two-fold: the subjectivity of interpretations of keywords and their application to the data sources; and the reliance on the availability, accuracy and framing of web-based information which varied between faculties in a single institution and excluded new research staff and new research projects and programmes which have since been posted to the website.

Table 5 Ratings of researchers by university faculty

Faculty	'High Profile' SDR	'Potential' SDR	'Links to SDR'	Total	
FADA	17	3	7	27	17.0 %
FASS	6	0	26	32	20.1 %
FBL	10	15	24	49	30.8 %
FHSCE	1	0	10	11	6.9 %
FSEC	20	1	19	40	25.2 %
Total	54	19	86	159	100.0 %
	34.0 %	11.9 %	54.1 %	100.0 %	

Source Authors

Table 6 A sample of research subjects relating to sustainable development

Biodiversity	Ethical	Nutrition
Cities	Ethics	Production
Climate change	Future	Resource
Consumption	Global	Society
Economics	Inequality	Sustainability
Economic growth	Infrastructure	The elderly
Employment	Innovation	Water
Energy	Justice	Well-being
Environmental	Land	Women
Equity	Marine	Work

Source Authors

Although the method was strongly based on the 'reading and interpretation' of the persons carrying out the analysis, the object of the study was to achieve a 'workable' rather than a 'definitive' benchmark, with the aim of identifying research and researchers to work with in future. In addition, the subjectivity of the analytical method can in part be mitigated in the next stage of the study by the principle of 'snowballing', that is finding the best access to data which can also inform who else in the university should be included in the data sample and analysis, with the aim of directing and signposting the research to as many current sustainable development researchers, research activities and initiatives as possible. The 'snowballing' approach to data collection is intended to be inclusive and offers the opportunity to learn about the university from the university staff themselves.

The data set and data collection process was based on publications available through the university websites. Therefore, the data is dependent on how well these websites and the associated researcher profiles have been updated with new information and publications. There were some instances of error message links between the main website and the faculty websites which meant that the profiles of some researchers were not available. In addition, the format of the university faculty websites and associated researcher profiles differ, leading to less detail being available for some researchers, this is particularly the case with the faculty of

Health, Social Care and Education which has a shared website with St George's University of London, and which may account for the apparent relative lack of researchers from this faculty featuring in the analysis. However, the research offers the possibility of identifying opportunities for the university to improve its communications with the public through various media, and to frame communications to target differing audiences who have differing awareness and values aligned with the sustainable development goals.

In terms of replicability, the limitations of time and resources available to undertake the benchmarking review are key considerations. The method adopted was based on the 'rapid assessment' of a large quantity of data and evidence to produce an initial benchmark, therefore a second stage study will consider the usefulness of capturing full details through a qualitative process with the key stakeholders identified.

12 Conclusions and Recommendations

The study identified the extent of research on sustainable development currently being carried out by the university, and demonstrated that many university educators are already contributing to this agenda, albeit under different discipline areas, offering the possibility to develop cross-faculty and cross-disciplinary collaborative research opportunities and improve communications and monitoring in this respect. The study has provided the sector with a quick and reliable, replicable methodology to establish a baseline of existing sustainable development research and to make an initial assessment of impact.

The study findings demonstrated that existing research which has potential to make a significant contribution to sustainable development can be either understated or categorised under other discipline areas. Whilst further research is required to clarify this finding, it appears to be because of a range of factors including: the relative immaturity of sustainable development as a recognised research area; the emerging methods to monitor engagement in this area; the many areas covered by the term sustainable development; and the sometimes interdisciplinary or cross-faculty nature of some research projects. The result appears to be a missed opportunity in such research being able to contribute to the development of the international sustainable development research agenda.

The next stages for this research include the sharing of initial findings from the desk based study with those researchers it identified, with the research office and support officers and with the communications teams in each faculty, with a request for feedback. The study outcomes potentially provide useful information for the Kingston University Research Support Office to: stimulate internal debate about cross-faculty working; and potentially, applications for grants; formation of specialist research groups; and support for the implementation of the University's research strategy more broadly.

An interview process will be developed which will involve qualitative interviews designed to allow the identification of potential synergies and networks both internally and externally and gather input from key stakeholders themselves on how they choose to communicate about their research and how engagement with sustainable development research can be measured. This will involve those researchers identified as having 'high profile' sustainable development research; those with 'potential' to develop a high profile, and those with links to sustainable development, as they all offer the possibility to develop cross-faculty and cross-disciplinary collaborative research opportunities, but may need different support. A further element of the next stage research will extend to the identification of research students working in sustainability related areas.

The current study supports the long-term aims of raising the profile of sustainable development research at the university, and the wider goal of embedding ethics and sustainability in research across the university. It is a first attempt to benchmark both the extent of sustainable development research and the profile of that research across disciplines.

References

Association of University Leaders for a Sustainable Future (AULSF). (1990). *The talloires declaration*. Available at: http://www.ulsf.org/programs_saq.html. Accessed April 16, 2016.

Association of University Leaders for a Sustainable Future (AULSF). (2009). *Sustainability assessment questionnaire* 2009. Available at: http://www.ulsf.org/programs_saq.html. Accessed April 16, 2016.

Bothwell, E. (2016). The world's most international universities 2016. *Times Higher Education*, January 14, 2016. Available at: https://www.timeshighereducation.com/world-university-rankings/kingston-university?ranking-dataset=133819. Accessed April 16, 2016.

Bullock, G., & Wilder, N. (2016). The comprehensiveness of competing higher education sustainability assessments. *International Journal of Sustainability in Higher Education, 17*(3), 282–304.

Centre for Climate Change Economics and Policy (CCCEP). (2016). *Centre for Climate Change Economics and Policy: About us*. Available at: http://www.cccep.ac.uk/about-us/. Accessed April 27, 2016.

Complete University Guide (CUG). (2016). *The Complete University Guide*. University League Table 2016. Available at: http://www.thecompleteuniversityguide.co.uk/league-tables/rankings . Accessed April 16, 2016.

Dawe, G., Jucker, R., & Martin, S. (2005). *Sustainable development in higher education: Current practice and future developments*. The Higher Education Academy.

Engineering and Physical Sciences Research Council (EPSRC). (2016). *Urban living partnership —pilot phase*. Available at: https://www.epsrc.ac.uk/funding/calls/ulppilotphase/. Accessed April 29, 2016.

Gibson, M. (2001). *A history of Kingston University*. Kingston: Kingston University Press.

Higher Education Funding Council for England (HEFCE). (2009). *Sustainable development in higher education—2008 update to strategic statement and action plan*. Available at www.hefce.ac.uk. Accessed April 27, 2016.

Higher Education Funding Council for England (HEFCE). (2014a). *HEFCE's sustainable development framework—Policy guide*. Available at: http://www.hefce.ac.uk/workprovide/Framework/. Accessed April 16, 2016.

Higher Education Funding Council for England (HEFCE). (2014b). *REF impact—Policy guide*. Available at: http://www.hefce.ac.uk/rsrch/REFimpact/. Accessed April 16, 2016.

Kingston University. (2012). *Kingston University's vision and strategy: Led by Learning*. Available at: http://www.kingston.ac.uk/aboutkingstonuniversity/howtheuniversityworks/universityplan/. Accessed April 28, 2016.

Kingston University. (2013). *Kingston University sustainability policy, institutional sustainability goals*. Available at: http://www.kingston.ac.uk/sustainability/governance/. Accessed April 16, 2016.

Kingston University. (2015a). *Research strategy 2015–2020*. Available at: http://www.kingston.ac.uk/research/about-research-at-kingston/research-strategy/. Accessed April 21, 2016.

Kingston University. (2015b). *Facts and figures: What you'll find at the University, Kingston University*, December 1, 2015. Available at: http://www.kingston.ac.uk/aboutkingstonuniversity/factsandfigures/. Accessed April 16, 2016.

Kingston University. (2016a). *Faculties and schools at Kingston University*. Available at: http://www.kingston.ac.uk/faculties/. Accessed April 16, 2016.

Kingston University. (2016b). *Find a researcher*. Available at: http://www.kingston.ac.uk/research/research-interests-and-expertise/research-interests-directory/. Accessed April 16, 2016.

Kordestani, A., Peighambari, K., & Foster, T. (2015). Emerging trends in sustainability research: A look back as we begin to look forward. *International Journal of Environment and Sustainable Development, 14*(2), 154–169.

Krippendorff, K. (2013). *Content analysis: An introduction to its methodology* (3rd ed.). California: SAGE Publications Inc.

Lang, D. J., Wiek, A., Bergmann, M., Stauffacher, M., Martens, P., Moll, P., Swilling, M., & Thomas, C.J. (2012). Transdisciplinary research in sustainability science: Practice, principles, and challenges. *Sustainability Science, 7*(Supplement 1), 25–43.

Leal Fihlo, W., & Davim, P. (Eds.). (2015). *Challenges in higher education for sustainability*. Berlin: Springer.

Lozano, R. (2006). A tool for a Graphical Assessment of Sustainability in Universities (GASU). *Journal of Cleaner Production, 14*, 963–972.

Orr, D. W. (2004). *Earth in mind: On education, environment, and the human prospect* (Tenth Anniversary Edition). Washington. Island Press.

Patton, M. Q. (2002). *Qualitative research & evaluation methods* (3rd ed.). London: Sage Publications.

People and Planet. (2015). *2015 People & Planet's University League Table*. Available at: http://www.peopleandplanet.org/university-league/2015/tables. Accessed April 16, 2016.

Research Council UK. (2014). *Research Council UK Submission to House of Commons Environmental Audit Committee Inquiry into Sustainable Development Goals (SDGs)*, December 2014. Available at: www.rcuk.ac.uk/documents/submissions/eacsustainabledevelopmentgoalsrcukresponse-pdf. Accessed April 21, 2016.

Research Excellence Framework (REF). (2014). *Sector impact assessment of the REF*. Available at: http://www.ref.ac.uk/about/background/. Accessed April 21, 2016.

Stern, N. (2007). *The economics of climate change: The Stern review*. Cambridge: Cambridge University Press.

Sustainability Knowledge Alliance (SKA). (2016). *Sustainability knowledge alliance: About us*. Available at: http://www.exeter.ac.uk/ska/about/. Accessed April 27, 2016.

Sterling, S., Maxey, L., & Luna, H. (Eds.). (2013). *The sustainable university—Progress and prospects*. Abingdon, Oxon: Routledge.

Taylor, R. (2013). Bottoms up for sustainability: The Kingston experience. In S. Sterling, L. Maxey, & H. Luna (Eds.), *The sustainable university—Progress and prospects*. Abingdon, Oxon: Routledge.

United Nations Environment Programme. (1977). *The Tblisi declaration*. Available from http://www.gdrc.org/uem/ee/tbilisi.html. Accessed April 28, 2016.

United Nations Framework Convention on Climate Change (UNFCCC). (1997). *The Kyoto protocol.* Available from http://unfccc.int/kyoto_protocol/items/2830.php. Accessed April 28, 2016.

United Nations, General Assembly. (2015). Transforming our world: The 2030 agenda for sustainable development, A/RES/70/1 (25 September 2015). Available from undocs.org/A/RES/70/1.

Universities that Count (UTC). (2010). *Universities that Count Annual Report 2009/10: Transforming environmental and social performance in universities*. Available at: http://www.eauc.org.uk/search?q=UTC+Annual+Report+2009-10. Accessed April 28, 2016.

White, R. (2013). Sustainability research: A novel mode of knowledge generation to explore alternative ways for people and planet. In S. Sterling, L. Maxey, & H. Luna (Eds.), *The sustainable university—Progress and prospects*. Abingdon, Oxon: Routledge.

Winter, J., Cotton, D., Grant, V., & Hopkinson, P. (2015). The university as a site for transformation around sustainability. *International Journal of Innovation and Sustainable Development, 9*(3/4), 303–320.

Wright, T. S. A. (2002). Definitions and frameworks for environmental sustainability in higher education. *International Journal of Sustainability in Higher Education*, *3*(3), 203–220.

Education for Sustainable Development and the Eco-school Initiative in Two Primary and Two Early Years Settings in the North East of England

A. Chatzifotiou and K. Tait

Abstract

Eco-school is an international initiative that offers schools the opportunity to develop practices on education for sustainable development (EfSD). Such practices need to focus on nine areas, namely: energy, water, biodiversity, school grounds, healthy living, transport, litter, waste and global citizenship. Acquiring the green flag status is the ultimate stage (silver and bronze are the other two) that is awarded by a committee external to the school and it lasts for two years. Our project focused on two such primary schools and early years settings that had acquired the green flag status. The project aimed to describe how teachers perceive sustainability through the eco-school agenda. We focused on the settings' approach of becoming an eco-school and the practitioners' role in promoting the values and principles of such endeavours. Sustainability is a term mentioned in the eco-school literature in a number of different instances. Thus, we chose eco-schools because this gave a straightforward way to identify a setting with an interest in EfSD. Our interest in this project and the conscious choice we made not to use explicitly the term sustainability to invite the settings to our project are due to other scholars' work in the field such as Green and Somerville (Environ Educ Res, 2014), Davies (Environ Educ Res 15(2):227–241, 2009), Gayford (Can J Environ Educ 8:129–142, 2003) who have highlighted issues that teachers and early years practitioners face when it comes to EfSD (e.g. lack of confidence, skills, knowledge, etc.). This is a qualitative project that used a multiple case study design to focus on the practices of four educational settings to gain a green flag status. A semi-structured interview was

A. Chatzifotiou (✉) · K. Tait
Department of Social Sciences, Faculty of Education and Society,
Sunderland University, 6 Green Terrace, Sunderland SR1 3PZ, UK
e-mail: athanasia.chatzifotiou@sunderland.ac.uk

K. Tait
e-mail: Karen.tait@sunderland.ac.uk

W. Leal Filho (ed.), *Sustainable Development Research at Universities in the United Kingdom*, World Sustainability Series,
DOI 10.1007/978-3-319-47883-8_4

used with the lead teachers/practitioners of the schools; an audit was also conducted as part of a tour of the settings' premises highlighting the initiatives, curriculum work, projects undertaken as well as resources available to school. Interview transcripts were analysed with the creation of response categories by the two researchers first working individually and then collaboratively; the findings of the project reflected issues that concerned: (a) pupils' cognitive, physical and socio-emotional development, (b) the wider community and (c) the lead practitioners' role and status in school. In relation to EfSD, our findings indicated that its impact upon these settings was rather minimal; a mismatch was identified between the eco-school practices and a holistic understanding of issues that EfSD aims to achieve. This mismatch between eco-school practices and EfSD is discussed with regard to: (a) pupils' understanding of the sustainability dimension in the topics they addressed; and (b) teachers' knowledge of sustainability and willingness to keep on such work in schools.

Keywords
Eco-schools · Environment · Primary schools · Early years · Sustainability education

1 Introduction: Education for Sustainable Development and Eco-schools

Education for sustainable development (EfSD) is 20 years old; over the last two decades it has gained a prominent status in the international and national literature linking environmental, social and economic dimensions. In England, EfSD has been included in the different versions of the National Curriculum thus far; internationally, it has started to appear as a dimension that needs to be included in early years settings too (Arlemalm-Hasger and Sandberg 2011; Davies 2009; Cutter-Mackenzie and Edwards 2013; Reynamo and Suomela 2013). Although EfSD is put forward in rhetoric, in reality things maybe different; for instance a UNESCO brief policy report (2013) discussing EfSD in the UK context highlighted that the Teaching Agency in England which is responsible for the curriculum for school teachers does not have any interest in sustainability; while studies such as that of Barrett (2007) showed that practices with a focus on environmental education activities reflect mainly individual teachers' interests. Here lies one of the current project's interests in the role of the lead teacher/practitioner in the implementation of the eco-school initiative and consequently the implementation of EfSD.

Schools address EfSD in different ways (e.g. project work, topics, cross-curricular approaches, etc.). 'Eco-schools' is an initiative that reflects one such approach towards achieving practices relevant to EfSD. The eco-schools initiative is an international initiative that is managed in Britain by the organisation 'Keep

Britain Tidy'. This initiative aims to inspire and help schools achieve different levels of environmental and sustainable practices. There are three such levels—namely, silver, bronze and green flag status—which schools can achieve based on different actions. Schools that decide to become an eco-school need to follow a number of steps where they have to register, form an eco-committee, conduct an environmental review and draft an action plan. The environmental review needs to address environmental topics identified by the Eco-school agenda. There are nine such environmental topics and depending on the kind of flag a school aims to apply for, they need to review either all nine or a number of these at different stages. These topics include: energy, water, biodiversity, school grounds, healthy living, transport, litter, waste and global citizenship.

These nine topics resemble the eight 'doorways' of the National Framework for sustainable schools in England (DCSF 2009). These 'doorways' include: Food and drink, Energy and water, Travel and traffic, Purchasing and waste, Buildings and grounds, Inclusion and participation, local well-being and global dimension. The overlap between the eco-school themes and the national framework 'doorways' is evident as similar themes and terminology is used in both cases; This parallel between the eco-school initiative and the National Framework for sustainable schools has been drawn because the current project—while focusing on the eco-school agenda—aims to investigate sustainable education practices in these settings. We chose to do so via the eco schools initiative because it gave a straightforward way to identify settings with an interest in EfSD. This interest in sustainability was surmised based on information provided by the eco-school initiative; for instance, in the eco-school webpage (http://www.eco-schools.org.uk/aboutecoschools/theprogramme) we read that the eco-school initiative aims to: "… *guide schools on their sustainable journey…*", that it provides "*…a simple framework to help make sustainability an integral part of school life.*", and its mission is to "*… help make every school in the country sustainable…*". The reason we did not explicitly use the term sustainability when we approached the settings was due to a concern that practitioners might have declined to talk to us. Scholars in the field such as Green and Somerville (2014), Davies (2009) and Gayford (2003) highlighted barriers to EfSD that relate to practitioners' training, knowledge, skills, etc.

Furthermore, this interest in EfSD via the eco-school initiative may provide a limited context within which EfSD is viewed. Scott (2013, p. 185) argued that: "*… the fragmented view of sustainability which eco-schools present, the way that success is possible without the whole-hearted involvement of the entire school, along with the relative ease with which such flags are obtained, mean that this will not, in and of itself, suffice. Neither will any of the increasing number of awards that are readily available for UK schools to collect.*" Scott raises here an interesting point that relates to the way that initiatives/policies attempt to 'attract' schools' and teachers' interests in embracing particular educational aspects, EfSD in this instance.

2 Methodology and Conceptual Framework

This is a qualitative study that used semi-structured interviews to discuss with particular practitioner(s) the issues concerning their schools' interest and practices in the eco-school, within a multiple case study context. This strategy is interested in examining a phenomenon within a real-life context (Demetriou 2013) aiming to describe the case(s) as accurately as possible; our case studies were instrumental/exploratory case studies (Stake 1998) where a case is examined to clarify further an issue; our case studies helped to gain insights as to how practitioner(s)' work has contributed to changes in a setting, in this instance pertaining to an eco-school and EfSD. The focus in all the settings we visited (two primary schools and two early years settings) were on the practitioner(s) who have started and developed the eco-school initiative. There seems to be a dearth of studies that focus on teachers as McNaughton (2012) has highlighted that the 'voices' of teachers who develop and implement EfSD topics have not been heard as much in the literature. Thus, this study seeks to describe and explain why/how practitioners/teachers become interested and involved in initiatives relevant to environmental and sustainability issues.

The main method used to collect data was semi-structured interviews; the researchers were also shown around the school premises where the practitioners/teachers demonstrated the different resources they had acquired for their setting in relation to the eco-school status. All settings chosen are located in areas within the north east of England (referred to as School A and B, early years setting A and B onwards).

School A is in an underprivileged area while School B is in an affluent area of the North East of England. In School A the lead practitioner was a high level teaching assistant known to one of the researchers; six years before she had studied for a foundation degree in the university where the researcher worked. In School B we talked to a qualified primary school teacher. Even though, our participants had different teaching qualifications, they were the 'lead person' in their schools when it came to the eco-school initiative. Similarly, the Early years setting A was a private day care provision situated on a city centre university campus, newly constructed and purpose built with energy efficiency features (e.g. living roof). The lead practitioner who talked to us was one of the managerial staff. The Early Years setting B was a local authority, community nursery school situated in an underprivileged area. The lead practitioner who talked to us was the head-teacher.

The transcribed interviews were read through thoroughly by the two researchers separately; we followed an inductive approach where the focus was on the content the lead practitioners wanted to communicate. A descriptive/narrative analysis of these transcripts (guided by the data, hence exploratory) led to three emergent topics, namely (a) how the setting became interested in the initiative and the lead practitioner/teacher's role in that, (b) the impact upon pupils and (c) the impact upon the wider community. The narratives with the emergent topics from each interview were compared between the two researchers and there were no significant differences. Finally, one narrative for each setting was produced which described and further contextualized the above three topics.

3 Results and Analysis

The results from the two schools will be presented together and then similarly the results from the early years settings. The results will be presented under the three emergent topics mentioned already.

4 Schools A and B

4.1 Topic 1: *How the Schools Became Interested in the Eco-school Initiative—The Lead Practitioner/Teacher's Role*

In school A the participant was a teaching assistant and her starting point emanated from a realization that the grounds of the school were not good enough for pupils' learning. She felt competent and autonomous to look for funding in order to start buying equipment and changing the schoolyard (e.g. nature garden, something to climb on, a picnic table). She was introduced to the eco-school initiative and she started pursuing its agenda. In school B the participant was a qualified teacher, a Key stage 1 teacher. The teacher and school became interested in the eco-school initiative via an email or letter that alerted the school about the initiative. The teacher clearly explained that she took the eco-school project under her supervision without though clarifying that she did so because of any environmental or other particular interests.

In school A even though parental involvement was rather passive (e.g. parents not complaining going outside in mud) and the practitioner found difficult to 'recruit' parents and governors to be part of the eco-ambassadors' team and participate in the meetings, her vision and determination was not diminished. She talked in the first person highlighting her dynamic role in all this.

In school B, the teacher did not remember how the school became involved in the initiative; she believed they had received an email or letter alerting them to the initiative. In terms of support from the rest of the school, the teacher highlighted that their school's ethos is such that if people are asked to help in something, they usually and readily become available—this happened with the eco-school agenda and other staff members (teachers, head-teacher, dinner-ladies) have been briefed and contributed one way or another. The senior management seemed to be also interested as long as the initiative 'paid back'.

According to self-determination theory (Ryan and Deci 2000) competence, autonomy and relatedness are three necessary elements for intrinsic motivation to occur. In school A, the participant had a sense of competence, she felt autonomous enough within the school to pursue her goals but relatedness that refers to the support given by the wider context was not really present. The principal of the school was not against her but leadership is an important factor for introducing and further developing initiatives that can contribute to effective teaching and learning (Kadji-Beltran et al. 2013). In a Cypriot study (ibid.) it was revealed that there were a number of different constraining factors for successful sustainable schools like: "*Principals' reported lack of confidence in administrative skills for sustainable schools, limited willingness to challenge the status quo and limited frequency of engaging in actions important for supporting ESD [education for sustainable development] activities such encouraging networking with external groups…*" (p. 318). In our participant's case (from School A), she mentioned the principal had been helpful without giving any details of particular actions; however, as the principal's record of detentions indicated fewer detentions since the start of the eco-school activities, his interest in our lead practitioner's work was mostly linked to the management of his school rather than to assisting the practitioner in her endeavours for outdoor learning within the eco-school agenda. Similarly, with school B senior management was on board as long as the endeavour 'paid back'.

4.2 Topic 2: Impact upon Pupils

4.2.1 Emotional Development

In school A the lead practitioner discussed how/why the new school yard premises might have affected pupils' behaviour—she argued that: "*…I've done millions of playground duties and I only ever had one playtime when I thought there hasn't been a little incident happened; there's always something but it is how you can approach it and what you can divert that child to, that is important.*" She makes an interesting point that reflects how inclusion and participation—one of the eight 'doorways' from the National framework for Sustainable schools—can be facilitated with equipment and activities that can become relevant to children's interests and needs. The school's revamped outdoor area—one of the nine eco-school agenda topics 'school grounds'—provided teachers and pupils alike with a context that were able to take advantage of and use it both for learning and for leisure. As a result children felt more comfortable and able to play and work in an environment they enjoyed. The literature also shows how play in natural environments contributes to more diverse and creative play activities for children (Fjortoft and Sageie 2000). The changes that occurred in the school's outside area included: an orchard (where pupils planted trees with their parents), a butterfly garden, two sets of benches with matching seating and a bigger 'wildlife' garden with a bridge and containers with plants and flowers. The physical environment was used both as a resource—education IN the environment—and as an object—education ABOUT the environment according to the distinction made by the environmental education remit.

In school B there were not any statements clearly indicating how the eco-school agenda the school followed actually impacted upon pupils' emotional development as there were with our practitioner in School A. There were mostly on cognitive development.

4.2.2 Cognitive Development-Sustainability Issues

In school A, the practitioner mentioned that the use of the outdoor environment can contribute significantly on teaching academic aspects of the curriculum. During our conversation around the benefits of the outdoors in the teaching of the school subjects, the term 'sustainability' was mentioned (almost half way through our overall conversation); it was mentioned by a colleague of our lead practitioner (a teacher who joined our conversation for a short time) in an instance that she was explaining about the kind of visitors they had at school and the kind of invitations they had received to participate in 'sustainability networks'. At this point, we asked them to talk to us about sustainability in terms of what they think it is, how they understand it and how they take it under consideration when planning the curriculum. They acknowledged that sustainability is what it is all about; Our participant said: "*Well, that's what the nine out of ten are…this is what you are trying to do, promote sustainability within…*". They went on to describe sustainability as something that needs to be meaningful and on-going. Our participant claimed: "*It's also got to be something that you can keep on doing because you can't take the children out for one week and then think that's it. So, you've got to have all your ideas, you've got to feed of how you develop the school grounds and see the opportunities that you can take the learning further out there.*" It seems that both our participant and her colleague thought of sustainability as something on-going and mainly realized in the outdoors. Here are some examples of the sustainability dimensions they mentioned.

They were aiming to cover issues about the rainforest, deforestation, engendered animals, etc. When we tried to probe further and more about the way they plan these activities, our participant's colleague said: "*Because of the type of topics we do and now across the school really, there are such a vast range of topics that are being done across the school, its…is almost not planned and because we are so used using our school grounds, it becomes part of what you do, in the same way…. that you don't say that I have to teach maths…we know that we are going to take children on visits and trips and things…it just happens because we are so used working in that way*". This reply highlights the use of 'hands-on, experiential learning approaches as widespread and successful in their school fitting as well the eco-school agenda. There was nothing more explicit said that related to how particular sustainability dimensions emerged and taught respectively under the topics; the weight seems to fall more on the 'hands-on' pedagogy which is important for pupils' learning but not clearly linked to how it contributes to the teaching of sustainability dimensions.

In our effort to investigate further aspects of sustainability that may be taught, we asked how children, especially older children in Year 6, understand links they make to community, to potential employment and to the wider context of their life. Our participant claimed that all the work they do in the community certainly helps and makes children feel part of this community; also, she commented on children's enthusiasm to become part of the eco-team, she stressed how children ask her almost every day if they can be an eco-ambassador (an essential part of the eco-school agenda).

Another example relates to a project they did, entitled 'Living Streets'; this is a walk to school initiative. This project tied really well with the Travel and Traffic theme (again common theme between eco-schools and eight 'doorways'). Under this project our participant described to us how they updated the school's travel plan. Our participant formed a questionnaire that included questions about how children and adults travel between home and school. The questionnaire was administered by one of her eco-ambassadors—she said: "*…even though I wrote the questionnaires out for her she went around other classes and she gave that questionnaire to teachers, to children….to identify some…sort of issues about how when you walk to school, what you like, what you don't like, had any ideas how to make the journey better…*". When we asked her whether the children understood why is better to come to school on foot or by bike, our participant said that during assembly all children have heard about the importance of the 'walk to school' initiative and if asked, children should be able to "*…give you the healthy answer*"—that is, it is good for one's health to exercise. When we further prompted her with questions about issues like traffic, pollution she did not give a straight answer—she talked about the man from the 'walk to school' initiative and how he contributed to their travel/traffic theme. The sustainability dimensions of this topic seem to have been introduced to pupils via an assembly (lecture-like format) and the prominent issue projected to children was the one that related mostly to them (be healthy) rather than taking into consideration a more holistic approach (humans, environment, pollution, economy, etc.). The term 'sustainability' is presented as the ultimate goal of all the topics and the work they do for the eco-school agenda but we never really got a clear picture of what and how they perceive 'sustainability' to be.

Another interesting issue concerns pupils' active participation which may not be as 'active'; our lead practitioner mentioned that she wrote the questionnaire in the 'walk to school' project which pupils then administered. In another project on the 'green procurement policy' that the school needs to have for the eco-school agenda, the practitioner explained how she wrote again that policy in 'child speak' after she talked about it with the children in a meeting they had. This is not unusual; Katsenou et al. (2013, p. 244) argued that: "*…pupils become involved, either in participatory actions while continuously guided by teachers, or activities planned solely by teachers.*" While it is not within the remit of this project to evaluate the active or not participation of pupils, it becomes relevant to ask whether the practitioner's sense of competence and confidence may have, unwillingly compromised pupils' active participation.

In school B, they work with different topics through the year and for the eco-school initiative the school as an institution seems to have a 'priority' over making sure that all pupils engage with these topics. When asked whether the eco-school related activities are mapped against the curriculum, the teacher said that they do so in retrospect; that is, there is no specific planning and cross-referencing because this is how the school works anyway. Similar to School's A practices as well.

During our conversation over the energy awareness week the school had, we had the opportunity to initiate a discussion around the notion of 'sustainability'. We asked how they identify and link the sustainability dimensions of the topics they do, energy in this instance. The teacher said: "*Sustainability was a bit harder, that was all about energy*". That is, she differentiated between sustainability and energy as two different topics with sustainability being the harder to do. The teacher was aware of the term 'sustainability' as an element that was mentioned in the eco-school website; she clearly stated that there were a number of things teachers could do in relation to 'sustainability'. She claimed: "*…I asked where sustainability was happening. Each year group had a sheet to fill in, what curriculum area it was and what activity it was that they did.*" The aforementioned quote indicates that sustainability is viewed as an activity/element that takes place in a particular instance rather than as an overall idea that can permeate different activities. Further on, the teacher identified as well the three curriculum areas that these sustainability activities/instances took place, namely: Design and Technology, Science and Art. She did highlight that the 'sustainability ideas' were taken by the eco-school website—they were not devised by the teachers in the school. She actually said that: "*…they were there for ideas if people hadn't achieved sustainability across the year, that they could then 'Oh I'll do that idea' and quickly put that in, so it could be ideas for them to work on.*" At this point we asked how teachers feel about the notion of sustainability (since it was mentioned that they found it hard to implement); the teacher said that they were all on board with it because they want to offer pupils more than just the curriculum.

With the environmental review that the school has to do as part of the eco-school agenda, the teacher very clearly stated that she takes leadership here. They take the questions from the eco-school website, she makes them more child-friendly and then pupils start asking the prescribed questions and along with the help of the teacher they work on an action plan. The teacher's 'presence' is very 'prominent' in most steps of the initiative just like it was in School A. She actually said: "*…I created an action plan and then we shared it with the Eco-friends [the eco-committee], we've shared it in a staff meeting, so the other members of staff could add to it.*"

While we had the chance to see the questionnaires and action plan used for the conservation area topic, we prompted with another question as to how much understanding pupils have about conservation and why it is important. Her reply was that pupils do not question any of these; they accept the eco-school activities as something they have to do as part of their school engagements. She actually said: "*I think they just accept that it's part of learning. If we said...we are going to learn about electricity, they don't say 'why are we learning about electricity? Like we do PE, we do eco-schools.*" Thus, the teacher seems to assign 'sustainability' a status similar to the status of the rest curriculum subjects.

Overall, it seems that activities are mostly focused on educating pupils ABOUT the environment (e.g. recycling paper, learning about the water/energy and how to be sensible consumers, etc.) and IN the environment (e.g. being outside the class, going to the lighthouse, visiting local centres, etc.). Activities that highlight the connections between the choices humans make and the implications these have on the planet have not been readily available. For instance, when the teacher (from School B) talked about the school's use of local produce, she did not highlight how such an approach can be linked to issues of carbon emissions, consumer patterns, etc. Similarly, when the same teacher mentioned pupils' knowledge about recycling, healthy eating, energy conservation, globalisation there was no indication that pupils learn to value something inherent in these activities; they learn and do these things (recycling, cycling to school, etc.) as they learn anything else in school.

4.3 Topic 3: Impact upon the Community

In school A, the participant highlighted in a number of different instances how she turned to the community that the school belonged for help. She had links with a local country park that she took pupils over for activities. Other initiatives she took included contacting a landscape gardener, a landscape architect—all of whom were quite expensive to use but they did offer her ideas about the way she could address the school's outdoor area. When she talked about a local resource they were using with pupils, she referred to a partnership they had created. This resource was a local community centre and at some point they were inviting schools to visit and do all sorts of activities, e.g. gardening. Our participant said that every time they were invited she made sure they went and their latest activity just a week before our discussion, was to plant an orchard which they named 'Whispering trees' as part of a name competition. She explained that part of the eco-school agenda is to explore and create community partnerships (also reflecting the eight 'doorways'). She described a number of activities pupils did in the community centre (e.g. cooking with chefs, talking to the community police, etc.) and highlighted how all these activities not only raised the school's profile and partnerships but also made children feel that they are part of a community. Inclusion and participation (one of the eight 'doorways' of the National framework for sustainable schools) is an important aspect that schools with sustainability interests need to promote.

In school B, the eco-school initiative has certainly given the school opportunities to open up to the community. The teacher mentioned a number of such instances; for instance, the school organized the eco-festival the summer before. Another activity which fostered closer links with the community emerged from Northumbrian Water; they came to school, they gave water saving kits and they talked to pupils about looking after the water and how to take care drains and sewers.

5 Early Years Settings A and B

5.1 Topic 1: *How the Settings Became Interested in the Eco-school Initiative—The Lead Practitioner/Teacher's Role*

In setting A, the lead practitioner had worked at the setting for many years and explained that she began to act upon her personal interest in environmental issues 12 years ago when she introduced recycling to the setting. Over time her interest developed, staff became more involved and this was gradually becoming embedded into nursery routine. She eventually recognised that their practices need to be highlighted. Through her leadership and active modelling, staff became more conscious of their practice. Underpinning the practitioner's development of practice and pursuit of the Eco-school status was her question to staff and children, "*What do you think we could do better to support children and look and sustain, well, sustain life, really*?" This kind of question is at the heart of sustainability education but this was the only instance a term relevant to sustainability was used explicitly within the conversation at this setting.

In setting B, the head teacher talked to us; there was not one 'lead' person like in the other settings. They were always interested in such things namely '*outdoorsy*', '*nature park*', '*nature*', '*do the best for the world*', '*environment and how to look after* it'. The journey to the eco-school started with a litter check in the yard conducted by someone from Keep Britain Tidy. The person who started the initiative is no longer at the setting but their rational to go after the Eco-school initiative reflected: "*…something we could do and it was just going to be really recognition of what we were trying to do already*." So it started as an activity that could further add value to children's experiences.

5.2 Topic 2: Impact on Pupils

5.2.1 Emotional Development

In setting A, the ethos and staff commitment to environment and sustainability education are embedded into everyday routines and practice. The 'voice of the child' and active engagement is thoughtfully promoted by the staff. In considering the development of the outdoor environment the lead practitioner explained, "*…we look at what the children want so… we have the children draw plans*". Through asking their opinions, engaging the children in conversation and looking at their drawings, the children were able to convey what they "*…wanted to see in the garden*." The practitioner explained their practice is to give children responsibility at the "*right level*", to "*make it fun*" by using "*small steps*". Thus, empowering children's agency is one of the main approaches, in this setting, towards the eco-school agenda.

In setting B, there was not much discussion and reference to children's emotional development.

5.2.2 Cognitive Development-Sustainability Issues

In setting A, the lead practitioner highlighted a number of examples explaining how their pedagogical approach and use of resources/materials helps children develop their understanding of environment and sustainability issues. She noted the importance of helping children's imagination to flourish by using reusable materials, books and involving children in their own story creations. She linked children's imagination with helping children to develop their thinking skills; she said "*it's basically all about getting children to use their imagination. I think that's the main thing. Getting them to think about things*." In this way children acquire not only a sense of responsibility for tasks within the setting but also an understanding of the importance of these tasks (switching off lights, water etc.). Children's involvement in an environmental review highlighted the importance of focusing not only on particular learning outcomes, but more crucially, on the process that contributes to their thinking. She said, "*It's the process and getting them to think that... they have to look after and they have to save energy and we have to look after the planet.*"

Staff is conscious to embed these principles and this pedagogy across the curriculum citing examples such as when they collect the recycling from each room, saying "*They find the numbers or they write the numbers and they stick them onto how many bags they've collected*." Using recyclable materials to create junk model dragons for Chinese New Year and using stories such as Loony Little illustrate this application to developing knowledge and understanding of the world. Thus, by the time children leave this nursery they hopefully have emergent scientific understandings of 'change' as they have experienced for instance waste products breaking down and being used in a different way for another purpose.

In setting B, their overall philosophy is to have all different aspects of the eco-school initiative embedded in their everyday life and classroom learning. Through their regular staff meetings, their medium term planning, their whole school assemblies they plan to have everyone involved in the different activities. From a point onwards this holistic approach becomes sort of a 'given' in the sense that it is not easy to talk about their activities separately. She said: "*Its sometimes quite hard within our setting to think "well, this is healthy eating, this is eco-schools, this is early years" because it's actually just all part and parcel of our ethos*." While such a statement does reflect a cross-curricular, holistic pedagogical approach, at the same time it does highlight a feature that may be problematic—where are the distinctions between the eco-school and the healthy eating activity? Such distinctions could have been helpful to identify particular features relevant to bigger issues like sustainability for instance.

'Sustainability' as a term was not discussed in any particular way by the practitioners; the practitioners did not make any reference to the term. We brought it up when we asked how the term 'sustainability' that is found in the eco-school literature is implicitly or explicitly introduced to young pupils. Their response indicates a rather weak and limited understanding of the term. They said: "*…I certainly think the nursery promise about not breaking sticks off trees, and we've got very definite rules in the yard of what things can be picked where and why. That's* *sustainability*...[the practitioner's name] *has put big smiley faces, which you can see for areas that they can pick things for the mud kitchen.*"

Another example where the 'weak' link to sustainability can be shown is when we were discussing the gardening and planting activities; the practitioners talked about the joy that young learners experienced when they dig up the potatoes for instance; they said how they give to pupils information about the digging circumstances (e.g. temperature, etc.) needed for the vegetables to grow but there was no mention about the 'seasonality' or the 'locality' of the vegetables used. Similarly, when discussing composting, the practitioner described the session they had with pupils as one where they shared fruit and put their peels in the bin in the yards without making any other point about the activity.

Overall, the topics they worked on included: recycling, composting, healthy living, packed lunches, bringing to school only water for drinks, school grounds (with mud kitchen, bug holes, mini beasts, etc.) and biodiversity. The topics they found harder to work on because of the age of pupils were energy and water for which they try to do as much and as best they can. These topics they found easier to work on were also topics that were further supported by activities relevant to Forest school. A number of the staff had training in Forest school activities and so they made the best of these by using both the school grounds and a nearby park.

The practice of eco-school activities were mainly adult-led; the practitioners mentioned how for instance, they tried to involve children in the environmental review. They had a list with pictures and they were asking children to identify the things they did in the school grounds; thus children were able to identify that they did recycle paper but no bottles. Such an approach is interesting because it can be challenging to involve pre-literate children in such activities. At this point, the practitioners did highlight that a number of the eco-school activities seem to be geared towards older children and they need to tailor them to their settings needs.

5.3 Topic 3: Impact upon the Community

In setting A, developing parental involvement is a point the lead practitioner and staff have reflected and acted upon. The lead practitioner acknowledged that parents "*... haven't got the time to come and offer the support they would like to...*". But she explained that over time they have grown to make use of mascots, props and story-books used in the setting as a vehicle for informing parents and including parents in the environment and sustainability 'message'. They make effective use of 'Handa' a snail hand puppet, 'Garbage' and 'Scoop', mascots made from recyclables as well as story-books such 'Loony Little' as vehicles to enable the children to talk about what they are learning and doing in the setting. Parental involvement may include collecting recyclables to bring to nursery and some families have begun recycling at home as a result of the children talking about how and why they do it in nursery.

Similarly, in setting B, the practitioners talked about mainly the involvement of parents. Even though they have only one parent in their eco-committee, they have involved more parents in a number of different activities they did like cooking and gardening activities. They described the parents' group as a very lively and dynamic group comprising both local Geordies and people from other nationalities like Iraqis and Iranians who are also very proud of their school having acquired the Green Flag.

6 Discussion and Conclusion

6.1 Practitioners and Settings

6.1.1 Eco-school Is Seen as an 'Add-on' Rather Than a 'Built-in' Activity

All practitioners had different starting points but they all seem to view the eco-school agenda as an 'add-on' rather than a 'built-in' aspect of the curriculum; practitioners in both schools tried to capitalise on the eco-school initiative as something that would bring added value on pupils' learning. Practitioners in both the schools and early years settings had outdoor interests in general and they wanted to give something extra to their pupils. None of the practitioners had a strong, inherent interest in environmental/sustainability issues; they were mostly interested in their pupils' learning and experiences, especially in the outdoors.

Scott (2013) when discussing how sustainable schools can contribute to UK sustainable development, talked about different stages that a sustainable school may go through. He described four stages starting from stage zero to stage four. Within these stages one can see the role that different people/professionals can play within an organization. Drawing a parallel between the eco- and sustainable schools that Scott (2013) described, we can argue that all settings in this project may be found somewhere between stage one and stage two. Stage one "*…is characterized by the work of individuals, with isolated curriculum inputs…school leaders…are reasonable tolerant…*" (Scott 2013, p. 186); while stage two is "*…where the school leadership has accepted the idea that a broad view of sustainability needs to be taken seriously in relation to school's curriculum and supports the opportunities that exist for mutually beneficial links with the local community… providing active leadership…*" (ibid.). Practitioners from all settings had their work acknowledged by the principal and other staff of the school, they had their moral support and support for pursuing further developments but not in a dynamic manner where more concerted efforts could be planned to contribute financially, structurally and educationally both for pupils and the other teachers.

6.2 Sustainability

Knowledge ABOUT the environment (e.g. recycling, planting activities, learning about energy, etc.*) and working IN the environment (e.g. being outside the class,* *visiting local centres,* etc.*) are more prominent features within the topic work approach that settings used than activities that highlight the connections between the choices humans make and the implication these have on planet.*

In terms of what 'sustainability' is, all participants seem to understand 'sustainability' as something that is 'on-going', as the ultimate goal, that takes place mainly in the outdoors, highlighting 'hands-on' approaches; links and references to the National Curriculum subjects or early years learning areas are not necessarily planned out, while links made between society, economy and the environment are rather difficult to detect.

Lead practitioners were not in a position to clearly and explicitly discuss sustainability dimensions in the curriculum. They were able to discuss the topics of the eco-school agenda in relation to the knowledge imparted to pupils (education ABOUT the environment/potential content for sustainable development), in relation to the pedagogies used (hands-on, cross-curricular, integrated approaches—education IN the environment) but less so in relation to values and principles that should permeate a sustainable school (education FOR the environment/a commitment to care).

A school's job is first and foremost to educate pupils rather than save the environment and the planet (Scott 2013). This latter aspect is certainly harder to achieve; in this instance, the obstacles against EfSD were due to: this whole endeavour being mainly one person's ambition, vision and work (hence, 'added on' rather than 'built in'); lack of pertinent knowledge around sustainability; pupils' active participation being restrained to a reactive approach; lack of leadership for sustainability from senior management and community's passive support. On that last element (community support) it is worth mentioning that Green and Somerville (2014) in their study of sustainability education in primary schools in Australia noted that: "*The layering of webs of connection between schools and their local community members and organisations produce an active school ecology of place that underpins sustainability education practice.*" (p. 12).

Mapping these against the notions of Education for sustainable development 1 and 2 (ESD 1–ESD 2) (Vare and Scott 2007) we can claim that all practitioners in their settings have promoted changes in pupils' behaviour and knowledge about environmental issues; but in terms of ESD 2 which is characterized by building a capacity to think critically and explore contradictions inherent in sustainable living, they do not seem to have succeeded. This is because they have highlighted learning more as an outcome rather than as a process via which such outcomes may come about. In order for practitioners to be able to develop and focus on the process, knowledge/skills on sustainability need to be enhanced and understood before they are able to implement these in their pedagogical approaches.

Nevertheless, one also needs to highlight positive aspects/seeds for developing a systemic view of the local and global space/environment. These included the value they posed upon outdoor, experiential learning, the 'tangible' links they made with the local community and the work towards issues that go beyond the traditional learning aiming to enhance social cohesion.

7 Limitations

The nature of an exploratory case study and the sheer number of these (four in this project) cannot allow us to claim generalizability of our results. However, these four case studies may reflect similar settings in the UK in terms of practitioners' training; training that does not necessarily take into consideration EfSD. Practitioners with different levels of engagement with environmental/sustainability issues and training can lead to varied results in schools. Finally, our main source of data came mostly from the practitioners' input via the interviews and we did not have the chance to observe some of these activities when they were taking place to further enhance/complete or illuminate different aspects of our findings.

References

Arlemalm-Hasger, E., & Sandberg, A. (2011). Sustainable development in early childhood education: In-service students' comprehension of the concept. *Environmental Education Research, 17*(2), 187–200.

Barrett, M. J. (2007). Homework and fieldwork: Investigations into the rhetoric-reality gap in environmental education research and pedagogy. *Environmental Education Research, 13*(2), 209–223.

Cutter-Mackenzie, A., & Edwards, S. (2013). Toward a model for early childhood environmental education: Foregrounding, developing and connecting knowledge through play-based learning. *The Journal of Environmental Education, 44*(3), 195–213.

Davies, J. (2009). Revealing the research 'hole' of early childhood education for sustainability: A preliminary survey of the literature. *Environmental Education Research, 15*(2), 227–241.

DCSF. (2009). *National framework for sustainable schools.* Available at: http://webarchive.nationalarchives.gov.uk/20090608173126/http://publications.teachernet.gov.uk/eOrderingDownload/Sustainable%20Schools%20Framework%20Poster.pdf. Accessed November 14, 2015.

Demetriou, H. (2013). The case study. In E. Wilson (Ed.), *School-based research—A guide for education students* (pp. 256–269). Beverly Hills: Sage.

Fjortoft, I., & Sageie, J. (2000). The natural environment as a playground for children: Landscape description and analyses of a natural playscape. *Landscape and Urban Planning, 48*(1–2), 83–97.

Gayford, C. (2003). Participatory methods and reflective practice applied to research in education for sustainability. *Canadian Journal of Environmental Education, 8*, 129–142.

Green, M., & Somerville, M. (2014). Sustainability education: Researching practice in primary schools. *Environmental Education Research.* doi:10.1080/13504622.2014.923382

Kadji-Beltran, C., Zachariou, A., & Stevenson, B. (2013). Leading sustainable schools: Exploring the role of primary school principals. *Environmental Education Research, 19*(3), 303–323.

Katsenou, C., Flogaitis, E., & Liarakou, G. (2013). Exploring pupil participation within a sustainable school. *Cambridge Journal of Education, 43*(2), 243–258.

McNaughton, M. J. (2012). Implementing education for sustainable development in schools: Learning from teachers' reflections. *Environmental Education Research, 18*(6), 765–782.

Reynamo, L., & Suomela, L. (2013). Education for sustainable development in early childhood education in Finland. *Journal of Teacher Education for Sustainability, 15*(2), 91–102.

Ryan, R., & Deci, E. (2000). Intrinsic and extrinsic motivations: Classic definitions and new directions. *Contemporary Educational Psychology, 25*, 54–67.

Scott, W. (2013). Developing the sustainable school: Thinking the issues through. *The Curriculum Journal, 24*(2), 181–205.
Stake, R. (1998). Case studies. In N. Denzin & Y. Lincoln (Eds.), *Strategies of qualitative inquiry* (pp. 86–109). Beverly Hills: Sage.
UNESCO. (2013). *Education for sustainable development (ESD) in the UK—Current status, best practice and opportunities for the future*. UK National Commission for UNSECO.
Vare, P., & Scott, W. (2007). Learning for a change: Exploring the relationship between education and sustainable development. *Journal of Education for Sustainable development, 1*(2), 191–198.

Authors Biography

Athanasia Chatzifotiou has gained her Ph.D. from Durham university in the UK. She examined primary school teachers' knowledge and awareness of environmental education in two European countries, namely England and Greece. Her subsequent work addressed issues concerning the status of education for sustainable development in the National Curriculum in England and Greece. She teaches in the BA Hons Childhood Studies degree in Sunderland University where she is a Senior Lecturer in the Department of Social Sciences.

Karen Tait (B.Ed. and M.Ed.) studied in USA where she worked for years in specialist residential education for deaf, blind and profoundly multiply disabled pupils. She was part of toddler services sensory development team. Range of experience in UK includes service development and teaching across a range of SEN contexts including the maintained, independent and voluntary sectors. She has years experience in supporting staff from a range of early years settings, particularly with regard to staff development of inclusive practice and pedagogies. Additionally, she has taught a wide range of subjects and modules in a number of further education settings and contexts. She leads child development modules and teach on the BA Hons Childhood Studies degree. She has been heavily involved in both teaching and assessment on the PGCE/Early Years Professional degree and vocational qualification.

Education for Sustainable Development for Employees—A Route to Behavioural Change

Alexandra Mifsud

Abstract
The UN Decade of Education for Sustainable Development (2005–2014) has undoubtedly raised the discourse on the principles of ESD and provided a platform for healthy debates on infusing ESD in curricula and ways to overcome the barriers that exist to implementation programmes. Furthermore, the decade has also strengthened community based ESD activities and initiatives. This paper addresses a gap in research within the field of ESD by exploring the potential of ESD for employees at their workplace through a research study carried out with a set of employees in a UK higher education institution. The findings suggests that a design process for employee programmes on ESD should be needs based and context specific. Whilst it is acknowledged that employees have an important role to play in driving the organisation's sustainability strategy forward, the study has found that not only are ESD training programmes for employees non-existent, but neither are employees effectively invited to participate and engage in shaping the sustainability strategy of the organisation. The paper will present the potential of infusing thinking skills into ESD training programmes to assist employees feel adequately empowered to engage in needs based ESD training programmes relevant to their role at work and to their life beyond the workplace. The study highlights the role thinking has in cultivating a thinking culture within an organisation as part of its response to the challenges of sustainable development today. Results from the study indicate that employees are of the opinion that bespoke ESD training for employees would most likely lead to behavioural change.

A. Mifsud (✉)
Centre for Environmental Education and Research, University of Malta, Msida MSD 2080, Malta
e-mail: alexandra.mifsud@um.edu.mt

W. Leal Filho (ed.), *Sustainable Development Research at Universities in the United Kingdom*, World Sustainability Series,
DOI 10.1007/978-3-319-47883-8_5

Keywords
Education · Sustainable development · Employee/staff training · Thinking skills · Organisations · Behavioural change

1 Introduction

a few thousand words from Rachel Carson and the world took a new direction

Linda Lear cites an American editorial writer in her afterword of Carson's book, *Silent Spring* (1999: 258), a book that played an influential role on the beginnings of the environmental movement in the 1960s. The significance of Carson's book to this study is that she put into question the status quo which, despite prompting great resistance and even attempts to silence her work, opened the gateway to a new era for behavioural change and social transformation. Half a century on and there is increasing evidence (McKeown et al. in Chalkley et al. 2009, and more recently in UNESCO's Roadmap for Implementing the Global Action Programme on ESD, 2014a) for the need for more work to assist individuals in making changes in their decisions and actions for the benefit of the future of our planet and future generations around the world. Carson's stance to question the status quo remains valid today. Individuals need to be equipped with the necessary skills (cognitive, practical, thinking, decision-making and problem-solving) to feel adequately confident and sufficiently motivated to engage in a more sustainable lifestyle.

This research study took place at the start of the UN Decade of Education for Sustainable Development 2005–2014 (UNDESD). On the global arena there were high expectations and great enthusiasm. Higgit (2009: 3) captures this enthusiasm when he says "the Decade for Education for Sustainable Development is widely held to offer the best opportunity to date to implement lasting and radical changes to educational programmes." Education for Sustainable Development (ESD) practitioners and researchers around the globe greeted this decade with a mix of cautious excitement such as Cloud, in Chalkley et al. (2009) and scepticism as expressed by Jickling and Wals (2008) and Sauve and Berryman in Chalkley et al. (2009). A lively debate on ESD ensued to tease out the issues around it and bring in fresh ideas to the field.

The researcher was keen to find ways of how best to assist sectors of society to become sufficiently engaged in making behavioural changes so that they can lead a more sustainable lifestyle that would be of benefit to our planet and to future generations. Being an ESD practitioner, the researcher had prior experience of experimental work on infusing De Bono's thinking programmes into ESD training. The training received by the researcher in De Bono thinking skills helped to discover that if people are taught thinking skills then with practice, they are able to apply them to all areas of their life both personal and professional. De Bono (1991: 18) claims that "on a personal level, people have to do more thinking and take more decisions than ever before" whilst on a socio-political level "some people will undoubtedly do more thinking than others … but it is desirable that the rest should at least do enough thinking to decide for themselves whether the special thinkers make sense or not." De Bono (ibid.)

Interestingly, De Bono (ibid.) continues to argue that "in a complex society political decisions and pressures depend very much on individual thinking. If that thinking can see only narrow self-interest, or only an immediate future, then society becomes a power struggle for self-interest." When such statements are applied to the pressing issues of sustainable development the world is facing today, it immediately becomes clear as to why there is the need to infuse thinking skills in ESD. In so doing, individuals are not only equipped with the skills to lead a sustainable lifestyle but also with ones that provide them with the self-confidence needed to question the status quo, to become actively engaged in decisions, actions and initiatives on sustainable development at their place of work, at home and within their community.

The study explored the design process for employee training that has the potential to be flexible enough to provide bespoke training to employees in a given organisation without causing undue disruption to the operations of the entity, yet be sufficiently adaptable that would make it possible for it to be implemented by different organisations and in different settings. Ultimately, the study explored ways of how to instigate change by involving people in forming part of the design process of ESD training. This would make the training more relevant, interesting and inspirational to those attending the training.

The study examined how ESD could be designed and introduced when it is situated within the professional lives of people—employees. Within the realms of human resources and development there exists a great deal of literature on employee training. For example, Noe (2013: 11) describes the employee training design process as comprising of seven parts. These are: (a) Conducting needs assessment; (b) Ensuring employee readiness for training; (c) Creating a learning environment; (d) Ensuring transfer of learning; (e) Developing an evaluation plan; (e) Selecting training method; and (f) Monitoring and evaluating the programme. Furthermore, he states that "to fully benefit from employee knowledge requires a management style that focuses on engaging employees…employees who are engaged in their work and committed to their companies give those companies a competitive advantage, including higher productivity, better customer service, and lower turnover." (ibid: 19.)

Data was collected from a higher education institution in the UK, a sector which has recently witnessed an increase in ESD activity through infusing ESD into university curricula and greening campuses initiatives. Yet, it is imperative to point out that this study did not address these curricular initiatives at universities but focussed exclusively on how employees can be trained in a manner that would help them become sensitised to sustainable development concepts and take action to lead a more sustainable lifestyle. Nonetheless, the fact that the research participants were employees at a university in the UK makes it easier for other universities to replicate the work carried out in this study with their employees.

2 Significance of the Study

The reasons for undertaking this research study were:

- To gain a better understanding of an ESD practitioner and the role of a change agent.
- To ensure that the work carried out with the selected set of employees would serve as a platform to contribute to organisational change.
- To make a contribution to the development of educational theory in ESD. (McNiff and Whitehead 2010: 242).

Documented evidence of work and/or research undertaken to apply ESD principles in employee training within organisations with the infusion of thinking skills is absent. There is however, significant work to infusing ESD in higher education curricula by encouraging and supporting academics to include ESD concepts in their study units or modules across the courses and programmes offered at the higher education institution. In the UK, recent reports (HEA/QAA 2014; EAUC/NUS/UCU/AoC/CDN 2015) are evidence of this growing movement for a cross-curricular and interdisciplinary approach to ESD in higher education institutions.

The exploratory work conducted with a set of employees sheds light onto a relatively untouched area within ESD. Thus, the findings and conclusions of the study provide guidance on how organisations can set about institutional change through a process in the design of employee training that best fits the needs of the employees and the overall sustainability goals of the organisation.

3 The Research Context

Data was obtained from De Montfort University in Leicester. An active sustainability team promotes sustainability through various programmes for students and staff. These are supported by operational measures to reduce the organisation's environmental impact whilst improving its sustainability performance. http://www.dmu.ac.uk/about-dmu/sustainability/sustainability-strategy.aspx (2015).

The inception of assisting the private sector towards sustainable development through education and training dates back to the 1992 publication of Changing Course: A Global Business Perspective on Development and the Environment, by the Business Council for Sustainable Development (Schmidheiny 1992). Yet, Melhmann and Pometun (2013: 84–85) express concern that 'greenwash' organisations "whose wish for behaviour change is driven not by a longing for the immensity of sustainability, but rather by a desire to call their organization or product 'green' or 'socially responsible'" can pose an educational problem in that they succeed in "convincing their stakeholders that they are following a sustainable path.". It follows then to explore how education can help organisations in becoming accountable for and achieving a return on their investment not only financially but also in the social and environmental dimensions of their operations.

This research work conducted at De Montfort University (DMU) explored the level of contribution education for sustainable development training programmes for employees within an organisation could possibly have to create a shift for more corporations becoming socially and environmentally responsible thus giving them a strong foothold on attaining their sustainable goals. Furthermore it provided valuable insight into the views, needs and expectations of a set of employees in relation to making changes in their behaviour that would lead them to adopt a more sustainable lifestyle. Human resource management and development departments provide a learning environment in organisations through staff training and development. The study explored how such a learning environment in an organisation can be strengthened (or created if it is not present) in such a manner for it to infuse ESD in all agendas, programmes and activities that promote sustainable development. The concept of 'education' or 'training' provision carries the principal goal of offering a 'learning' environment for employees that would assist them to become sensitised towards sustainable development concepts. Key elements in the sensitisation process of employees are employee engagement and behavioural change.

4 Performance and Reporting Tools

DMU (Leicester, UK) is a typical organisation that strives to excel in environmental management systems and is indeed performing well in this respect (http://dmu.ac.uk/dmu-staff/hot-topics/2015/august-2015/dmu-recognised-for-green-credentials.aspx 2015). Performance based tools in industry such as the ISO 14000 series are based on environmental management principles. Improving the environmental performance of an organisation does not necessarily bring about an increase in the level of environmental literacy of its employees. Neither does it empower the employees to become proactive in working towards sustainable living. Melhmann and Pometun (2013: 84) capture this underlying concept inherent in this research study so eloquently that it merits a full citation of their view:

> Many businesses as well as public agencies and NGOs have ambitions and programs that go beyond 'business as usual'. They may start with some kind of certification, for example ISO 14001….the initial focus is often technical or administrate investments and innovations. But sooner or later – assuming that the sustainability ambitions are genuine – it becomes apparent that the full potential of such investments can only be reached when the hearts, minds and hands of employees are engaged in adopting new behaviours and, finally, also in innovation: in the design of more sustainable ways to work.

5 Social Responsibility of Organisations

Businesses are becoming increasingly aware of their impact on the environment and society leading them to take responsibility of managing sustainability. This is "sometimes referred to as the "triple bottom line" or "people, planet, and profits."" Jackson et al. (2012: 15). By the time this study unfolded, UNESCO (2014a, b: 31) reported the gains from an increase in awareness and activity made through education and training within organisations to strengthen their ability and capacity to respond to sustainable development. Indeed, UNESCO (ibid.) states that "In many cases, education, training and awareness raising efforts are leading to the adoption of sustainability as a business strategy." This shift is encouraging and further strengthens the need to address how education and training is taking place in organisations and if it is not yet taking place, what are the barriers. It is evident that there is no 'one size fits all' model for organisations to follow on their journey towards sustainability. Similarly, ESD programmes for employees should reflect this diverse reality by modelling elements into the programme that are truly needs-based and relevant to the context within which they are being implemented.

6 The United Nations Decade of Education for Sustainable Development

With The United Nations (UN) launching its Decade of Education for Sustainable Development (UNDESD) in 2005 it signalled a recognition that education at all levels is key for changes in social attitudes required to protect future generations. Fien (2006) states that the UNDESD "has been established to help build commitment and skills across the world's education system so that human society can develop an enhanced understanding of what it means to work for a sustainable future, a sense of responsibility for future generations, and a spirit of optimism and hope for a sustainable future." Tilbury (2006) was optimistic whilst warning that "it will depend on how meaningfully stakeholders engage with national efforts and on whether they reflect upon the experiences of ESD to date." Contrastingly, Jickling (2006) argues that if policy makers expect that work be "bent to the sustainable development agenda, then this Decade will be little more than an annoying distraction for many environmental educators." Fien (2006) resists this notion by referring to the International Implementation Scheme for the Decade and argues that "education for sustainable development is not a global imposition on countries and education systems but an invitation for them to explore the themes and issues, the objectives and the pedagogies that can make education locally relevant and culturally appropriate in the search for a better world for all."

At the end of the UNDESD and following on from the publication of the report, *Shaping the Future We Want* (UNESCO 2014a, b), UNESCO drafted a roadmap for implementing the global action programme on ESD (2014) which largely deals with the points raised by Fien (ibid.). The GAP (Global Action Programme 2014) has two overall objectives:

- To reorient education and learning enabling everyone to have the opportunity to acquire the knowledge, skills, values and attitudes that empower them to contribute to sustainable development.
- To strengthen education and learning in all agendas, programmes and activities that promote sustainable development.

The spirit of this study was to ensure ESD for employees has the facility to be adaptable to various contexts and organisational cultures whilst simultaneously encapsulate a flexible element to be able to fit around the day to day operations of the organisation.

The declaration from UNESCO (2014a, b: 28) at the end of the UNDESD falls in line with the stance taken in this research study because its context and focus (a) addressed employee training, which is a sector in ESD that is often missed; and (b) explored ways to assist employees become sufficiently empowered and engaged in a process whereby they are better equipped to lead a sustainable lifestyle. Despite the conceptual differences between EE and ESD highlighted in UNESCO'S international policy documents (Tbilisi declaration 1977) and (DESD 2006), Pavlova (2012: 667) rightly points out that these differences are not always visible and present in the realm of practice across the globe.

Upon careful consideration of the EE principles listed in the Tbilisi Declaration (UNESCO-UNEP 1977: 27) and UNESCO's International Implementation Scheme for ESD (UNESCO 2005: 30–31) it is evident that both EE and ESD share the following pedagogical features:

- An emphasis on life-long learning and inclusion of formal and non-formal education.
- Interdisciplinarity.
- Inclusion of social, environmental and economic realms.
- The use of a variety of pedagogical techniques that promote participatory learning, first-hand learning and development of higher order thinking skills (referred to as problem solving and critical thinking in the Tbilisi declaration).

Figure 1 captures the interconnectedness and overlap of EE and ESD concepts that this study adopted in the work on the design of ESD training programmes for employees with the input of thinking skills.

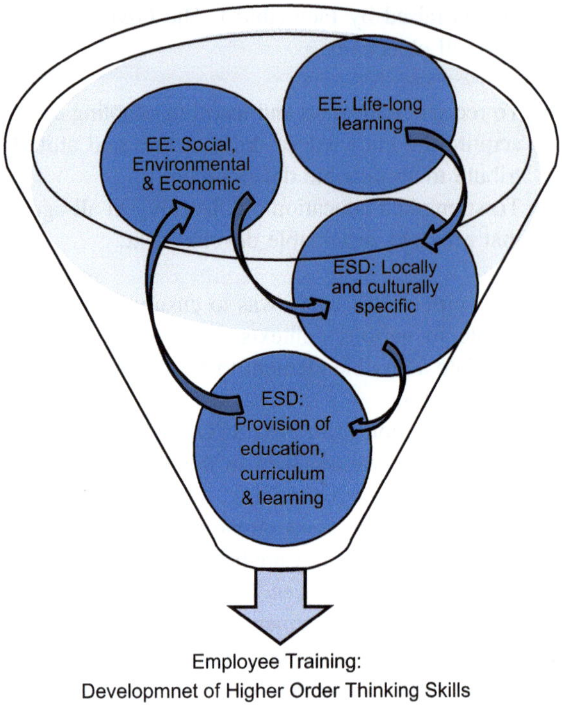

Fig. 1 Pedagogical and conceptual elements of EE and ESD that feed into employee training

7 Current Trends in Research on Education for Sustainable Development

Research in the field of education for sustainable development can be found within the spectrum of 'sustainability literacy', 'sustainability curriculum', 'environmental sustainability' and 'sustainable education'. The International Journal of Environment and Sustainable Development serves as a good showcase of research carried out in the field of sustainable development in its broadest scope. Indeed, in his editorial of the first issue of the journal, Leal Filho (2002) articulates this research need when he says that the journal aims at "fostering the cause of sustainable development by means of the publication of scholarly research, studies and projects taking place." Significant work has been carried out in the formal education sector at all levels of education. For example, London South Bank University (LSBU) has been running post-graduate courses for over 21 years' and is a shining example of ESD presence in the formal education sector. Parker and Wade (2008: 2) claim that the programme at LSBU together with its associated research activities and its alumni can offer an important contribution to the UNDESD. Albeit, literature indicates that very little attention if any has been given to ESD for employees.

Lessons can be drawn from the way the formal education sector has responded to ESD since research has been conducted in higher education settings ranging from higher education pedagogy to campus wide initiatives for organisational change. Yet, even within the formal higher education curricula there is a lack of consistency in the implementation of ESD across all courses for students. Thomas and Nicita (2002) and Sherren (2006) claim that in Australian universities, ESD is often limited to disciplines that have an environmental component rather than across the entire range of disciplines. Indeed, "education for sustainability continues to be accessible only by those most directly involved in environmentally focussed education courses, such as environmental sciences." (Thomas and Nicita 2002: 477). Such evidence suggests that, aside from academics and members of staff directly involved in an environmentally-related discipline, those responsible for designing course programmes and study modules in higher education lack the vision, belief and competency to see the need to integrate ESD within their study modules. A case study by Qian (2013) focussed on the development of educational change for sustainability at the University of South Australia and the impact of such development "on an area that has long been resistant to the sustainability initiative—accounting." Qian (ibid: 90) concludes that "a well-designed change strategy needs to be built within an institutional environment where capability and cultural support can be developed to formalize and stabilize sustainability values during the change."

Hence current research on ESD in the formal education sector demonstrates that whilst there is broad agreement on ESD and its place within formal educational curricula, uptake of ESD implementation plans and initiatives is slow and sporadic at all levels of the formal education sector. Evidence from this study at DMU found a similar scenario with an inadequate presence of undergraduate courses that are infusing ESD principles (http://www.dmu.ac.uk/about-dmu/sustainability/teaching.aspx, 2015). Moore et al. (2005) and Thomas (2004), in Qian (2013: 90) attribute such a disparity to "the lack of effective mechanisms to engage with staff in non-environmental disciplines and to institutionalize change." Action research to explore the factors influencing academic staff engagement in ESD and their views and visions on ESD was carried out at the University of Southampton (Cebrian et al. 2015). The research study concludes by appealing for "the creation of professional development programmes for staff" and to "engage and empower academics in their journey to embed ESD in the curriculum" (ibid: 85).

This reinforces the importance of designing ESD training programmes for employees working in a formal education institution or in industry. The need is coherent across all organisations since employees, regardless of what sector they work in, are increasingly being expected to integrate sustainable development practices in their day to day work as part of their organisation's way of meeting its sustainability targets, requirements and legal obligations.

8 Education for Sustainable Development Within Organisations

A study by Eboli and Mancini (2012) on corporate education systems in fostering the competences necessary for companies to face the challenges of sustainability in their management processes concluded that "the importance of corporate education systems in the management process of sustainability becoming protagonists in developing competences for sustainability is high, due to its nature of creation of knowledge, skills and values." Szovics et al. (2011: 92, 93) identify the need for the development of not only technological expertise but also communication skills whilst suggesting a revision of curriculum development for existing students and the re-training of professionals and blue collar workers.

Whilst a top-down approach may partially lead to some improvement in the environmental and sustainable performance of the organisation, the change in behaviour by employees to have the capacity to embrace and adopt meaningful sustainability values is dubious. Qian (2013: 90) argues that "a strategic approach needs to embrace a top-down initiation for change and bottom-up capability building to develop institutional commitments that can sustain the change." This suggests a two-pronged approach that promotes senior management supported ESD for employees, that is interwoven at its core. Hence there is requirement for a learning framework that is (a) able to be sensitive to the ever changing needs of society and the environment and; (b) equipped with the skills to respond appropriately to these needs. The challenge for education particular that of business management and employee training and development is to shift from a focus purely on environmental performance to embedding sustainable development concepts throughout the entire operations of the organisation. ESD that is needs based and context specific with elements of flexibility and adaptability, has a role to play in developing action strategies for sustainable living amongst the employees of an organisation.

Recent research by Zibarras and Coan (2015) focussed on the human resources management (HRM) practices in UK organisations used to promote pro-environmental behaviour. Their study highlights a gap between research and practice about HRM's role in supporting the attainment of sustainability because their findings show that the use of HRM practices is not being used sufficiently. Most relevant to this study are the recommendations made by Zibarras and Coan (ibid.: 2136) that "training should be made available to all employees, including management, which focuses on improving environmental knowledge, awareness and skills … [and] findings imply that organizations need to empower employees to take ownership of some of the environmentally related issues and/or initiatives themselves, for example, including employees in the design and implementation of any new environmental change initiative …" Similarly, Epstein (2008: 52) identifies employee training as one of the internal actions to be taken by organisations when driving a sustainability strategy through an organisation.

Redmond and Walker (2009: 126) express concern when they state that "Serious issues are raised for educators seeking to design and deliver environmental education for small businesses who are a disparate group in need of individualised and specific programs." A key point raised in their study that is of great significance to the one carried out at DMU is that "Starting from where learners are at is not a new idea in education, however, this appears to have been forgotten by many who develop programs that are not contextually-specific and have not been developed in consultation with small businesses." (ibid.). Indeed, Noe (2013: 113) states that any effective training for employees involves the use of a training design process which starts with a needs assessment. Amongst reasons cited by Noe (ibid.: 114) for conducting a needs assessment are (i) training programs may have the wrong content, objectives, or methods; (ii) training will not deliver the expected learning, behaviour change, or financial results that the organisation expects; and (iii) money is spent on training that is unnecessary due to being unrelated to the organisation's corporate strategy.

If ESD within a sector that is lacking both in activity implementation and in research is not introduced effectively and appropriately, there is the risk for ESD to lose support from such an important sector. Organisations need to have some reassurance that any ESD initiatives for employees will meet their expectations as a route to creating behavioural and organisational change for sustainable development.

9 Research Methodology and Research Methods

The research approach and process adopted in this study reflect the values held by the researcher and is thus largely within the interpretivist framework. The interpretivist paradigm explains that human behaviour is situation specific and results from research cannot be used to predict similar human behaviour at a different time and situation. The social researcher is fully aware that personal values and beliefs can "intrude at any or all of a number of points in the process of social research: choice of research area; formulation of research question; choice of method; formulation of research design and data collection techniques …" (Bryman 2004). The ontological position for this study with employees at DMU is not too distant from Grbich's explanation (2013: 7) of constructionism (constructivism) when she states that the researcher's knowledge is a constructed understanding and interpretation based on life experiences and subjectivity. The study stems from the researcher's understanding of the selected area of interest as well as the experiences in the field of ESD as a practitioner. It was established that the researcher was working "within a certain view of the social world (ontology) and how to generate knowledge of it (epistemology)" (Mason 1997: 18). Therefore the study is aligned to the interpretative ontological position with a constructivist epistemological approach. Furthermore, it features the basic tenets of action research identified by O'Leary (2004: 139–140) and Denscombe (2010: 126) namely it: (i) addressed a practical problem—ESD programmes for employees; (ii) generated knowledge—design needs-based ESD programmes for employees; (iii) enacted change—employees leading an improved sustainable lifestyle; (iv) was participatory—through seeking views of employees on their own ESD training needs; and (v) relied on a cyclical process—observations at DMU.

Methodological triangulation was adopted by using multiple data collection methods, namely content analysis of documents, interviews, casual conversations and observations, to study a single problem. NVivo qualitative analysis software was used for the coding of data and for content analysis of documents. Covert integration of De Bono's thinking tools helped to word the questions for the in-depth interviews and casual conversations. The observations at DMU were unstructured as defined by Punch (2005: 179–180) in Bell (2010: 193) since they were more suited for this study.

Gaining access to research participants for the purposes of the collection of data from employees at DMU posed the greatest challenge in the research study. Despite the assistance of a gate keeper within DMU and numerous attempts and requests for employees to volunteer for participation in a 30–45 min interview, the total number of interviewees was seven. These were supplemented by two casual conversations that assisted the researcher to gain information not captured through the seven interviews. Nonetheless, the small sample size for the in-depth interviews provided a wealth of data that contributed significantly to the research findings. Data was collected from participants in research, academic or management roles at DMU. The study would have been richer if employees in other roles would have been available. This is considered as the main limitation to the study because it was not possible within the timescales of the study to obtain the viewpoints of other employee sectors at DMU. In line with the interpretivist research paradigm the aim of the study was to gain insight into the views of a set of employees at a given context, therefore the researcher did not tap into employees from another organisation as this would have created variables inherent to a different research context which would have impacted the research findings of the study.

The findings of the study are described and discussed in the subsequent sections and fall within the broad research aims identified by the researcher at the outset of the study.

A. **Provision of Thinking Skills Training at DMU and Applicability of De Bono's thinking programmes to the design of employee training**

None of the research participants were familiar with De Bono's thinking programmes but a few had heard of his work hence this theme was adapted to encapsulate soft skills training.

- Soft Skills Training Provision: Mostly soft skills (not thinking skills). Four interviewees were aware of this provision and have attended whilst three stated that there is no provision at DMU.

- The Design of Employee Training: Mixed views from the interviews about the potential benefits of thinking skills (not specifically deBono's) to employees' participation in designing their own training programmes. Only two interviewees saw the link of how being equipped with thinking skills, employees would be in a better position to feel able to participate in the design of their own training.

A trained thinker has the advantage of viewing information and situations more comprehensively with the ability to take a more appropriate line of action. "Thinking is the operating skill through which innate intelligence is put into action." (De Bono 1991) This study explored how thinking skills could equip employees with what they need in order to engage in identifying their own training needs on sustainability and participate in the design of their own training programme. It was surprising to discover the low level of awareness on thinking skills amongst the research participants and their apparent inability to pick out the potential of a trained thinker in becoming engaged in the design of staff training programmes. The data gathered suggests DMU employees interviewed were broadly convinced any staff training is not their remit but that of the HR department. Some expressed the view that it would be presumptuous to intervene in what is viewed as the responsibility of the HR department, whilst others shared the sentiment that they did not wish to be involved as they already had a demanding work load. Mehlmann and Pometun (2013: 84) highlight this point by arguing that "there is a need for an adequate pedagogy: methods and tools to convey the vision and engage the creativity of employees in the search for ways to move towards the vision." Furthermore, they make a valid point when they say that in this respect "a workplace is no different from any other educational arena." (ibid.) It is therefore useful for HR departments to consider ESD pedagogies being used within the formal education sector when developing employee learning and training programmes for their organisation. The data highlights the notion that employees do not feel confident or empowered to become engaged in their training needs and that they are too busy to take charge of their own professional development. Yet UNESCO (2011a: 8) explains that ESD learning refers to "learning to ask critical questions; learning to clarify one's own values; learning to envision more positive and sustainable futures; learning to think systemically; learning to respond through applied learning; and, learning to explore the dialectic between tradition and innovation." Infusing thinking skills such as those by De Bono would only aid in the ESD learning that can take place during employee training programmes.

B. **Applicability of De Bono's thinking programmes to a sustainable lifestyle**

Whilst the interviewees had only a vague awareness of De Bono's work the responses for this theme suggest that they found it easier to make a connection between improved thinking ability and leading a sustainable lifestyle.

- Sustainable Lifestyle: Two interview participants admitted they could not "*see the relevance*" (INT002) and were "*struggling to make that connection.*" (INT006). Whilst the remaining were of the opinion, some more clearly than others, that being equipped with thinking skills would assist employees in their decisions and actions to leading a more sustainable lifestyle "*Yes they would benefit from that because if they were given training on decision making* etc. *they would obviously be more actively practising it rather than just using their common sense.*" (INT007).

UNESCO (2014a, b: 12) identifies one of the four dimensions of ESD as "Stimulating learning and promoting core competencies, such as critical and systematic thinking, collaborative decision-making, and taking responsibility for present and future generations." Morris and Martin in Stibbe (2009) had made similar assertions "Our contention is that learners cannot deal with the wicked problems of sustainability without learning to think and act systematically." The responses from DMU employees taking part in the study supported by the theoretical background give strength to the proposed suggestion that integrating thinking skills in ESD programmes would not only enhance the training for employees but would also meet the requirements on ESD as a way of bringing about behaviour change.

C. **Factors when designing ESD for DMU staff**

The interviewees were asked to identify what factors needed to be kept in mind when designing ESD training for DMU staff. This question was phrased in line with another of deBono's thinking tools, CAF (Consider All Factors) which "is the process of exploring all factors in a situation … The CAF is the prime information input tool." De Bono (1997). The key factors highlighted by the respondents (Fig. 2) were that the training programme needed to be relevant to the audience and their role at DMU. This was followed by: (i) it ought to be needs based; (ii) the duration of the programme; and (iii) it ought to include messages on moral ethics and responsibility.

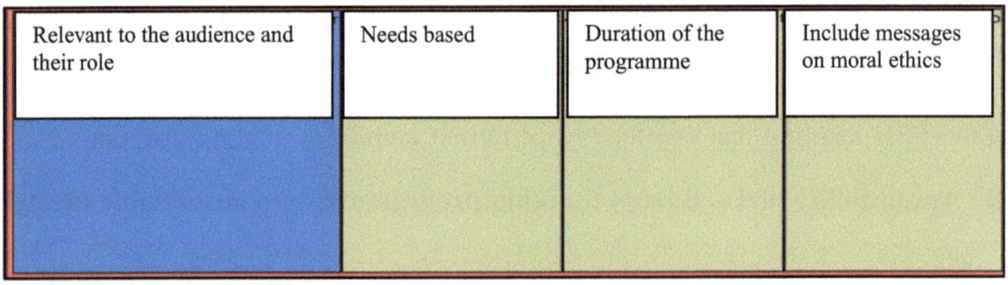

Fig. 2 Factors to be considered when designing ESD for staff at DMU

- Relevant to audience and their role: "*it needs to be relevant to the people who are going to listen otherwise they will not listen.*" INT003. "*relevance and context*" (INT006).
- Needs based: "*I would organise these sort of cafes where members of staff from different faculties and departments can do brainstorming sessions and see what their needs are*" (INT001).

It was encouraging to note how strongly the respondents felt about having a programme that adopts a needs-based approach by being relevant to employees at DMU. These findings are in line with arguments made by Epstein (2008) and Redmond and Walker (2009) on the importance of individualised and tailor-made programmes for employees. They also concur with Pavlova's definition that "ESD prioritizes embedding learning into locally and culturally appropriate contexts" (2012: 667). Results further support the concept that a needs-based training programme would strongly benefit the employees and can be made possible when employees form part of the design process and are given engagement tools to participate in such a process. The findings make it clear that the employees value the concept of needs-based and context specific training programmes on sustainable development. The next step would be to provide them with the skills to feel sufficiently confident in recognising that they are a crucial element in the design of such a programme. Their input is what would inherently make the programme design relevant to their needs and their role within the organisation.

D. **The design process of needs-based employee training**

If employees are involved in identifying their own training needs and are given the opportunity to become sufficiently engaged in the training programme design, it is more likely to provide the knowledge, attitude, skills and behaviour required for employees to make changes to their day to day decisions and actions. However, the difficulty arises when determining the 'how' element within the ESD training programme for employees. This point was therefore addressed in the study with results (Fig. 3) identifying staff forums and senior management commitment and relevance to role as key elements for a successful process of the design of the ESD programmes.

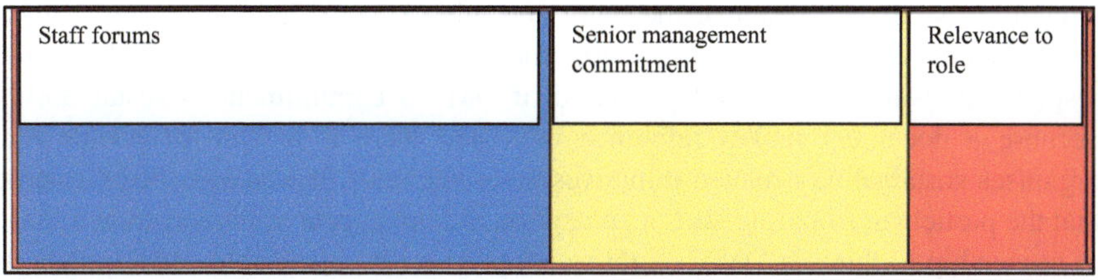

Fig. 3 The process for the design of needs based employee training

- Staff forums: "*there is a staff survey as a platform to seek the viewpoint of members of staff at DMU.*" (CONV008); "*I would try to do some sort of brainstorming session.*" (INT001); "*Through a staff forum, well that sort of thing, then probably yes but time is so precious that … I've been involved in a couple of staff forum things and we met once and it never happened again.*" (INT003).
- Senior management commitment: "*We also need to know that that's not at odds with top management and what they want. So there has to be a continuation right the way up the scale and that does not always happen. There is a sense of somehow we're above and beyond all that when you're up at the top management level. So vertically and horizontally it's got to be spread throughout the organisation.*" (INT004).

Evidence from the data suggests the research participants are of the opinion that the most appropriate manner with which to assist DMU staff to engage in training on ESD that leads to action and change is by involving them through staff forums. DMU does have a staff survey system in place (CONV008) however from the responses one would conclude that it may be underutilised or not viewed as a valuable means to engage in identifying one's training needs on ESD. The possibility to explore successful strategies adopted by Eco Schools through their Eco-Schools Committees and whether these could be adjusted and adopted in workplace settings is recommended. Furthermore, senior management commitment was also identified as key to implementing ESD programmes leading to behavioural and institutional change. Research carried out at organisations and their commitment to sustainable development points to similar findings (Legis and Collerette 2006; Herold and Fedor 2008; Karp and Helgo 2008 in Qian 2013: 78). Once again, the importance of the training programme being relevant to the role of DMU staff was mentioned. Clearly, the relevance of the training is a recurring theme throughout the findings of this study.

E. **DMU's commitment to ESD**

As can be seen in Fig. 4, research participants attribute DMU's commitment to sustainable development to the fact that it is a theme in the corporate strategy for the period within which this research study took place (DMU Strategic Plan 2011–2015). Another factor as to why they believe DMU is actively working on sustainability issues is a result of the work and initiatives carried out by the sustainability team within the Estates department. The general sentiment from the respondents was that they felt reassured of DMU's commitment to sustainability because a document makes reference to it and there is a team of people that organises sustainability related initiatives across campus. It was of concern to note that the participants holding such a perception did not appear to demonstrate a sense of ownership to the role DMU ought to adopt towards sustainable development as they seemed to believe that someone else was taking care of that.

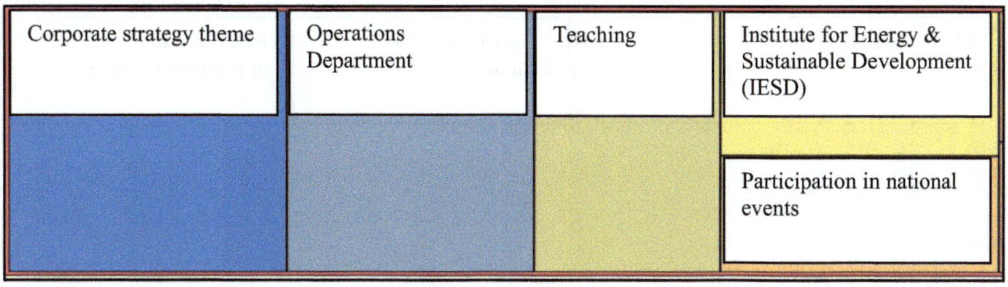

Fig. 4 DMU's commitment to ESD

- Corporate strategy theme: "*Well DMU is really committed towards environmental sustainability. If you see its strategic plan, we have one of the main themes for the University to be a top leader on environmental sustainability in the higher education sector of the UK. So it's one of the 4/5 key themes in the University's corporate strategy.*" (INT001); and "*I know DMU has a commitment to sustainability.*" (INT005).

Interestingly, two interviewees gave insightful responses and almost expressed frustration that sustainable development is not embraced across the board at DMU. Such on the ground realities are recorded in literature (Qian 2013: 79) (Redmond and Walker 2009: 126) on organisations and their commitment to sustainability so it comes as no surprise that this was the case at DMU too.

- "*the University as with all other organisations, seems to jump on the bandwagon of sustainability*" INT003); and "*As an organisation overall I think it probably pays lip service to it but doesn't really fully integrate it.*" (INT004).

ESD is only mentioned in DMU's Environmental Report for 2012–2013 in the context of students learning. The subsequent report for 2013–2014 gives more prominence to staff (and student) engagement with a target stating DMU "will deliver at least one environmental behaviour change project per year for staff (Green Impact or similar) to 2016/2017.", DMU Environmental Report 2013/2014 (2015: 3). There is mention of teaching and research but it steers away from the term ESD notwithstanding that this report was compiled within the UNDESD. On a more positive note, it was encouraging that in DMU's 2013/14 environmental report the semantic is more in line with the principles of ESD when it makes reference to "environmental behaviour change".

F. **Factors to be Considered by DMU Before Committing to ESD for Employees**

Data was derived from the interview question which was worded by using deBono's CAF tool (Consider All Factors) as it prompts respondents to scan as widely and comprehensively as possible when providing their response. The most

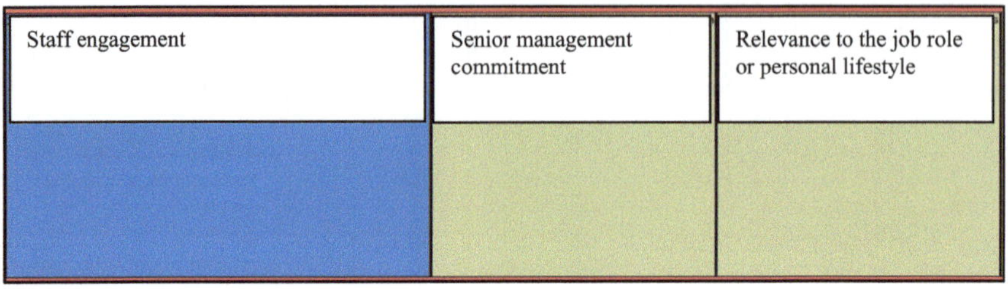

Fig. 5 Strands in ESD: factors to be considered by DMU

common factors (Fig. 5) listed by the respondents were: (i) staff engagement; (ii) senior management commitment; and (iii) relevance to job role or personal lifestyle.

- Staff engagement: "*we need staff engagement in the wider context at the University*" (INT001); "*the challenge is to engage them in a way where they are keen to get involved, to make it fun and make it not feel like one of those fairly useless empty gestures but something real.*" (INT004); and "*We need to engage more people in everyday activity somehow*" (INT006).
- Senior management commitment: "*we need top level commitment. The University senior leadership commitment is really important to drive any change so senior management and leadership commitment is key.*" (INT001); "*if you seem to be imposing stuff from above, there's so many different ways in which you can get people's backs up and it becomes utterly counter-productive.*" (INT004); and "*a clear message from the top and secondly clear and sustained top-level involvement. So if they are genuinely saying we need to do this, it has to be more than just printing a poster and forgetting about it for a year.*" (INT006).
- Relevance to job role or personal lifestyle: "*So making people aware and understand to make them realise why it is so important to achieve sustainability and what their role is.*" (INT001); "*something to relate it to the subject that is being taught in any particular module rather than leaving it up to the module leader by saying here is a new topic, dump it into your module. Well no, it doesn't work like that. You have to work at this and figure out its relevance or which pieces of it are relevant.*" (INT003); "*I think there's far too many staff that see it as something that: 'has nothing to do with me, it does't affect me.' How do we help people understand that it matters to them personally?*" (INT006); and "*the members of staff will have to keep in mind obviously that their daily duties, their daily work that they do, is not affected in any way by making that (ESD) commitment.*" (INT007).

Respondents felt helpless and unable to take charge of how DMU could engage in sustainable development across the entire organisation. This could potentially lead to feelings of inadequacy amongst employees, several of whom had already expressed concerns about their unrealistic work load. Indeed it is unfortunate that adopting a sustainable lifestyle by making behavioural changes is seen as a burdensome initiative. Reflecting upon the data gathered at DMU and upon careful consideration of UNESCO's report (2014a, b), it is believed that providing employees training on ESD with elements of thinking skills would give them the confidence needed to not only make small yet effective changes but also become convinced of the benefits attached to making those changes. Their efforts are key to driving DMU forward in its endeavours to attain the sustainability goals as set out in its documents. Indeed, UNESCO (ibid: 185) reports that "scaling up these efforts will require one of the most important success factors identified during the DESD, that of leadership. Put simply, leadership within and across education systems will be essential to sustain efforts and ensure ESD objectives are adopted and put into action." Thus the journey for DMU will require leadership commitment to enable a sustained educational commitment across the board in all its learning and teaching, operations and staff development work.

From the conversations and interviews carried out at DMU it would be safe to say that other than the work being done by the sustainability team at the Estates department and the research and teaching carried out by the staff at the IESD, rigorous strategic commitment to sustainable development and ESD at DMU is rather lacking. Such a scenario is reported and highlighted by Epstein (2008), Redmond and Walker (2009), Qian (2013), and Zibarras and Coan (2015). The value of the work by the sustainability team and the IESD is not fully and wholly recognised. Moreover, institutional support particularly from senior management needs to be seen and felt more strongly if DMU is to honour its pledge to sustainability as stated in its very own corporate strategy where it states: "We will make a significant contribution to global efforts to achieve environmental sustainability." (DMU Strategic Plan 2011–2015, 2011: 29–31).

On the basis of information obtained from the data sources through interviews and casual conversations it is evident that gaps exist between what is stated in DMU's publications and the viewpoints of the research participants. This may be attributed to complex factors however it is broadly due to (a) low employee engagement resulting in a lack of awareness of, not taking an interest in, and not taking ownership of sustainability at DMU; and/or (b) DMU as an organisation only pays lip service to sustainability resulting in weak support to employee engagement in sustainability.

If employees have not been given training on sustainable development, then it is rather presumptuous to assume that the promotion of sustainability in a manner that creates behavioural and institutional change would be achieved solely through the endeavours of the teams at the Estates department and the IESD. That scenario presents a near to impossible task for these teams especially since a considerable part of their work focusses on student engagement and operational aspects for achieving sustainability targets. This is similar to the research at the University of Southampton on academic staff engagement in ESD at a higher education institution highlights the need for universities "to provide a clear vision and strategy in ESD and build sustainability awareness through clear dissemination and communication strategies and the creation of professional development programmes for staff." Cebrian et al. (2015: 85).

G. **Staff development programmes and staff initiatives on Sustainable Development at DMU**

As can be seen in Fig. 6 none of the research participants in the study were aware of any ESD training at DMU for its members of staff. This scenario echoes similar findings by the UN Global Compact published in its *2013 Corporate Sustainability Report* which states that despite 65 % of its signatories are committing to sustainability at CEO level, "only 35 % are training managers to integrate sustainability into strategy and operations" (UNGC 2013: 7).

If no formal ESD training is provided at DMU for its employees then it was hoped that some informal ESD was taking place through other initiatives and/or events. All research participants (Fig. 7) were able to mention DMU's participation in Green Impact which clearly indicates that this initiative is given effective prominence across the organisation. Nonetheless from other responses elsewhere in the interviews, it was revealed that there are pockets of departments and buildings at DMU that do not participate in Green Impact so coverage is still not at desirable levels. Most interviewees lamented that any invitations to sustainable development initiatives were done through a blanket invitation via email. Some respondents also complained that as employees they feel they receive initiative overload and most times due to time pressure or a belief that the initiative is not of relevance or interest, they would discard the invitation as soon as it reaches them.

Once again, this is alarming because opportunities are being missed in gathering more support in inspiring employees to "act for sustainability." UNESCO (2014a, b: 12) Hence this study supports Lenglet (2014: 124) when he states that "more and sounder research is needed on how ESD-inspired content and learning methods can make a real difference in getting people to move onto paths of sustainability." More recent findings in the UNESCO's report, (2014a, b: 151) claim experts have suggested that "on the basis of the last 10 years of work on private sector education, short experiential training events, which focus on systems thinking and practical decision-making, and which challenge participants from different sectors to co-create solutions to real problems, are most effective." This suggests that there is

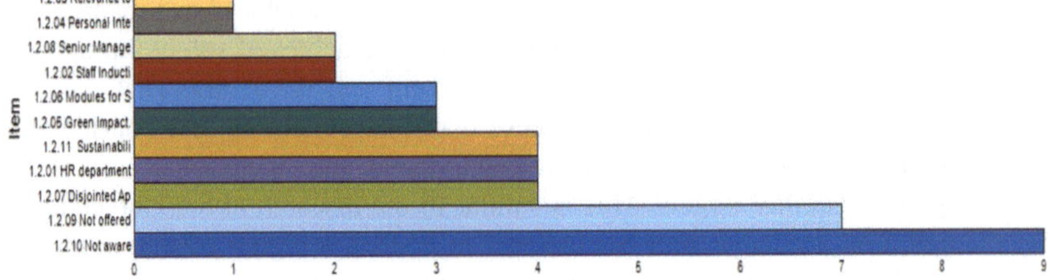

Fig. 6 Provision of ESD for staff at DMU

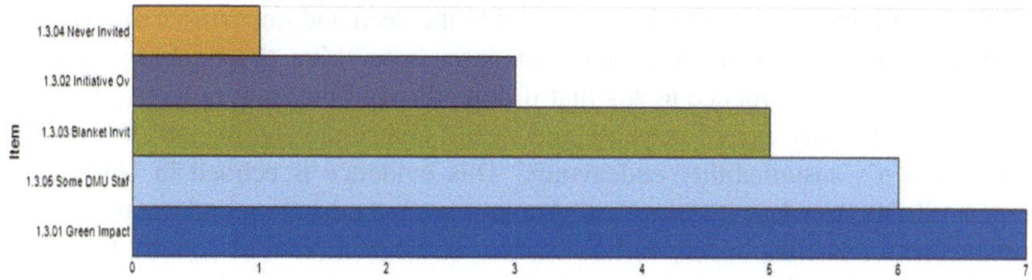

Fig. 7 Provision of initiatives on sustainable development for staff at DMU

a need and a demand for ESD in the work setting with a growing body of leadership teams and senior management teams in organisations recognising this need. Yet they are requesting "a greater focus on scaling up business action as it relates to their own companies or industries and greater experiential learning" (UNESCO 2014a, b: 151). This concurs with two key findings in the study conducted at DMU that ESD for employees requires (a) senior management commitment and (b) needs to be relevant and context specific. Furthermore, there is an increasing awareness that "technical 'know-how' will not be sufficient. Skills and capacities for whole-systems approaches, critical thinking and collaborative problem-solving will also be needed for private sector transformation." (ibid.: 152). This gives strength to the work undertaken with employees at DMU on exploring ways how thinking skills would enhance behavioural change if they are infused into ESD for employees.

H. **Skills required for empowerment of staff**

Findings shown in Fig. 8 highlighted three top empowerment skills: (i) provision of knowledge and understanding of ESD principles to staff; (ii) provide staff with motivation; and (iii) provision of soft skills and thinking skills training to staff. "*Make employees aware of the issues on sustainable development by providing training.*" (CONV009); "*Motivation means 'I really want to do this'*" (INT002); and "*I think to have a thinking mind, the inquisitive mind not to just accept things. Not to be risk averse.*" (INT006).

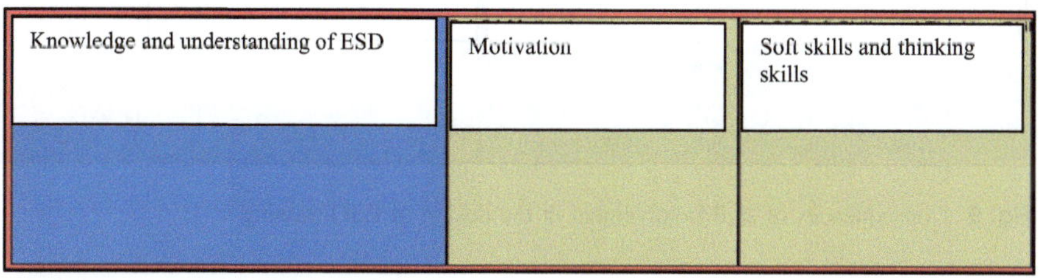

Fig. 8 Empowerment skills required by employees

Such findings are encouraging and point to the need and importance of providing training to employees if an organisation intends to drive sustainability forward. There is sufficient evidence to say that if done through the right process, employee training on ESD is indeed possible and would contribute significantly towards an organisation's sustainability endeavours. This evidence is echoed in UNESCO's report (2014a, b) when it states that "Another type of education and training for the private sector is customized, in-house training. Such initiatives contribute considerably to ESD in the workplace, including training staff to implement sustainable business models in corporate environmental management, corporate social responsibility and support for local sustainable development initiatives."

I. **Consequences of Staff Involvement in the Design of Needs Based ESD for DMU Staff**

This interview question was worded by using another of deBono's thinking tools called Consequence and Sequel (C&S) since it is a prime evaluation tool and directs us to "run things forward in our minds" (De Bono 1997).The two most commonly cited consequences were positive ones: (i) Positive initiative and needs-based; and (ii) Higher buy-in and a sense of ownership. "*The consequences would be that through the engagement, through the workshops and feeding into the design, it would personalise the messages.*" (INT002). INT005 was able to highlight the impact of such employee engagement in the design of ESD training and link it to behavioural change by the employees attending the training: "*They would be more interesting for the people who have to take part in them and I think that the people who were involved would be more engaged in the process afterwards.*" Whilst INT007 sums it up well when saying: "*It would be more efficient because if I was asked how I should be trained I would tell them with this thing in mind by stating my training and learning needs. Then they can train me up more accurately rather than somebody just sitting in an office and deciding that ... So if people were asked and their opinions were taken into account I think it would be more efficient and effective* (Figs. 9 and 10)."

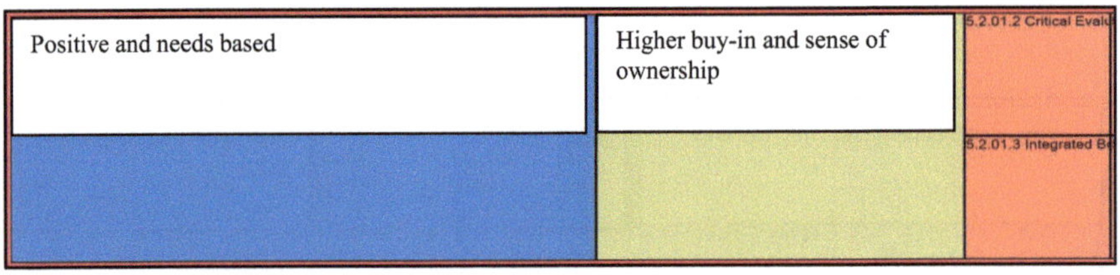

Fig. 9 Consequences of staff involvement in the design of ESD training

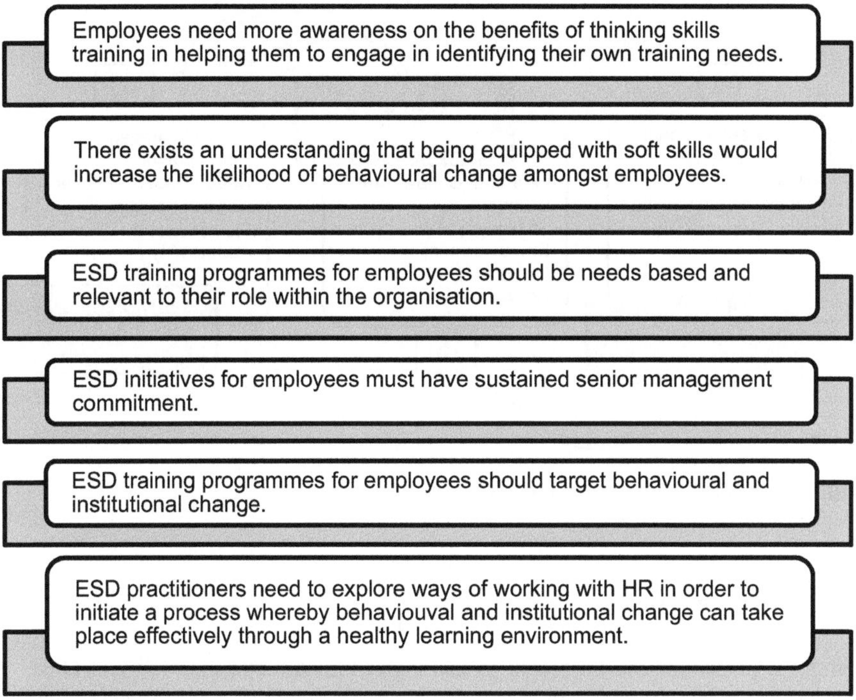

Fig. 10 Key findings

10 Summary of Findings

A summary of the key findings of the study is presented below. These were used to formulate the diagram (Fig. 11) that the researcher has designed to depict the role ESD can play in employee learning and training both as a *driver* for change and an *enabler* to deal with change.

The three key elements proposed in the diagram below are a reflection of the findings form the study at DMU and reflect the literature reviewed. These are: (i) Learning—for behavioural change; (ii) Instruction—for technical know-how; and (iii) Thinking Skills—for empowerment and engagement.

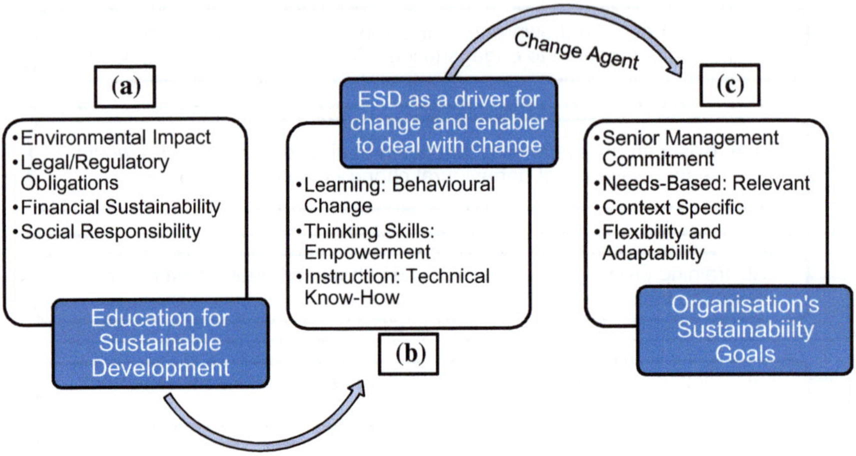

Fig. 11 The role of ESD in employee learning and training

11 Conclusion

Findings presented and discussed in this paper indicate that the reality at DMU is not in line with, or working towards objective 1 of the GAP programme (UNESCO 2014a, b: 14) which states that it is "to reorient education and learning so that everyone has the opportunity to acquire the knowledge, skills, values and attitudes that empower them to contribute to sustainable development." This is because employees are not offered ESD training despite research participants highlighting its benefits. UNESCO (2011a: 8) points out that key sustainable development issues are to be integrated into teaching and learning. Yet it warns that as these issues are "characterized by uncertainty, complexity and a high degree of systemic inter-connection, ESD requires participatory teaching and learning methods like critical thinking, imagining future scenarios and making decisions in a collaborative way in order to empower learners to take action for sustainable development." Findings from the study at DMU on exploring the process for such ESD learning and teaching to take place in an organisation are evidence to the dire need for employees to become empowered to participate and take action towards sustainable development. The respondents have voiced their opinions on the benefits of staff involvement in the design of needs based ESD for employees yet they are dubious of the level of senior management commitment to see any of these ESD initiatives through.

In view of these findings organisations will need to carefully consider the learning environment and employee engagement that exists and possibly explore adapting and implementing recommendations the researcher has compiled for the study at DMU which are presented in the final section of this paper below.

12 Recommendations

The implications and recommendations that can be drawn from this research study are:

1. Rigorous and consistent training to all employees in soft skills, though thinking skills training is preferable.
2. A participative approach enabling employees to identify their own training needs and engage in the design of their training programmes. This can be done by introducing a scheme to enable employees to identify their own training needs and participate in the design of their training and staff development. Such scheme could be spearheaded by the HR and Staff Development Department with the support of the expertise held at the IESD.
3. Revamp or market the staff survey scheme amongst DMU staff and tap into it as a vehicle to engage with employees in order to assist them to list their training needs on ESD and to consult with employees throughout the design process of the training programme to ensure it is based on relevance to DMU staff and context.
4. Infuse thinking skills in ESD training programmes.
5. Senior management need to be seen committed to sustainable development at DMU and ought to ensure that it is a strong and sustained commitment.
6. Publicise information and initiatives on sustainable development amongst DMU staff effectively and efficiently through staff forum, staff café sessions and team meetings within every department across the organisation.
7. Phase in a compulsory ESD staff training session to all members of staff, the content of which takes into account the specific needs and context of target groups.
8. ESD training needs to place a strong focus on the ease and simplicity of taking charge of one's decisions, actions and behaviours in a manner that identifies those that can be altered or changed to lead a more sustainable lifestyle.
9. A strategy for employee engagement and training on sustainability that is infused in the corporate strategy of the organisation supported by a system for periodic review.

Whilst it is acknowledged that the study took place with a sample of employees at a higher education institution in the UK and reflects the realities within that context at a given time when the data was collected, the findings would serve as a good platform for debate and springboard for action by other organisations, particularly higher education institutions. The issues that surfaced through this study resonate amongst many other organisations with lessons to be learnt on how ESD can be introduced in employee training as a route to behavioural change within organisations.

References

Bell, J. (2010). *Doing your research project*. Berkshire, UK: Open University Press.

Bryman, A. (2004). *Social research methods* (2nd ed.). Oxford, UK: Oxford University Press.

Cebrian, G., Grace, M., & Humphris, D. (2015). Academic staff engagement in education for sustainable development. *Journal of Cleaner Production, 106* (Elsevier Ltd.).

Chalkley, B., Haigh, M., & Higgit, D. (Eds.). (2009). *Education for sustainable development, papers in honour of the United Nations decade of education for sustainable development (2005–2014)*. Oxon, UK: Routledge.

De Bono, E. (1991). *Teaching thinking*. London: Penguin Books.

De Bono, E. (1997). *Direct attention thinking tools manual. Advanced Practical Thinking Training*. Inc. U.S.A.: The McQuaig Group Inc.

De Montfort University, (2011). *Strategic plan 2011-2015*. Leicester, UK: De Montfort University.

Denscombe, M. (2010). *The good research guide* (4th ed.). Berkshire, UK: Open University Press.

EAUC. (2015). Sustainability in education 2015. http://www.eauc.org.uk/the_state_of_sustainability_in_tertiary_educati. Last accessed January 11, 2016.

Eboli, M., & Mancini, S. (2012). Corporate education for sustainability. *International Journal of Environment and Sustainable Development, 11*(4), 339–354.

Epstein, M. J. (2008). Improving social and financial performance in global corporations. In *Making sustainability work*. Sheffield, UK: Greenleaf Publishing Limited.

Fien, J. (2006). A letter from the future: UNESCO and the decade of education for sustainable development. *Australian Journal of Environmental Education*, *22*(1), 63–70.

Grbich, C. (2013). *Qualitative data analysis—An introduction* (2nd ed.). London: Sage.

HEA/QAA. (2014). Education for sustainable development: Guidance for UK higher education providers. http://www.qaa.ac.uk/en/Publications/Documents/Education-sustainable-development-Guidance-June-14.pdf. Last accessed January 11, 2016.

Higgit, D. (2009). In Chalkley, B., Haigh, M., and Higgit, D. (Eds.). *Education for sustainable development, papers in honour of the United Nations Decade of Education for Sustainable Development (2005-2014)*. Oxon, UK: Routledge.

http://dmu.ac.uk/dmu-staff/hot-topics/2015/august-2015/dmu-recognised-for-green-credentials.aspx. Last accessed September 02, 2015.

http://www.dmu.ac.uk/about-dmu/sustainability/sustainability-strategy.aspx. Last accessed August 14, 2015.

Jackson, S. E., Ones, D. S., & Dilchert, S. (Eds.). (2012). *Managing human resources for environmental sustainability*. California, USA: Jossey-Bass.

Jickling, B. (2006). The decade of education for sustainable development: A useful platform? Or an annoying distraction? A Canadian perspective. *Australian Journal of Environmental Education*, *22*(1), 99–104.

Jickling, B., & Wals, A. (2008). Globalization and environmental education: Looking beyond sustainable development. *Journal of Curriculum Studies, 40*(1), 2008.

Leal Filho, W. (2002). Editorial. In *International Journal of Environment and Sustainable Development*. Switzerland: Inderscience Enterprises Ltd.

Lear, L. (1999). In R. Carson (Ed.), *Silent spring* (reprinted with a new afterword). London, UK: Penguin Books.

Lenglet, F. (2014). Can ESD reach the year 2020? *Journal of Education for Sustainable Development, 8*(2). London, UK: Sage Publications.

Mason, J. (1997). *Qualitative researching*. London, UK: Sage.

Mcniff, J., & Whitehead, J. (2010). *You and your action research project* (Vol. 3). Oxon, UK: Routledge.

Mehlmann, M., & Pometun, O. (2013). *ESD dialogues-practical approaches to education for sustainable development by and for educators*. Germany: Global Action Plan International, Books on Demand.

Morris, D. & Martin, S. (2009). Complexity, systems thinking and practice. In A. Stibbe (Ed.), *The handbook of sustainability literacy: Skills for a changing world*. Devon, UK: Green Books Ltd.
Noe, R. A. (2013). *Employee training and development* (Vol. 6). New York, USA: McGraw-Hill.
O'Leary, Z. (2004). *The essential guide to doing research*. London, UK: Sage.
Parker, J., & Wade, R. (Eds.). (2008). *Journeys around education for sustainability*. London, UK: London South Bank University.
Pavlova, M. (2012). Towards using transformative education as a benchmark for clarifying differences and similarities between environmental education and education for sustainable development. *Environmental Education Research, 19*(5), 656–672.
Qian, W. (2013). Embracing the paradox in educational change for sustainable development. *Journal of Education for Sustainable Development, 7*(1). London, UK: Sage Publications.
Redmond, J., & Walker, E. (2009). Environmental education in small business: The owner-manager's perspective. *Australian Journal of Environmental Education, 25*.
Schmidheiny, S. (1992). *Changing course: A global business perspective on development and the environment*. Cambridge: MIT Press.
Sherren, K. (2006). Core issues: Reflections on sustainability in Australian University coursework programs. *International Journal of Sustainability in Higher Education, 7*(4), 400–413.
Szovics, P., Tessaring, M., Walmsley, C., & McGarth, J. (2011). *Future skill needs for the green economy*. UK: Dictus Publishing.
Thomas, I., & Nicita, J. (2002). Education for sustainability and Australian universities. *Environmental Education Research, 8*(4), 475–492.
Tilbury, D. (2006). Australia's response to a UN decade in education for sustainable development. *Australian Journal of Environmental Education, 22*(1), 77–81.
UNESCO. (2005). *International implementation scheme for ESD*. Paris, France: United Nations Educational, Scientific and Cultural Organization.
UNESCO. (2014a). *Roadmap for implementing the global action programme on education for sustainable development*. Paris, France: United Nations Educational, Scientific and Cultural Organization.
UNESCO. (2014b). *Shaping the future we want*. Paris, France: United Nations Educational, Scientific and Cultural Organization.
UNESCO-UNEP. (1977). Intergovernmental conference of environmental education: Tbilisi declaration. Tbilisi (USSR). http://unesdoc.unesco.org/images/0003/000327/032763eo.pdf. Last accessed January 11, 2016.
UNGC, (2013). *Global corporate sustainability report 2013*. New York: United Nations Global Compact.
Zibarras, L. D., & Coan, P. (2015). HRM practices used to promote pro-environmental behaviour: A UK survey. *The International Journal of Human Resource Management, 26*(16) (UK: Routledge).

Embedding Education for Sustainable Development (ESD) Within the Curriculum of UK Higher Educational Institutions (HEIs): Strategic Priorities

Obehi Frances Sule and Alison Greig

Abstract

Higher Education is recognised as having a significant role to play in achieving sustainable development, through its teaching and research, its business operations and community engagement and through the sustainability of its buildings (HEFCE in Sustainable Development in Higher Education 2014). The UK Government, e.g. through its funding councils and the United Nations e.g. through the Decade of Education for Sustainable Development (DESD), have provided considerable encouragement and support for Higher Education Institutions to embrace this agenda. This paper provides a critical commentary on the strategic importance that higher education institutions themselves have placed on sustainability in recent years. It builds on Sterling and Scott's (Environmental Education Research 14(4):386–398, 2008) paper which noted that although good progress had been made in promoting sustainability within campus management activities and to some extent research, very little had been done to re-orientate HEIs curriculum. They suggest that the curriculum dimension suffered from a lack of incentive to engage, inadequate leadership from UK HEIs main funding body (HEFCE) and the autonomous nature of teaching and learning. Most importantly however, their review identified that sustainability principles had largely not permeated institutions visions, ethos and practice, and suggest that without this, real transformation of all aspects of HEIs operations and practices may not be possible. This study investigates whether this vital shift is now taking place. The research involved a qualitative document analysis of 128 UK HEIs strategic or corporate plans to explore which and how aspects of sustainability are explicitly mentioned and to what extent ESD is being prioritised across institutions. Findings show that the main focus across

O.F. Sule (✉) · A. Greig
Global Sustainability Institute, Anglia Ruskin University, East Road, Cambridge, UK
e-mail: obehi.sule@anglia.ac.uk

W. Leal Filho (ed.), *Sustainable Development Research at Universities in the United Kingdom*, World Sustainability Series,
DOI 10.1007/978-3-319-47883-8_6

HEIs still remains campus greening and the embedding of sustainability within the curriculum is often still not included as an explicit part of HEI's strategic priorities. In short, the potential for higher education to contribute to sustainable development is not yet being fully realised. The study also identified what appears to be several existing missed opportunities to strategically advance sustainability in HEIs.

Keywords
Education for sustainable development · Sustainability · Curriculum · Strategic priorities · Higher education

1 Introduction

The vital role of education in attaining sustainable development has long been recognised on the world stage. A whole chapter (Chap. 36) of Agenda 21 (1992) presented at the first United Nations Conference on Environment and Development (The 'Rio' Earth Summit) was dedicated to 'Promoting Education, Public Awareness and Training'. Also, in 2002, the United Nations General Assembly declared the years 2005–2014 as a Decade for Education for Sustainable Development (DESD). The main thrust of the DESD was to aid the re-orientation of the world's populace towards a more sustainable future and assist nations make progress on their Sustainable Development (SD) agendas 'through the integration of the principles, values and practices of sustainable development in all aspects of education and learning' (UNESCO 2016).

The United Nations Education Social and Cultural Organisation (UNESCO) was lead agency for the DESD and developed an International Implementation Strategy (UNESCO 2005), which addresses both formal and informal education which nations could tailor to their own peculiar needs. During the DESD, Higher Educational Institutions (HEIs) faced a wave of pressure, driven by national and international policy and the apparent environmental, economic and societal challenges, to incorporate ESD in their programmes for students to develop the knowledge, skills and values necessarily for the shift towards a sustainable pathway (Martin et al. 2014). About 15 million students annually attend institutions of higher learning, including a disproportionate number who go on to become future world leaders, inventors, employers and employees. The UK has a particularly impressive record of educating the world future decision makers, for example, 55 past and current world leaders (Presidents, Prime Ministers and monarchs) from 51 countries graduated from UK HEIs and out of the 245 current heads of state, 26 attended a UK university (HEPI 2015). HEIs are therefore increasingly being looked to, as vital to enabling sustainability (UNESCO 2014).

UNESCO's end of DESD Report (2014) notes that within the decade, higher education increased its efforts to support sustainable development and that globally, the number of people engaged in efforts to infuse sustainability principles in curricula across all levels of education in both public and private sectors has increased to thousands. It highlights the significant increase in efforts to address the sustainability of campus operations (typically involving initiatives to reduce carbon and waste), sustainability research and the building of networks of HEIs and scholars for the sharing of tools, reporting frameworks and good practice. However, while acknowledging that there has been a rise in stand-alone specialist courses on sustainability and that wider pedagogic and curricular innovation and changes in policies and practice have taken place, they assert that systemic progress in the reorientation of learning and teaching practices has not taken place. They point to the need for strong political leadership in order to scale up effective programmes and translate policy commitments across operations, teaching and curricula both in formal systems and in non-formal learning (UNESCO 2014).

Leal Filho (2010) categorises the level of implementation of sustainability in higher education systems into three incremental stages; (1) those who lack understanding of the concept, make no substantial effort to promote it in their institutional operations and are not engaging in systematic projects to promote sustainability. (2) Institutions who understand the ideal and are significantly promoting sustainability in their practices, are engaging in projects and a programme of research and extension. (3) those who in addition to stage 2, exhibit long-term commitments to the SD agenda, as they make it core to their very existence/way of thinking, principles and practices, have sustainability policies and senior staff overseeing implementation efforts. Sterling and Scott (2008) emphasise that real transformation of all aspects of HEIs operations and practices towards SD may not be possible if sustainability principles has not permeated their visions, ethos and practices.

The aim of this paper is to investigate how far UK HEIs have travelled along this journey to make SD part of their core purpose and practice, that is, it explores the strategic importance being placed on Education for Sustainable Development.

2 UK Higher Education

In the UK, Sterling and Scott (2008) noted midway through the DESD that although good progress had been made in promoting sustainability within UK HEIs campus management operations and to some extent research, very little had been done to re-orientate HEIs curriculum. They suggest that the curriculum dimension suffered from a lack of incentive to engage, inadequate leadership from UK HEIs main funding body (Higher Education Funding Council for England) and the autonomous nature of teaching and learning. Most importantly however, their review identified that sustainability principles had largely not permeated institutions visions, ethos and practice, and they stress that without this, real transformation of all aspects of HEIs operations and practices may not be possible.

Arguably, UK political leadership for sustainability was strong at the start of DESD, with key policy documents such as its sustainability strategy "Securing the future" (2005), and other key policy initiatives, including its "Low Carbon

Transition Plan" and "Low Carbon Skills Challenge" highlighting the importance of ESD (BIS 2010; DECC 2010). Also, indicative of the importance being given to both education and sustainability, earlier in 1998, the UK government established a Sustainable Development Education Panel (SDEP) to facilitate action on embedding ESD in all education sectors throughout the UK. But progress faltered as the decade continued, as a result of changing political priorities and the re-organisation of the policy framework for higher education, including, in 1999 devolution of educational strategy to individual countries (England, Scotland, Wales and Northern Ireland).

In England, higher education's chief regulator is the Higher Education Funding Council for England (HEFCE). As Sterling and Scott (2008, p 387), note "key to understanding the current progress and prospect of ESD in HE in England is the policy of HEFCE and its relationship with HEIs". As a non-departmental public body of the Department for Business Innovation and Skills (BIS) HEFCE is a direct conduit between central government strategy and HEI policy which is crystallised each year in a letter (The Grant Letter) from the Secretary of State for Business Innovation and Skills to HEFCE confirming funding allocations and priorities for HEFCE and higher education the forthcoming year. Although in most years the Grant Letter includes specific mention of support for sustainability, this support invariably focused on supporting institutions in their efforts to improve *their* sustainability. It was therefore little surprise that in 2014 HEFCE's Sustainable Development Framework focused principally on reducing the environmental impacts of the sector. Curriculum support was included, but limited to the reorientation of existing disciplines to supply graduates, and in particular science, technology, engineering and mathematics (STEM) graduates to emerging low carbon or 'green' sectors of the economy.

HEFCE has, however, also taken action to raise awareness and increase engagement in ESD within the curriculum, albeit through a different route. The Higher Education Academy was founded in 2004 as a British professional institution to promote excellence in higher education learning and teaching. It is funded directly and indirectly by HEFCE and the other national funding councils (Scottish Funding Council, Higher Education Funding Council for Wales and the Department for Employment and Learning Northern Ireland) via grants and institutional subscriptions.

Although principally working within traditional subject disciplines, until 2014 HEA supported a number of cross-disciplinary themes including ESD. In 2011 it established its Green Academy change programme, to assist institutions of learning develop their ESD agenda, aiming mainly to incorporate ESD in 'student experience' primarily through the curriculum (McCoshan and Martin 2012). From 2011 to 2013 it worked with 18 HEIs to support strategic change in university curricula and processes and develop the leadership capacity for ESD (Luna et al. 2012; McCoshan and Martin 2012; Martin et al. 2014). Additionally the Green Academy became an informal ESD change management network where participants exchange ideas and resources (Kemp 2012). The programme was well regarded by those who participated as well as by other stakeholders and was instrumental in working collaboratively in the creation of two important publications. With the Quality Assurance Agency (QAA), HEIs academic standards regulator, it developed practical guidance for HEIs on how to work with students to foster their knowledge, understanding and skills in the area of sustainable development. This guidance was specifically intended to complement Chapter B3 of the UK Quality Code for Higher Education (QAA 2014) which sets out the expectations that all providers of UK higher education are required to meet and is used in QAA review processes. Secondly the HEA has, since 2011, worked with the National Union of Students (NUS), to carry out a national annual student survey assessing students' attitudes towards, and expectations on, sustainability (Bone and Agombar 2011; Drayson et al. 2012, 2013; Drayson 2015). These surveys now form a unique, large (around 20,000 respondents) and continuous data set of student attitudes to sustainability. Which has over the years consisted indicated students expect their institutions of higher learning to provide them with the skills, values and knowledge to live and work in a sustainable way.

HEI leaders have in various summits and conferences declared their support for the sustainable development and its education agenda. They have however been constantly criticised for signing declarations to show visible commitment, but mostly not taking any action or doing enough to sufficiently bring about the desired changes (Bekessy et al. 2007). Within varying political and regulatory leadership conditions and implementation challenges, there are indications HEIs have within the DESD been embedding the principles of sustainability in their operations and programmes (UK commission for UNESCO 2013; QAA 2014).

3 Methodology

This study's methodology consist of a background analysis reviewing literatures (including key reports) also utilises empirical evidence gathered from UK HEIs public strategic documents. Strategic documents normally called Corporate Plans (CPs) represent the organisations goals, objectives and future priorities. CPs show what senior management are prioritising and are seen as a useful tool for planning, implementing and monitoring efforts to achieve their strategic goals (HEFCE 2000). CPs therefore provide a rich source of data to address the research aims. Which is to explore if sustainability principles have permeated institutions visions, ethos and practices. In particular, the strategic importance being placed on Education for Sustainable Development.

Of the 148 HEIs in the UK with degree awarding powers (GOV.UK 2015), only 128 had published corporate plans during the study period—June 2015 to March 2016 (Table 1). These 128 CPs were analysed qualitatively.

The qualitative content analysis approach applied in this study, "is a research method for the subjective interpretation of the content of text data through the systematic classification process of coding and identifying themes or patterns" (Hsieh and Shannon 2005, p 1278). A conventional approach to content analysis

Table 1 UK HEIs included in the study by devolved administration/country

Administration	Published corporate plan/total
England	102/122
Scotland	15/15
Wales	8/8
Northern Ireland	3/3
Total No	128/148

was applied as described by Hsieh and Shannon (2005). Steps taken were familiarisation with the documents. Reading the documents word for word and understanding the documents as a whole. Enabling understanding of the context words were used, which aided in identifying words used within sustainability context as well as generally relevant to social, economy and environmental issues. Coding ensued without predetermined categories for classification. Codes were sorted into themes based on their relationship. Themes are defined and their links and implications are discussed and presented with theme exemplars mostly illustrated in the form of descriptive statistics used with simple graphics to summarise and quantitatively describe the data (Trochim 2002; Hsieh and Shannon 2005). This was by no means a linear process, rather it was iterative, and a computer aided qualitative data analysis software 'NVivo' was utilised for organising, managing and coding the data.

Though this approach has unique advantages including generating knowledge grounded in the actual data without the influence of preconceived categories. It is important to acknowledge there are potential pitfalls which authors describe as "credibility within the naturalistic paradigm of trustworthiness or internal validity within a paradigm of reliability and validity (Hsieh and Shannon 2005, p 1280)". This is pertaining to researchers failing to identify key categories, haven failed "to develop a complete understanding of the context, thus resulting in findings that do not accurately represent the data" (Hsieh and Shannon 2005, p 1280). In a bid to overcome this challenge and establish credibility, this study applied, persistent observation, prolonged engagement and carried out negative case analysis. However, were funding is available, future studies may want to consider having multiple coders and method triangulation to confirm findings.

3.1 Clarification of Terms

In this study, the term 'Sustainability' is used generally in relation to the pillars of sustainable development, that is, the environment, society and economy. Where there is need to distinguish particular dimensions (for instance environmental sustainability) or indicate all aspects are integrated (sustainable development) it is clearly stated. Similarly, 'Sustainability Education' is used to denote education geared towards sustainable development. Where distinction needs to be made, the particular dimension (for instance environmental education) or that it incorporates all aspects (education for sustainable development used interchangeably with education for sustainability) is clearly stated.

4 Results

4.1 Sustainability Practices

77 % (99/128) of UK HEIs' state they are engaging in sustainability in their corporate plans, but for 66 % of these it is mentioned in the context of campus greening (see Fig. 1).

Noticeably, campus greening (CG) was the most frequently mentioned aspect of sustainability throughout the devolved administrations. Welsh HEIs' were most likely to be engaging with the broader sustainable development agenda (Fig. 2), which may be attributed to stronger Welsh government leadership in this area (UK National Commission for UNESCO 2013).

Amongst the Institutions, sustainability relevant practices (see Fig. 3) were identified in all except one HEI's corporate plan.

The findings indicates that for majority of institutions with sustainability agendas, their sustainability practices tended to cut across all three aspects of

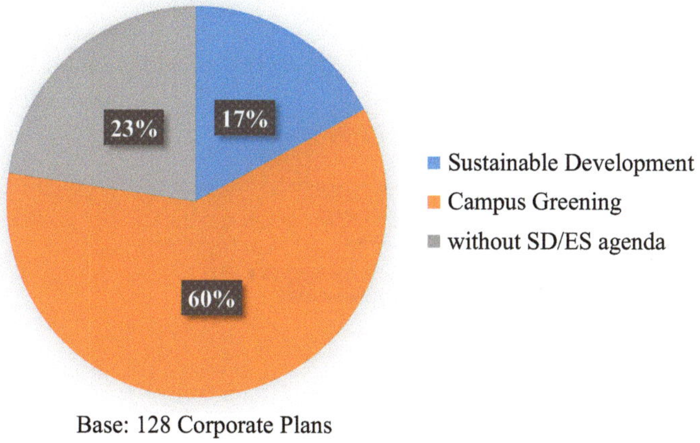

Fig. 1 Sustainability in UK HEIs' corporate plans

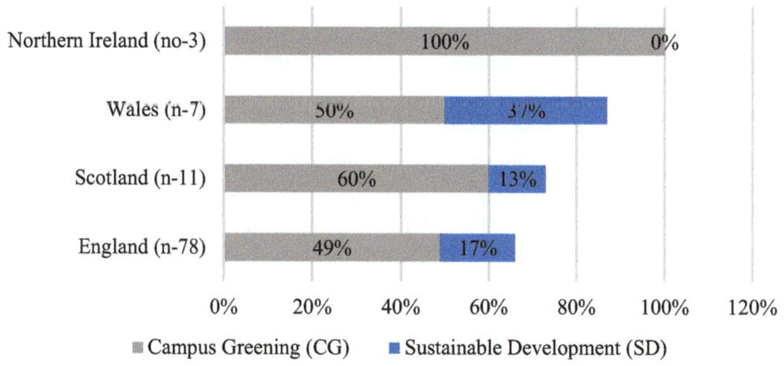

Fig. 2 UK HEI's with sustainability agendas by devolved administration

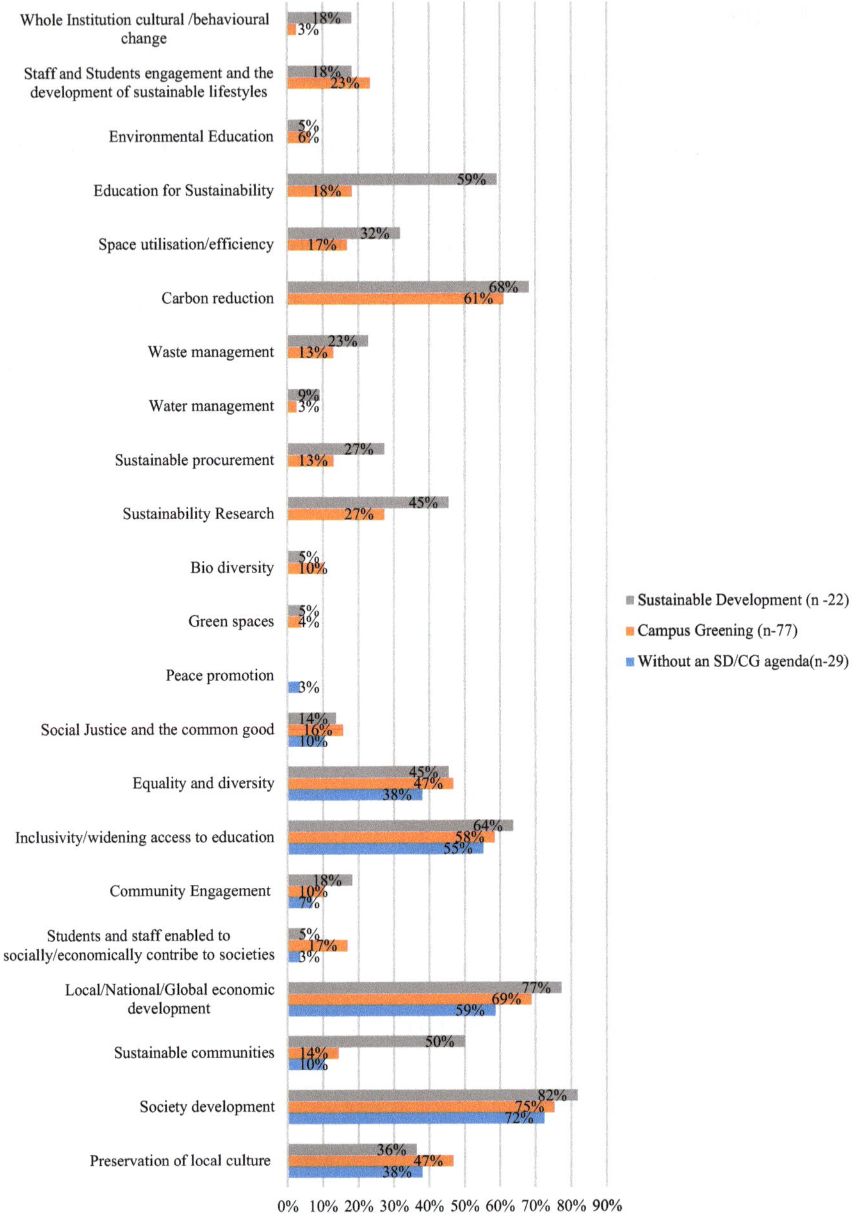

Fig. 3 Sustainability practices identified in HEIs' strategic plans

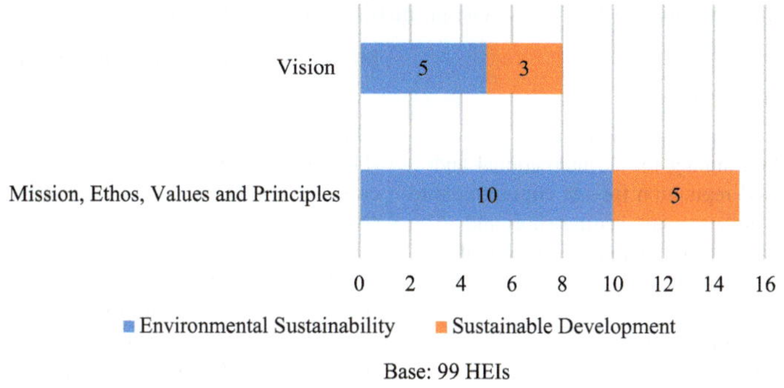

Fig. 4 Sustainability in HEIs' vision, mission and ethos

sustainability. HEIs' focusing on the environmental aspects of sustainability, often had the economic and social dimensions of sustainability as part of their strategic direction but do not link these to their sustainability agenda. The findings also indicate that institutions without any stated environmental or sustainable development agendas did actually include aspects which could be considered as relating to sustainability. In these institutions, it could be argued that aspects of sustainability are being considered at a strategic level but that they are not calling it sustainability.

4.2 Sustainability in Visions and Ethos

Only a handful of HEIs' have included SD or Environmental Sustainability in their Vision, Mission and Values. It most often appears as part of their aspirations around ethos (Fig. 4 and Table 2).

4.3 Sustainability in Institutions Curricular

4.3.1 Environmental Education

Only five percent of HEIs (6 institutions) specifically mention environmental education (EE) in their strategic documents (Table 3) but none of these had set targets or key performance indicators relating to EE. Interestingly 3 of the institutions appear to be using this environmental focus to attempt a whole institution approach to sustainability (connecting curriculum & pedagogy, research, campus operations and community). They make no mention of the economic or social aspects of sustainability (Fig. 5).

Table 2 Sustainability in UK HEIs' vision, mission, ethos, values and principles

Vision	Mission and ethos, values and principles
Sustainable development	
"We shall continue our emphasis on sustainability, aiming for an international and 'best in class' reputation for our commitment to sustainable development, setting short and long-term targets to integrate all aspects of sustainability into our daily operations" (Bangor University) "committed to operating in a sustainable manner" (University of Bristol) "We will strive to be a sustainable and responsible organisation which contributes to positive environmental, social and economic futures across the communities we serve" (University of Gloucestershire)	"Sustainability planned sustainable development (financially, socially and environmentally) is crucially important to securing our future" (Brunel University) "Innovation our contribution to the sustainable development of communities, organisations and society is built on our ability to innovate through research, enterprise and our own practice" (University of Bedfordshire) "Sustainable development through a system-based approach to delivering meaningful and relevant educational pathways we will promote learning and social responsibility that supports "development that meets the needs of the present without compromising the ability of future generations to meet their own needs" (Brundtland Commission 1987) (University of Wales Trinity Saint David) "A commitment to health, well-being, sustainability and sustainable development" (University of Central Lancashire) "Sustainability through education and research we are aware of the ecological limits of the planet and promote the careful use of resources" (University of Exeter)
Environmental sustainability	
"We want Brighton staff and students to be known for their commitment to impact, community and sustainability in their chosen field" (University of Brighton). "Our vision is to be recognised around the world for our signature contributions, especially in global food security, energy and sustainability, and health" (University of Nottingham) "Keele vision we will be a leading campus-based university that stands out due to our unique community, our world leading research and our broad-based education that produces graduates who have a genuine positive impact across the globe. Our research will be transformational in higher education and across society more broadly and we will be internationally recognised for our professionalism, collegiality and environmental sustainability" (Keele University) "Make a significant contribution to global efforts to achieve environmental sustainability" (De Montfort University) "The University of Surrey is committed to being a leading national and international university. Our high quality teaching, learning, research and	"We aim to achieve this through high-quality education, research and enterprise activities. Success is demonstrated by significant cultural, economic, environmental and social contributions at local, national and international scales" (University of Greenwich) "Through teaching, learning, research and innovation we work in partnership with our students, staff, community, business and the professions to drive social inclusion, economic prosperity and sustainability in Plymouth, across the nation and throughout the world" (Plymouth University) "We are committed to the twin principles of sustainability and social responsibility as foundations for all our activities" (Aberdeen University) "Environmental and financial sustainability we will exploit our strengths in research and education to achieve progressive social, environmental and economic benefits, locally, nationally and internationally. We will manage resources to deliver a sustainable and long-term future for the University" (Durham University)

(continued)

Table 2 (continued)

Vision	Mission and ethos, values and principles
enterprise, will be delivered in a financially and environmentally sustainable manner, within an academic community that values collegiality and professionalism, providing our students with skills that allow them to maximise their potential"	"Maintain and develop a campus that is both of outstanding quality and sustainable" (University of East Anglia) "Commitment to sustainability" (Teesside University) "We will be environmentally and financially sustainable and resilient" (Edinburgh Napier University) "We will be committed to environmental sustainability, setting and meeting the highest possible standards across the full range of our activities" (University of Manchester) "Respect for the environment: we will manage the school's resources in ways that meet the needs of the present without compromising the options of future generations" (London School of Economics and Political Science) "The development of our staff, estate and physical resources, as the bedrock of a professional and supportive academic community, and with equality, diversity and environmental sustainability to the fore" (Norwich University of the Arts) "We are committed to (…) Financial and environmental sustainability" (London School of Hygiene and Tropical Medicine)

Table 3 Environmental education (EE) in UK HEIs' by administration

Administration	Incoporating EE	Percentage (%)
England	6/102	6
Scotland	0/15	0
Wales	0/8	0
Northern Ireland	0/3	0
Total HEIs	6/128	5

4.4 Education for Sustainable Development

21 % of HEIs' across three administrations (England, Scotland and Wales) stated in their corporate plans that they are embedding Education for Sustainable Development (ESD) in their institutions (Table 4).

Only some institutions provided information on how this was happening, 10 % talk only very generally about ESD being included in their curricula and pedagogy. A further 10 % indicated they are embedding ESD across all types of curricula and/or all disciplines. For example, Anglia Ruskin University states it "*continue*[s] *to incorporate sustainability across the curriculum and embed it generally in student life and activities*". Several universities also explicitly mention that they will support staff in this process.

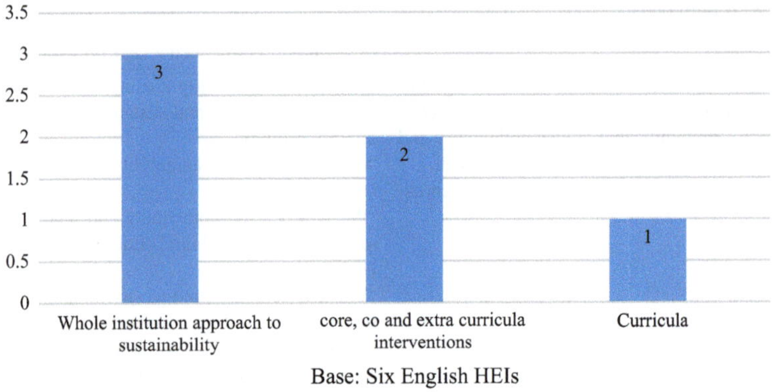

Fig. 5 HEIs' approach to incorporating environmental education in their institutions

Table 4 Education for sustainable development in UK HEIs' by administration

Administration	Incoporating ESD	Percentage (%)
England	20/102	20
Scotland	3/15	20
Wales	4/8	50
Northern Ireland	0/3	0
Total HEIs	27/128	21

Only one university has chosen to strategically prioritise the use of specialist courses on sustainability although a number of universities have developed specialist sustainability courses as part of their more general efforts on embedding sustainability into the curricula (Fig. 6).

Amongst the 21 % of HEI's incorporating ESD, only a handful (3 %) indicated they are strategically prioritising the inclusion of ESD as part of their academic processes (1 %) or have set ESD key performance indicators/targets (2 %). University of Gloucestershire stated it "embed[s] sustainability into the design and delivery of teaching programmes as a required component of initial course approval and revalidation and review for all programmes". Institutions with key performance indicators/targets in their strategic documents are University of Wales Trinity Saint David who appears to have an annual target to "*complete curriculum audits and develop the curriculum with due regard to the emerging sustainability agenda*". University of Bedfordshire amongst its key measures of success is their sustainable development strategy focusing "on cultural as well as practical change and on the development of future generations of leaders with a firm understanding of, and commitment to, sustainability". Anglia Ruskin University with a 41 % baseline on "percentage of students who say that sustainability has been a feature of their experience", aims to achieve an increasing annual target of 50 % in 2015, 60 % in 2016 and 70 % in 2017.

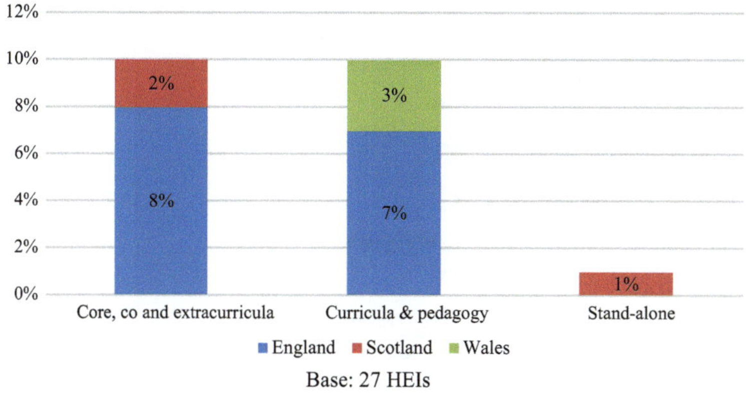

Fig. 6 HEIs' approach to incorporating ESD in their institutions

4.5 Sustainability Education Themes

Though only 33 HEIs' explicitly stated their engagement with sustainability education (either ESD or EE), a further search of the strategic documents for sustainability education related a number of themes (Fig. 7) which relate closely to sustainability but are not explicitly linked. For example the University of Bolton as part of its' Learning, Teaching and Assessment practices, aims to "*help students develop, recognise and use their potential, and make positive contributions to society, developing as global citizens*".

5 Discussion

This study presents a critical commentary of ESD in HEIs, explores how aspects of sustainability are explicitly and implicitly included in strategic priorities and also assessed the extent ESD is being prioritised. This study found relatively and the HEA strong support from Government, education councils and the HEA in the early part of the DESD. However, this appears to have waned in the later part of the decade, as can be inferred from HEFCE's recent (2014) sustainability framework which focuses on STEM subjects only.

Sustainability principles have largely not permeated HEIs visions and ethos and for most sustainability practicing institutions, their activities still mostly focus on campus greening activities. Although, other aspects of sustainability may be part of corporate visions they may not be explicitly linked to the sustainability agenda. For example, aspiration to make positive impact on local culture, communities, social and economic development was relatively common. The question this raises is therefore how to make these links explicit, and help 'mainstream' sustainability into

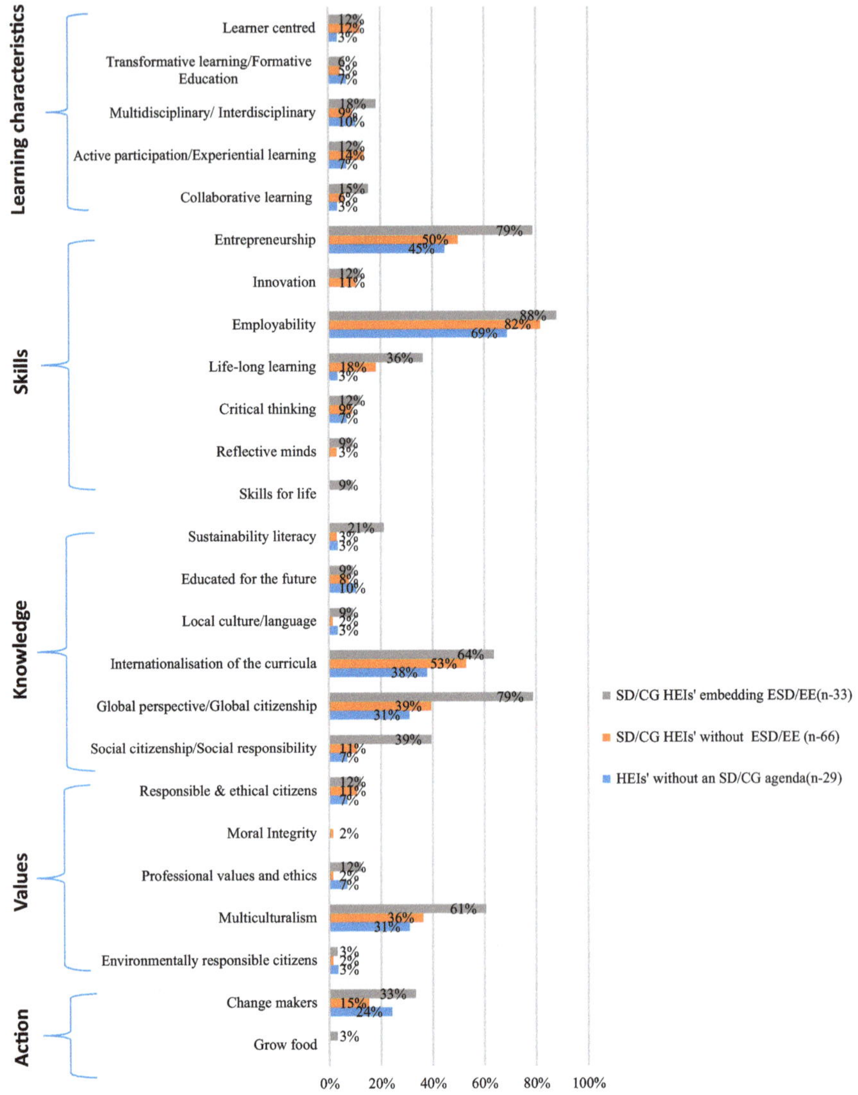

Fig. 7 Sustainability education themes in HEIs corporate plans

the university. One suggestion could be working with institutions possibly on a case by case basis, where the relevance of their existing priorities to the goal of incorporating sustainability in their Institutions is made clear and understanding around this is fostered.

Findings also indicate that, ESD has generally not penetrated HEIs strategic priorities; while 21 % indicated a general commitment to ESD only 2 % of these institutions have set targets/key performance indicators to monitor their progress.

Using the three level framework suggested by Leal Filho (2010), it appears most (around 63 %) of UK HEIs are in, or aspiring to be at level 2, while a further 14 % claim to be at least aspiring to level 3. The reasons for this lack of progress will take further investigation but may relate to the continued dominance of the environmental agenda. Though, the data revealed that a number of Universities include themes within their corporate plans (e.g. Global citizenship and Social citizenship) which have clear links to sustainability and sustainability education but are not making these links explicit. This indicates that there may be more happening than a document analysis would first indicate. It also raises the question on how to harness these existing but currently lost opportunities to advance the ESD agenda.

6 Conclusion

This paper builds on Sterling and Scott's (2008) mid DESD paper which noted that although good progress had been made in promoting sustainability within UK Higher Educational Institutions' campus management activities and to some extent research, very little had been done to re-orientate HEIs curriculum. Their review identified that sustainability principles had largely not permeated institutions visions and ethos. The findings of this study suggest some further progress, despite the reduction in government support. Specifically it suggests that sustainability within campus management activities has progressed even further but that ESD in the curricula is still largely contained within specific courses and programmes. ESD is still not a strategic priority in most UK HEIs, and sustainability has not permeated most institutions vision, ethos and values. Although there may be other sustainability and education for sustainability related activity happening within UK HEIs there is little evidence in CPs that transformation of all aspects of HEIs operations and practices to included sustainability, explicitly or implicitly has progressed significantly.

It is important to note the limitation of this study. Corporate Plans are aspirational documents intended to provide high level strategic direction. They demonstrate a 'willingness' by institutions but do not indicate their level of success, what actually happens or how. This work has been undertaken as part of a PhD study which is also investigating these other aspects, in particular the role of individual and collective leadership among staff and students in progressing sustainability in their organisations.

References

Bekessy, S. A., Samson, K., & Clarkson, R. E. (2007). The failure of non-binding declarations to achieve university sustainability: A need for accountability. *International Journal of Sustainability in Higher Education, 8*(3), 301–316.

BIS. (2010). *Skills for sustainable growth*. Department for Business, Innovation & Skills. www.bis.gov.uk/policies/further-education-skills/skillsfor-sustainable-growth. Last accessed October 18, 2015.

Bruntland Commission. (1987). *Our common future: The world commission on environment and development*. Oxford UK: Oxford University Press.

Bone, E., & Agombar, J. (2011). First-year student attitudes towards, and skills in, sustainable development. In *National Union of Students: A report for the Higher Education Academy*. https://www.heacademy.ac.uk/sites/default/files/firstyearattitiudes_finalreport.pdf. Last accessed March 9, 2016.

DECC. (2010). Meeting the Low Carbon Skills Challenge—A Government Response. Department of Energy and Climate Change. www.decc.gov.uk/assets/decc/Consultations/lowcarbonskills/1116-meeting-low-carbon-skillscons-response.pdf. Last accessed September 18, 2015.

Drayson, R. (2015). Student attitudes towards and skills for sustainable development. In *National Union of Students; A report for the Higher Education Academy*. https://www.heacademy.ac.uk/sites/default/files/executive-summary-students.pdf. Last accessed March 9, 2016.

Drayson, R., Bone, E., & Agombar, J. (2012). Student attitudes towards and skills for sustainable development. In *National Union of Students: A report for the Higher Education Academy*. https://www.heacademy.ac.uk/sites/default/files/student_attitudes_towards_and_skills_for_sustainable_development.pdf. Last accessed March 9, 2016.

Drayson, R., Bone, E., Agombar, J., & Kemp, S. (2013). Student attitudes towards and skills for sustainable development. In *National Union of Students; A report for the Higher Education Academy*. https://www.heacademy.ac.uk/sites/default/files/esd_student_attitudes_2013_v4.pdf. Last accessed March 9, 2016.

GOV.UK. (2015). *Check if a university or college is officially recognised*. GOV.UK. https://www.gov.uk/check-a-university-is-officially-recognised/recognised-bodies. Last accessed August 4, 2015.

HEFCE. (2000). *Strategic planning in higher education; A guide for heads of institutions, senior managers, and members of governing bodies*. Higher Education Funding Council England http://webarchive.nationalarchives.gov.uk/20100202100434/http://www.hefce.ac.uk/pubs/hefce/2000/00_24.pdf. Last accessed March 9, 2016.

HEFCE. (2014). *Sustainable development in higher education*. HEFCE's Role to date and a framework for its future actions. Higher Education Funding Council England http://www.hefce.ac.uk/media/hefce/content/pubs/2014/201430/HEFCE2014_30.pdf. Last accessed March 9, 2016.

HEPI. (2015). *Now that's what we call soft power: 55 world leaders educated in the UK*. Higher Education Policy Institute. http://www.hepi.ac.uk/2015/10/01/now-thats-call-soft-power-55-world-leaders-educated-uk/. Last accessed March 9, 2016.

Hsieh, H., & Shannon, S. (2005). *Three approaches to qualitative content analysis*. http://www.iisgcp.org/pdf/glssn/Supplemental_Reading_on_Coding_2.pdf. Last accessed April 18, 2016.

Kemp, S. (2012). *Green Academy: Curricula for tomorrow*. Higher Education Academy. https://www.heacademy.ac.uk/sites/default/files/green_academy_ptt_presentation.pdf. Last accessed April 18, 2016.

Leal Filho, W. (2010). *Sustainability at Universities: Opportunities, challenges and trends*. http://www.guninetwork.org/articles/sustainability-universities-opportunities-challenges-and-trends. Last accessed May 18, 2016.

Luna, H., Martin, S., Scott, W., Kemp, S., & Robertson, A. (2012). Universities and the green economy: graduates for the future. In *Higher Education Academy policy think tank report*. The Higher Education Academy. https://www.heacademy.ac.uk/sites/default/files/graduates_for_the_future_print_1308_12_1322.pdf. Accessed April 18, 2016.

Martin, S., McCoshan, A., & McEwen, L. (2014). Embedding Sustainability Into the Higher Education Curriculum: Lessons from the UK's Green Academy Change Programme. In *Proceeding of the 4th World Sustainability Forum Sciforum Electronic Conference Series*, 4.

McCoshan, A., & Martin, S. (2012). *Evaluation of the impact of the Green Academy programme and Case Studies*. The Higher Education Academy. https://www.heacademy.ac.uk/sites/default/files/green-academy-evaluation-case-studies_0.pdf. Last accessed April 18, 2016.

QAA. (2014). *Education for sustainable development: Guidance for UK higher education providers*. Quality Assurance Agency. http://www.qaa.ac.uk/en/Publications/Documents/Education-sustainabledevelopment-Guidance-June-14.pdf. Last accessed March 11, 2016.

Securing the Future. (2005). Delivering UK sustainable development strategy. In *The UK Government Sustainable Development Strategy*. UK Government. https://www.gov.uk/government/uploads/system/uploads/attachment_data/file/69412/pb10589-securing-the-future-050307.pdf. Last accessed December 10, 2015.

Sterling, S., & Scott, W. (2008). Higher education and ESD in England: A critical commentary on recent initiatives. *Environmental Education Research, 14*(4), 386–398.

Trochim, W. (2002). *Research methods knowledge base*. Atomic Dog Publishing. http://www.anatomyfacts.com/research/researchmethodsknowledgebase.pdf. Last accessed January 18, 2016.

UK National Commission for UNESCO. (2013). *Education for Sustainable Development (ESD) in the UK—Current status, best practice and opportunities for the future. Policy brief.* http://www.unesco.org.uk/wp-content/uploads/2015/03/Brief-9-ESD-March-2013.pdf. Last accessed March 19, 2016.

UNESCO. (2005). *United Nations Decade of Education for Sustainable Development (2005–2014) International Implementation Scheme*. United Nations Education, Scientific and Cultural Organisation. https://www.bibb.de/dokumente/pdf/a33_unesco_international_implementation_scheme.pdf. Last accessed April 18, 2016.

UNESCO. (2014). *Shaping the future we want; UN Decade of Education for Sustainable Development (2005–2014) final report*. United Nations Education, Scientific and Cultural Organisation. http://unesdoc.unesco.org/images/0023/002301/230171e.pdf. Last accessed June 3, 2015.

UNESCO. (2016). *Education for sustainable development*. United Nations Education, Scientific and Cultural Organisation. http://www.unesco.org/new/en/education/themes/leading-the-international-agenda/education-for-sustainable-development/mission/. Last accessed March 9, 2016.

Agenda 21. (1992). Rio de Janeiro, Brazil: United Nations Conference on Environment & Development 3–14 June 1992. United Nations. https://sustainabledevelopment.un.org/content/documents/Agenda21.pdf. Last accessed April 20, 2016.

Energy Management in UK Non-domestic Buildings—A New Perspective to Provoke Reduced Carbon Emissions

Kay Emblen-Perry and Les Duckers

Abstract

Politicians, industry and the public generally accept the need for energy consumption to be cut to deliver climate change mitigation measures essential for us to avoid climate disaster. For non-domestic fuel users current energy policy has attempted to drive this through rational economic responses to energy cost pressures. This reliance on voluntary action has created an "Energy Inconsistency", that is a marked difference between energy opportunities that have been proven technically viable, financially rational and retrofit feasible and those actually adopted. Other factors must therefore be involved to influence what appear to be simple carbon and cost saving opportunities. This paper presents a new approach to energy efficiency and consumption in non-domestic buildings, viewing attitudes and behaviours of building owners and tenants as the key driver of energy consumption. A new framework is proposed as a method to examine the impact of building ownership on the tenants' and owners' abilities to improve energy efficiency and consumption and identify opportunities to overcome the barriers inherent in these ownership structures.

Keywords

Energy efficiency · Energy policy · Owners · Reduced carbon emissions · Non-domestic buildings

K. Emblen-Perry (✉)
University of Worcester, Worcester WR1 3AS, UK
e-mail: k.emblenperry@worc.ac.uk

L. Duckers (✉)
Coventry University, Coventry CV1 5FB, UK
e-mail: L.Duckers@coventry.ac.uk

W. Leal Filho (ed.), *Sustainable Development Research at Universities in the United Kingdom*, World Sustainability Series,
DOI 10.1007/978-3-319-47883-8_7

1 Introduction

The role energy plays in the emission of carbon is widely accepted by UK Government, politicians and the public, with energy concerns high on the global political and business agendas. However, energy consumption reductions and adoption of lower carbon energy generation, targeted through existing conventional approaches of encouraging voluntary energy conservation actions, taxes and financial and non-financial incentives have fallen short of climate change requirements. Non-domestic buildings contribute 18 % of UK carbon emissions (Delay 2013).

Challenging UK carbon reduction targets have been set to drive mitigation of climate change. However, actions have fallen short of progress required to deliver mitigation measures essential for us to avoid climate change disaster. To date Government policy has been split between "carrots" and "sticks". Greater emphasis on taxes (sticks) has been applied to energy intensive, non-domestic building users in an attempt to follow the Polluter Pays Principle. This has ensured that at least some of the costs of pollution are borne by those responsible and to obtain political momentum for higher taxation levels (Environment Audit Committee 2011).

For less energy intensive building users a non-interventionist approach through incentives (carrots) such as low-rate loans, grants and the provision of good practice information have been the main policy approaches. These have lacked effectiveness and although they have applied a balance of "command and control" and encouragement they have proved unable to sufficiently lower energy consumption or change energy attitudes (Lyon and Maxwell 2002).

To date, research on building energy consumption and efficiency has largely focused on domestic properties and energy intensive commercial sectors and views building type (Schleich and Gruber 2008; Janda 2008; de Groot et al. 1999) or commercial sector (Janda 2014) as the driver of carbon emissions. This approach mirrors the structure of Government policies whereby organisations below the intensive energy user threshold of the Carbon Reduction Commitment Energy Efficiency Scheme have largely been excluded.

This paper presents a new approach to building energy consumption, efficiency and conservation. It views the attitudes and behaviours of non-domestic building owners and users, driven by the tenancy structure operated and barriers generated, as key drivers of energy consumption. This drives the "Owner-User Stalemate"[1] for investment in building energy consumption improvements.

The new Energy, Ownership and Impacts Framework is presented as a tool to facilitate evaluation of the economic and environmental impacts of these attitudes and behaviours. It subdivides the non-domestic building sector into a number of distinct groups based on ownership and tenancy structures and the energy characteristics attached to them. This framework is used to examine the impact of

[1]The Owner-User Stalemate considers the incentive to invest in building energy efficient technologies and building materials is significantly reduced as the benefits accrued are split between the building owner and the user, thus neither is willing to invest for the others' benefit.

building ownership on owners' and tenants' ability and willingness to improve building energy efficiency and conservation and identify opportunities to overcome barriers inherent in these relationships and structures.

2 The Challenge

Over the last 150 years market forces have driven continual improvements in the energy efficiency of electrical equipment. Edison's evolution of the electric light bulb into sales of electric lighting is an early example of this. The convenience and reliability of light provided by electricity offered a customer significant cost and lifestyle benefits over gas and so was adopted quickly by consumers. This increasing demand for lighting services drove Edison's development of more efficient distribution systems and longer operating lives of bulbs. This is just one of the many market driven energy consumption interventions that have created the current energy system and policy approach that reflect long-term historical forces (Unruh 2013). However, in spite of increases to energy efficiency these interventions have resulted in a vastly increased per capita consumption of energy (Warde 2010), creating an energy system and usage patterns that are now widely recognised as key contributors to carbon emissions and are therefore no longer sustainable.

Challenging Government targets have been agreed to mitigate climate change: Zero Carbon new non-domestic buildings by 2019 and 80 % reduction in carbon emissions over 1990 levels by 2050 with at least 35 % by 2020 (Committee on Climate Change 2015). However, despite extensive energy information campaigns and financial and motivational incentives most writers and energy analysts agree that energy improvements have not been adopted as expected (DeCanio 1993; de Groot et al. 1999; Janda 2009; Warde 2010). The research presented in this paper suggests that an "Energy Inconsistency"[2] exists. Energy opportunities have been proven economically viable, financially rational and retrofit feasible and yet are not widely adopted, therefore other factors must be involved to influence what appear to be economically rational, simple carbon and cost saving opportunities. This research demonstrates that non-domestic building owners and tenants are not acting as the economically rational players that the energy policy planners have expected, consequently financial and carbon saving are not being achieved. In an alternative approach this research hypothesises that the ownership of non-domestic buildings is the driver of this Energy Inconsistency and has contributed to the inability of energy policy to deliver effective carbon emissions reductions in non-domestic buildings, effectively creating the Owner-User Stalemate.

Government targets for zero carbon non-domestic buildings implemented through Building Regulations legislation have achieved greater sustainability in building design, which has delivered improvements in new buildings and large

[2]The 'Energy Inconsistency' is a gap between proven energy efficiency improvements and what is actually been implemented.

scale retrofits. However, new non-domestic buildings complying with these regulations represent less than 2 % of the present building stock, energy savings must also come from existing non-domestic or commercial buildings if targets are to be met. Retrofitting energy improvements within existing smaller non-domestic buildings remains a voluntary intervention.

Withdrawal of funding for the Green Deal and Home Improvement Cashback Scheme in response to low take-up rates will put further pressure on energy policy to create new solutions to provoke carbon emissions reductions. The stock of existing, largely energy in-efficient, non-domestic buildings could offer an opportunity to achieve this.

3 The Research

The current rate of progress on carbon emissions reductions and future emissions projections cast doubt on energy policy's ability to deliver UK targets (Committee on Climate Change 2015). Some progress to meet carbon reduction targets has been achieved to date but at a rate that will jeopardise meeting the 2020 and 2050 targets. Significant expansion of energy efficiency within the stock of non-domestic buildings, which contribute 18 % of carbon emissions annually, therefore provides a significant opportunity for carbon abatement in the UK (Delay 2013).

The evolution in non-domestic property ownership over the last 150 years has created a complex pattern of building ownership and occupation within the overall sector (Dixon 2009). This evolution has resulted in a legacy of diverse tenancy styles that have generated non-cooperative relationships between owners and tenants, the Owner-User Stalemate, which prevent the adoption of energy efficiency and conservation opportunities. This evolution of tenancy styles has also resulted in a number of different energy supply routes including tenants purchasing energy directly from the utility company, purchasing directly from the building's owner or receiving supplies within full-service contracts. The legacy of this evolving complexity will continue to prevent the adoption energy efficiency within the current stock of buildings unless an alternative approach is provided. The current stock of buildings is being replaced at only 1–2 % per year (McAllister et al. 2009) will therefore provide 70 % of the non-domestic buildings standing in 2050 (Kelly 2010).

Non-domestic building ownership as a driver of energy inefficiency and source of carbon emissions has received less attention within energy research and by policy planners. An alternative perspective to driving energy behaviour change is therefore required. This alternative driver of change could be provided through unlocking the Owner-User Stalemate. The challenges presented within the Owner-User Stalemate manifest themselves as both practical and attitudinal barriers such as lease clauses preventing changes to building fabric or the split of incentives for, and rewards from, investment.

The authors' experience of organisational energy behaviour suggests that participation rates in energy efficiency vary between SMEs due to varying levels of interest in energy management, willingness to invest in energy improvements and ability to control the changes to buildings necessary to achieve energy improvements. A review of literature indicates that building type (de Groot et al. 1999; Janda 2008; Peacock et al. 2008; Schleich and Gruber 2008; McAlllister et al. 2009) and levels of building awareness (Lorenzoni et al. 2007; Fawcett 2010; UK Green Building Council 2011; Axon et al. 2012) are well researched. However, a knowledge gap exists for the impact of building ownership. These have directed the research hypothesis described above.

4 Energy, Ownership and Impacts Framework Design

In response, a framework with which to examine the influence of non-domestic building owners and tenants on energy efficiency and conservation has been created. This is the Energy, Ownership and Impacts Framework (Fig. 1), which enables the impact of building ownership to be examined through the segmentation of the non-domestic building sector into homogenous groups that reflect the relationships likely to be experienced by non-domestic owners and tenants.

Unlike business sector used by Janda (2014), building type used by Peacock et al. (2008) and location used by Li and Runming (2009) to examine energy consumption within buildings, this research adopts a new criterion for segmentation; the shared characteristics of building ownership combined with the responsibility for the purchase of energy consumed.

The authors have conducted a pilot survey with SMEs to test the validity of the framework. This has corroborated the eight distinct "ownership segments" proposed i.e. categories on non-domestic building ownership with specific routes of energy purchase, each of which displays a common impact on energy management. Further research to examine the impacts of building ownership on the ability and willingness of non-domestic building owners and tenants is underway, the results of which will be published in the future.

This research establishes that investing in building energy efficient technologies and building materials generates different benefits for owners and tenants within different ownership segments, which consequently drives differing attitudes towards investment in them.

These differences emerge as shared energy attitudes, and common behaviours associated with them, within each ownership segment and have been found to influence participation rates in energy efficiency and conservation interventions. The research also finds that they differ between the homogenous sub-groups of owners and tenants established within the Energy, Ownership and Impacts Framework.

	Categories of non-domestic building ownership (energy bill payee shown in brackets)							
	a Building owner and tenant (tenant from utility)	b Building owner as the user (owner from utility)	c Building owner and franchisee (1) (owner from utility)	d Building owner and branch (head office from utility)	e Building owner, manager and tenant (tenant from utility)	f Building owner as the energy provider and user (tenant from landlord)	g Building owner as a commercial investor (tenant from utility)	h Building owner and franchisee (2) (franchisee from utility)
Low cost of energy as a business overhead	✓	✓	✓	✓	✓	✓	✓	✓
Split incentive for energy investment	✓		✓	✓	✓	✓	✓	✓
Lease terms restrict action possible	✓		✓	✓	✓	✓	✓	✓
Energy user does not see energy bill			✓	✓				
Competition for corporate investment funds	✓	✓		✓				
Corporate behavioural barriers		✓		✓				
Corporate financial barriers		✓		✓				
Lack of energy saving advocate				✓				
Disincentive to reduce energy use – energy provided to tenant			✓	✓	✓	✓		
Disincentive to reduce energy use – energy as labour replacement	✓	✓	✓	✓	✓	✓	✓	
Owner remote from day to day activities of building			✓	✓	✓		✓	
Access to energy efficiency information	✓	✓						✓

Fig. 1 Energy, ownership and impacts framework

In order to examine the homogenous ownership segments further each is presented in detail below. To distinguish the segments clearly, an example of each category's impact on the adoption of retrofitted energy efficient building insulation is included. Insulation is chosen, as it is a tried and tested building improvement with proven cost benefits.

(a) **Building owner and tenant**—The tenant occupies the building as a sole tenant in return for rental or lease payments.

- **Building owner**—Energy is purchased and consumed by tenants so the owner has no incentive to invest in energy efficient changes.

- **Tenant**—The tenant has sole responsibility for energy used within the building.
- **Impact**—The owner has responsibility for insulating the building. However, as they gain no benefit from doing so it is unlikely to be undertaken. Lease clauses may prevent the tenant from making changes to the fabric of the building.

(b) **Building owner as the user**—The owner is the user of the building and has sole responsibility for energy purchased and consumed within the building.

- **Impact**—The owner has responsibility for insulating the building and will financially benefit from doing so. Availability of capital and/or the payback period are likely to influence the decision to retrofit.

(c) **Building owner and franchisee (1)**—The user occupies the building rent free providing a service to clients on behalf of the building owner.

- **Building owner**—The owner manages energy supplies. Energy is provided free of charge to the franchisee.
- **Franchisee**—The franchisee has no responsibility for the payment of energy bills so has no incentive to reduce energy consumption.
- **Impact**—The building user has no incentive to conserve energy. The owner benefits financially from insulating the building if installation costs and payback periods are acceptable.

(d) **Building owner and branch**—The building (owned or leased) is a separately managed unit from within a larger, multisite organisation.

- **Building owner**—The organisation's owner manages the energy for a remote branch centrally hence there is little concern for their energy usage.
- **Branch**—Energy is supplied to the branch with no bills received by the building users. The tenants have no incentive to conserve energy.
- **Impact**—The financial benefits from insulating the building will remain with the central area rather than the branch. The cost of energy and benefit to invest in insulation are likely to get lost in the geographical gap between the two sites.

(e) **Building owner, manager and tenant**—An agent on behalf of the owner manages the tenancy.

- **Owner**—The building owner subcontracts the running of the building to the manager.
- **Manager**—The manager for the common areas provides Energy. The handling cost is covered by the owners' management fee.

- **Tenant**—Tenants have sole responsibility for the energy supply to their premises. The service charge recoups the tenant's share of energy consumed in common areas.
- **Impact**—The owner has responsibility for insulating the building. However, as they gain little benefit from doing so it is unlikely to be undertaken. Lease clauses may prevent the tenant from making changes to the fabric of the building.

(f) **Building owner as the energy provider and tenant**—The tenants purchase all energy from the building's owner.

- **Building owner**—As the building owner provides tenants' energy they have no incentive to improve the building's energy efficiency.
- **Tenant**—All energy is purchased from the building owner.
- **Impact**—The tenants benefit from reducing energy consumption. However, the owner has no incentive to retrofit insulation to the building as they will gain no return for their investment.

(g) **Building owner as a commercial investor**—The owners have no interest in the building other than as an investment.

- **Impact**—Unless a long-term benefit of a rental increase can be obtained the owner has no incentive to insulate the building.

(h) **Building owner and franchisee (2)**—The user occupies the building rent free providing a service to clients on behalf of the building owner.

- **Building owner**—Energy is purchased and consumed by franchisees so the owner has no incentive to invest in energy efficient changes.
- **Franchisee**—The franchisee has sole responsibility for the payment of energy bills so has no incentive to reduce energy consumption.
- **Impact**—The owner has responsibility for insulating the building. However, as they gain no direct benefit from doing so it is unlikely to be undertaken. Contract clauses may prevent the tenant from making changes to the fabric of the building.

The Energy, Ownership and Impacts Framework is used here to evaluate the economic and environmental impacts of ownership on energy efficiency and conservation. As each segment demonstrates distinct characteristics members can be measured and targeted to obtain the best energy consumption and carbon emissions reduction return on investment. This is examined in detail in the following section.

5 Applying the Energy, Ownership and Impacts Framework to Evaluate of the Economic and Environmental Impacts of Ownership on Energy Efficiency and Conservation

The Energy, Ownership and Impacts Framework is applied to a business simulation of a brewery to understand the behavioural trends and patterns of non-domestic building energy consumption and the likelihood of improvement actions driven by the styles of ownership currently observed in the UK non-domestic building sector. The responses of the company, placed within each of the ownership segments of the framework, will be considered in the context of current energy policy.

6 Case Study: Three Wise Men Brewery, Monks Castle, Shropshire

In this case study we will consider a fictitious brewery that uses large volumes of natural gas for both space heating (sh) and for its brewing processes (bp). We will initially use a Degree Day analysis technique to determine which demands the most gas (sh or bp), and then go on to explore how the different ownership segments would apply to the brewery. That is to say, how the ownership segment of the brewery might determine what action is taken to improve energy conservation and efficiency measures.

7 History

The Three Wise Men Brewery, located within the centre of Monks Castle, Shropshire, is one of the oldest working breweries in Britain dating back to 1659. Most of the monthly production of beer (250,000 L) is carried out in the original buildings, and sold within 100 km of Monks Castle to pubs, restaurants and local supermarkets. The factory (Fig. 2) still utilises the brewing tower (installed in 1888) and operates for 4000 h per year: 14 h per day Monday to Saturday.

The compact factory occupies buildings of historical importance. Whilst it is unable to extend due to its location, its Grade II listing prevents alteration to the external appearance. The brewery has invested little in environmentally sustainable technologies and has undertaken few energy improvements since the mid 1980s, resulting in high energy and water costs; energy represents about 10 % of annual turnover.

Fig. 2 Frontage of Three Wise Men Brewery

There is concern that both space heating of the buildings (sh) and brewing processes (bp) could be inefficient and costly with a total of 13,333 MWh of gas used each year. To establish the split of gas consumption between heating and process energy for brewing the "degree day" method will be applied within this case study. This is explored below. The brewery will then be used to illustrate the eight different scenarios of ownership contained within the Energy, Ownership and Impacts Framework described above.

8 Degree Day Method

In order to analyse the application of gas the monthly gas consumption for 12 months is plotted against the Degree Days (DD)[3] for each month. This is shown in Table 1 and Fig. 3, which provides a graphical representation of monthly gas consumption versus DD.

When displayed graphically and a trendline added, the intercept value on the vertical axis indicates the brewery's monthly process (bp) energy requirements.

Subtracting the requirement for process energy from the total gas used for each respective month gives the MWh of gas used for heating (sp).

To illustrate the use of the Energy, Ownership and Impacts Framework the Three Wise Men Brewery will be given three scenarios of energy consumption:

[3]Degree Days is a representation of the weather conditions using a base temperature of 15.5 °C and is used to establish heating demand.

Table 1 Gas consumption in case study scenarios

		Scenario I	Scenario II	Scenario III
	Degree days	MWh	MWh	MWh
January	375	1055	1990	2540
February	397	1059	2069	2663
March	358	1052	1929	2445
April	250	1030	1540	1840
May	148	1010	1173	1269
June	21	984	716	558
July	3	981	651	457
August	0	980	640	440
September	32	986	755	619
October	156	1011	1202	1314
November	267	1033	1601	1935
December	321	1044	1796	2238
Total		12,226	16,061	18,317
Intercept		980	640	440
Process		11,760	7680	5280
Heat		466	8381	13037
Gas usage		Process	Same	Heat

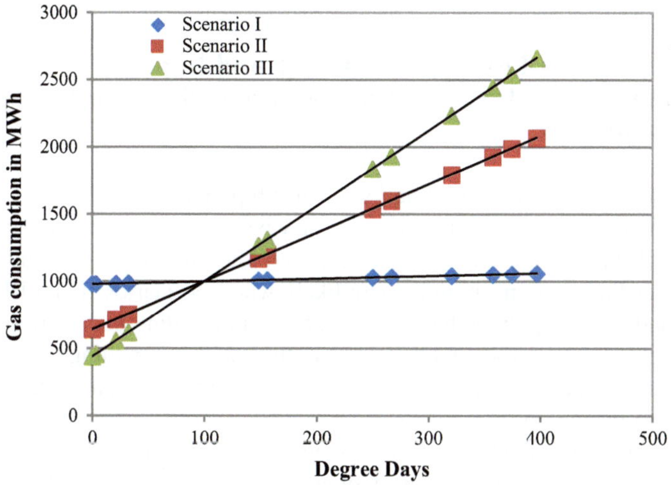

Fig. 3 Three wise men brewery gas consumption in case study scenarios

Scenario I: process gas (bp) is much greater than gas needed for heating (sp)
Scenario II: process gas (bp) is about the same as the gas needed for heating (sp)
Scenario III: heating (sp) requires three times as much gas as process gas (bp)

In Scenario I the operator of the brewery, whether an owner-occupier or a tenant, would find it valuable to assess the process and process equipment as the process gas consumption is so high that investment in new, efficient equipment could be economically viable and pay back the investment in a relatively short time.

In Scenario II gas consumed for processing is about the same as that for heating. An owner-occupier could choose to improve either or both the heating efficiency and process equipment. However, a tenant, with little long-term interest in the building, is likely to be reluctant to improve the fabric of the building fabric by insulating it or installing superior heating equipment, but may contemplate investment in superior process equipment.

Scenario III raises the issue of gas consumption dominated by the heating demand. Here, clearly, the owner-occupier might consider heating improvements. A tenant would need reassurance of long-term tenancy and business survival.

These scenarios are quantified in Table 1 and shown graphically in Fig. 3. They will also be discussed in exploring the impacts of owners' and tenants' ability and willingness to adopt energy improvement interventions through the application of the Energy, Ownership and Impacts Framework.

9 Applying the Energy, Ownership and Impacts Framework to Understand the Implications for Energy Management

If the brewery falls within the Energy, Ownership and Impacts Framework's segment 'a', building owner and tenant, there may be some lack of connection between the user of the energy i.e. the brewery and the owner of the building. The Owner-User Stalemate position is more likely where Scenarios II and III shown in Table 1 and Fig. 3 apply as benefits accrued from building energy improvements are split between the owner and the occupier thus neither is willing to invest for the others' benefit. Such a barrier is exacerbated if the lease or rental charges cannot be increased to recoup costs.

For the brewery situated within segment "a" of the model, any possibility to retrofit building energy efficient technologies is unlikely as leases frequently prevent tenants from making structural improvements (Bright 2010).

In segment 'a', where tenants manage their own energy, there is little financial incentive for the owner to invest for energy efficiency for the non-domestic building, as only the tenant will benefit in the short to medium term. This challenges the economically rational basis of current energy policy. This building owner and tenant relationship can therefore be expected to generate only small savings from building energy efficiency and conservation interventions such as those achieved through behavioural improvements e.g. energy housekeeping.

Current policies to encourage energy efficiency through provision of information and encouragement of voluntary energy conservation interventions will be of little use to tenants of segment 'a' as the owner is likely to be unwilling to invest in improvements.

If the brewery is situated within segment 'b', building owner as the user, the owner-occupiers of the building may be expected to readily invest in and implement energy efficiency actions, whether Scenario I, II or III applies (Table 1 and Fig. 3) as they will recoup all benefits from them. However, investment in energy efficiency must compete with general business investment so may not be the simple decision expected. Decisions vary in complexity according to the size, structure and attitudes of the company. Consequently the decision to adopt energy improvement interventions will be controlled by corporate culture and governance (DeCanio 1993). This challenges the basis of the effectiveness of the economically rational approach to energy policy.

Within the brewery there are potential points of friction that may prevent the adoption of economically viable and retrofit feasible investments. For example cultural disincentives such as compliance with hierarchical organisational decision-making structures may lead to economically rational decisions such as energy improvements being rejected (Sullivan and Sullivan 2009).

For smaller organisations the investment decisions may appear less bureaucratic. However, the financial aspects of energy investment may be more significant for them as energy efficiency investment decisions must be balanced against investment for business growth. Although energy savings can contribute directly to profits it is not always clear to the owner how to access energy efficiency. As with larger organisations, the attitudes of some small business owners and managers to energy efficiency will also be of major significance in whether energy reduction actions are taken (Department of Energy and Climate Change 2013; Johnson Controls 2013).

Within segment 'c', building owner and franchisee (1), the brewery would provide a service for a second organisation such as a sub-contracted brewing facility from one of their client's buildings provided free of charge. This category of building occupation combines a number of energy behaviours from other categories of ownership. For example, where the site is operated without payment of rental charges, energy bills are sent to the client so that the service provider is likely to have little or no incentive to reduce energy costs unless specifically included in the service contract. In this context the usage and cost of energy remains hidden from the service provider. Energy reduction may be a corporate policy for the building owner, but there is likely to be no direct financial or behavioural route for either party to intervene to deliver building energy savings for the brewery, whether it is within Scenario I, II or III (Table 1 and Fig. 3).

If the brewery used for this case study operates as a single site within a larger organisation it will be situated within segment 'd' of the framework, building owner and branch. It is remote from energy consumption or costs as energy purchases are controlled from a central point of the multi-site organisation. There is a barrier to energy efficiency whether Scenarios I, II or III applies (Table 1 and Fig. 3) as cost benefits are seen at the central level whilst site investments to achieve them are budgeted at the local level. Cost reduction targets set for the company will likely exclude reductions in energy use as there are more visible investments and savings that can be made. The split incentive of the Owner-User Stalemate experienced within category 'a' applies here with neither the building user nor head office recognising funding of building energy efficiency as offering a return on investment. Additionally corporate incentives for energy savings may be remote and not widely acted on if local management lacks an energy saving advocate to drive behavioural change (DeCanio 1993).

If the brewery occupies premises run by managing agents on behalf of the owner it will be fall into segment 'e', building owner, manager and tenant. Managing agents add another level of complexity to the owner-user relationship, which further discourages energy efficiency improvements, particularly if Scenario II or III applies (Table 1 and Fig. 3). Where they are responsible for selling energy to the tenants, the managing agent has a strong disincentive to encourage energy reduction. Where the brewery is located in a multi-occupancy building and their share of energy used in the communal areas is paid within the service charge there is little incentive for either owner or tenants to improve building energy efficiency.

If the brewery leases its premises and purchases energy from the building owner it will fall within category "f" of the Energy, Ownership and Impacts Framework, building owner as the energy provider and tenant. In this category of non-domestic building ownership the owner provides the tenant with their energy, consequently there is little or no financial incentive for the owner to invest for building energy efficiency and little motivation for the brewery to improve energy efficient technologies whichever Scenario applies. Financial incentives such as feed-in-tariffs and Renewable Heat Incentives may have payback periods that are too long to be attractive. This approach to energy efficiency challenges the economically rational response expected by energy policy.

If the brewery leases premises owned by a commercial investor they will be situated within category 'g'. In this category the building owners, whether privately owned companies or stockholding organisations, maximise stock value through favourable public perception of their energy behaviour (Green Building Council 2011). However, restrictive leases are a feature of this non-domestic building ownership category (Bright 2010; McAllister et al. 2009) and this owner is likely to be remote from the day-to-day activities of the building. Private owners maybe more open to energy investment and may be willing to engage in energy saving actions that benefit their tenants (Janda 2008). Once again the brewery situated within category 'g' is unlikely to invest to improve building energy efficiency if situated within Scenario II or III (Table 1 and Fig. 3) as they are usually bound by tenancy agreements which frequently prevent changes to the building fabric (Bright 2010; McAllister et al. 2009). Unless the owner or managing agent's corporate strategy favours energy efficiency there is little incentive for them to make the required investments.

Within segment 'h', building owner and franchisee (2), the brewery would provide a service for a second organisation such as a sub-contracted brewing facility from one of their client's buildings provided free of charge. The Owner-User Stalemate position applies whichever Scenario applies as neither party is likely to invest in insulation for the others' benefit.

Additionally there are generic barriers to reducing building energy consumption for non-domestic building owners and users that would impact the brewery and influence it's energy management policy. For example, within most organisations energy represents a small element of business overheads so it can rarely compete with other opportunities for attention. Energy cost is insufficient to drive behavioural change on its own (Bright 2010). Substituting lower cost energy for higher cost labour may support this and so remains a key incentive for commercial organisations to consume energy (Warde 2007). In addition, although utility companies are regulated to encourage energy efficiency they have a vested interest in maximising sales of energy. This regulatory responsibility, the Energy Company Obligation, requires the energy supplier to help improve the energy efficiency of their domestic customers' buildings (Department of Energy and Climate Change 2015). However, there is no requirement for such energy conservation relationships with commercial customers.

The researchers recognise that this preliminary study requires further research to validate its findings. However, applying the Energy, Ownership and Impacts Framework to the brewery case study highlights the need for policy to be redesigned to overcome barriers created by building ownership structure. The research has also recognised that neither owners nor tenants act as rational economic actors in response to energy price drivers as expected by energy policy planners. Barriers presented by the current system of building ownership and the Owner-User Stalemate generated by its complexity is not considered within energy policy. However, knowledge gained through the application of the framework can be used to more accurately and effectively engage owners and users in increasing building energy efficiency actions.

10 Using the Energy, Ownership and Impacts Framework to Inform Public Opinion

As described above current energy policies based on financial and non-financial incentives and disincentives have not delivered the economically rational responses to energy price manipulation expected or changes in energy behaviours planned. This research suggests that building ownership has introduced a number of barriers that have not been considered and therefore it is timely to consider whether an alternative policy approach should be the way forward.

Politicians dislike direct taxation as it forces people to be green (Ockwell et al. 2009). They fear that although it can drive greater energy efficiency, taxes will be unacceptable to the powerful business lobby. An alternative approach based on actions to overcome the barriers created by non-domestic ownership structures would be the initial step towards decoupling energy saving from political popularity, which Roodhouse (2007) considers is the key to future climate change mitigation. The research findings discussed here suggest that this alternative approach can be informed by the findings of the Energy, Ownership and Impacts Framework.

The initial findings of this research suggests that a series of bespoke, rational financial disincentives should be established linked to building ownership to overcome the barriers identified by the Energy, Ownership and Impacts Framework. When based on the segmentation of non-domestic buildings, bespoke actions can be designed to overcome the common attitudinal and behavioural responses to ownership constraints and applied more directly. The authors recommend that further research is undertaken on policies that could be targeted more effectively and are consequently more likely to deliver the behavioural change that is the key to long-term policy success and a permanent change in energy attitudes.

11 Conclusion

Although there has been a lengthy history of energy efficiency and conservation improvements and widespread recognition of the need for carbon savings through energy management, the UK has failed to overcome the Energy Inconsistency and deliver the energy efficiencies required to mitigate climate change. Successive governments have relied on politically attractive but voluntary energy efficiency actions which have been considered cost effective, rational investments. However, as the Owner-User Stalemate still persists alternative barriers to energy efficiency and conservation improvements must exist.

The research discussed seeks to break through these barriers by taking a different approach to energy management. This is based on two key factors identified within this research: firstly, attitudes and behaviours of non-domestic building owners and users drive the adoption building energy efficiency and conservation interventions and secondly, that these owners and users do not demonstrate rational economic responses to energy price controls. The alternative approach proposed is based on a segmentation of the ownership and occupation of buildings and identification of associated energy behaviours. This has led to a series of economic and environmental impacts being identified for each of these ownership categories. The authors recognise that further research is required to validate the initial findings presented here.

Using these impacts of ownership as the basis for energy improvements, a suggestion for more targeted approach to energy policy has been made. The authors recognise that directed financial disincentives may be politically disagreeable but propose that changes in attitudes towards energy efficiency and conservation within non-domestic buildings should be provoked through an alternative policy approach to reduce carbon emissions. A policy approach based on the initial findings of the

Energy, Ownership and Impacts Framework could introduce greater rationality into the policy landscape and assign energy responsibility to both owners and tenants.

References

Axon, C. J., Bright, S. J., Dixon, T. J., Janda, K. B., & Kolokotroni, M. (2012). Building communities: Reducing energy use in tenanted commercial property. *Building Research and Information 40*(4), 461–472. Available from http://www.tandfonline.com

Bright, S. (2010). Carbon reduction and commercial leases in the UK. *International Journal of Law in the Built Environment* [online] *2*(3), 218–231. Available from www.emeraldinsight.com 01 August 2013

Committee on Climate Change. (2015). *2010 to 2015 Government Policy: Greenhouse gas emissions* [online]. Available from November 30, 2015 https://www.gov.uk/government/publications/2010-to-2015-government-policy-greenhouse-gas-emissions/2010-to-2015-government-policy-greenhouse-gas-emissions

de Groot, H. L. F., Verhoef, E. T., and Nijkamp, P. (1999) *Energy Saving by Firms: Decision-Making, Barriers and Policies* [online]. Available from January 22, 2014 http://www.tinbergen.nl/uvatin/99031.pdf

DeCanio, S. J. (1993). Barriers within firms to energy-efficient investments. *Energy Policy* [online] September (pp. 906–914). Available from February 24, 2014 http://www.sciencedirect.com

Delay, T. (2013). *Energy efficiency in non-domestic buildings* [online]. Available from October 18, 2013 http://www.carbontrust.com

Department of Energy and Climate Change. (2013). *Attitudes towards and experiences of smart meters in the non-domestic SME market*. London: Department of Energy and Climate Change. Available from November 18, 2015 https://www.gov.uk/

Department of Energy and Climate Change. (2015). *Guidance, independent energy suppliers* [online]. Available from December 1, 2015 http://www.gov.uk/guidance/independent-energy-suppliers

Dixon, T. (2009). Urban land and property ownership patterns in the UK: Trends and forces for change. *Land use Policy* [online] *26S*, S43–S53. Available from May 22, 2013 http://www.elsevier.com

Environmental Audit Committee. (2011). *Budget 2011 and Environmental Taxes* [online]. Available from January 22, 2012 http://www.e-tid.com

Fawcett, J. (2010). *Department of Energy and Climate Change: Non Constrained Sector Research* [online]. Available from http://www.decc.gov.uk

Janda, K. (2008). *Implications of the ownership of buildings in the US and UK: An exploration of energy star buildings and the energy efficiency accreditation scheme* [online]. Available from May 15, 2013 http://www.eci.ox.ac.uk/publications/downloads/Janda08aceee.pdf

Janda, K. B. (2009). Buildings don't use energy: people do. *Architectural Science Review* [online] *54*, 15–22. Available from March 03, 2014 http://www.tandfonline.com

Janda, K. B. (2014). Building communities and social potential; Between and beyond organisations and individuals in commercial properties. *Energy Policy 67*, 48–55. Available from June 19, 2014 http://www.sciencedirect.com

Johnson Controls. (2013). *Energy efficiency indicator 2013 europe results*. USA: Institute of Building Efficiency. Available from March 12, 2015 http://www.institutebe.com

Kelly, M. J. (2010). Energy efficiency, resilience to future climates and long term sustainability: The role of the built environment. *Philisophical Transactions of the Royal Society* [online] *368*, 1083–1089. Available from February 07, 2015 http://rsta.royalsocietypublishing.org

Li, B., & Runming, Y. (2009). Urbanisation and its impact on building energy consumption and efficiency in China. *Renewable Energy* [online] *34*, 1994–1998. Available from October 18, 2015 http://www.researchgate.net

Lorenzoni, I., Nicholson-Cole, S., & Whitmarsh, L. (2007). Barriers perceived to engaging with climate change among the UK public and their policy implications. *Global Climate Change* [online] *17*, 445–459. Available from November 08, 2014 http://www.sciencedirect.com
Lyon, T. P., & Maxwell, J. W. (2002). Voluntary approaches to environmental regulation. In M. Franzini, & A. Nicita (Ed.), *Economics Institutions and Environmental Policy* (pp. 75–120). Vermont: Ashgate Publishing. Available from November 17, 2015 http://ostromworkshop.indiana.edu/papers/lyon031901.pdf
McAllister, I., Quartermaine, R., & McWilliams, F. (2009) *Costing Energy Efficiency Improvements in Existing Commercial Buildings* [online]. Available from March 12, 2015 http://www.cyrilsweett.com
Ockwell, D., Whitmarsh, L., & O'Neill, S. (2009). Reorienting climate change communication for effective mitigation, forcing people to be green or fostering grass-roots engagement?. *Science Communication* [online] *30*(3), 305–327. Available from May 18, 2015 http://scx.sagepub.com
Peacock, A., Banfill, P. F., Turan, S., Jenkins, D., Ahadzi, G., Bowles, G., et al. (Eds.). (2008). *5th International conference, improving energy efficiency in commercial buildings. Reducing CO2 emissions through refurbishment of non-domestic UK buildings held 10–11th April 2008 at Frankfurt*. Edinburgh: Herriot Watt University.
Roodhouse, P. (2007). *Rationing returns: A solution to global warming?* [online]. Available from June 19, 2014 http://historyandpolciy.org
Schleich, J., & Gruber, E. (2008). Beyond case studies: Barriers to energy efficiency in commerce and the services sector. *Energy Economics* [online] *30*, 449–464. Available from March 12, 2014 http://www.sciencedirect.com
Sullivan, M. J., & Sullivan, F. (2009). *Behavioral assumptions underlying energy efficiency programs for business* [online]. Available from June 19, 2014 http://www.calmac.org/publications/Behavioral_Assumptions_in_EE_Programs_for_Businesses_White_Paper.pdf
UK Green Building Council. (2011). *Carbon reductions in existing non-domestic buildings* [online]. Available from June 30, 2014 www.ukgbc.org
Unruh, G. (2013). The sweet spot of sustainability strategy. *MIT Sloan Management Review, 55* (1), 16–19. Available from November 15, 2015 http://search.proquest.com/docview/1438811686?accountid=15133
Warde, P. (2010). *Low carbon futures and high carbon pasts: Policy challenges in historical perspective* [online]. Available from June 18, 2013 http://historyandpolicy.org
Warde, P. (2007). *Facing the challenge of climate change: Energy efficiency and energy consumption* [online]. Available from June 9, 2014 http://www.historyandpolicy.org

Authors Biography

Dr Kay Emblen-Perry has several years' of senior environmental and ecology consultancy experience delivering consultancy projects in renewable energy technologies, contaminated land remediation, biodiversity offsetting and ecological assessment for UK organisations. She is qualified as an environmental and quality lead auditor, has implemented environmental management systems for both UK and multinational organisations and has trained environmental and quality assessors. In previous roles Kay gained senior project management and purchasing management experience in international automotive companies. She project managed the implementation of sustainable supply chain strategies, new vehicle projects and EU REACH Regulations. Kay's specialisation is in Sustainable Management including Environmental Management and Justice, Social Responsibility and Economic Sustainability.

Dr Les Duckers has been lecturing at Coventry University since 1975, initially in Physics and Engineering, later going on to establish the department of Environmental Sciences in 1992. Since then he has primarily taught masters course and supervised PhD students across a wide range of topics, reflecting his major research interest in environment and renewable energy, and especially wave energy. He is a visiting professor in Japan and Sri-Lanka and visiting lecturer to Reading and Loughborough Universities in the UK. He has experience of about 100 industrial environmental audits, is associate editor for The Renewable Energy Journal, and has acted as consultant to a number of companies.

Enhancing Our Social, Economic & Physical Environments by Embedding Sustainability into University Core Business

Jo Reed Johnson, Sandra Lee, Sarah Gretton and Derek Raine

Abstract

The rationale for this paper is to explore and share the ideological thinking behind the development of a University wide sustainable development strategy, with research at the heart. The University is the University of Leicester. Here we present an insight into the historical development of the sustainable pathways, explore the vision, key themes and activities for the 2015–2020 draft strategy. We then go on to present some of the successes, and lessons learnt particularly around collaboration, participation and voice. The aim of this new draft strategy (2015–2020), following the Environmental Sustainability Strategy (2010–2015), is to bring a new and coordinated approach across the University of Leicester, building on multi-disciplinary approaches. Previously sustainability had focused on the environmental aspect, thus led by an Estates function, as it has in many Universities. The new remit provides an opportunity to develop a more holistic approach across the curriculum, campus, community and culture in meeting the needs of the 21st century graduate and workplace. This is based on the thinking that: "*Discipline silos have no place in twenty first century education. The world today presents itself as a set of complex and interconnected problems and this is how learning institutions must help us solve those problems. Complexity is at the*

J.R. Johnson (✉) · S. Lee · S. Gretton · D. Raine
School of Education, Estates and Centre for Interdisciplinary Science,
University of Leicester, Leicester LE1 7RF, UK
e-mail: jarj1@le.ac.uk

S. Lee
e-mail: sll33@le.ac.uk

S. Gretton
e-mail: sng8@le.ac.uk

D. Raine
e-mail: jdr@leicester.ac.uk

W. Leal Filho (ed.), *Sustainable Development Research at Universities in the United Kingdom*, World Sustainability Series,
DOI 10.1007/978-3-319-47883-8_8

very core of sustainability, and dealing with complexity an essential sustainability literacy skill (Fagan 2009)". Whilst the University of Leicester are early in their journey to becoming a Sustainable University, the University of Leicester has top-down commitment coupled with lots of grassroots ideas in making this successful.

Keywords
Education for sustainable development · Participatory situated pedagogies · Interdisciplinary · Sustainability · Curriculum · Strategy · Higher education

1 Introduction: Historical Development of the Sustainable Pathways

The role of the University in society has evolved over time. In the early years Universities were set up as institutions for training in theology, medicine and the law, and then they developed into ivory towers of ancient learning, into seats for the generation of pure knowledge, then engines of economic development and, currently, underpinnings for the knowledge economy. They have been both idle luxuries and economic drivers, mechanisms of selection of dominant elites and instruments of social mobility. Their role is universal and local and they are both for a public and private good. Universities are defined by how they handle these contradictions (Castells 2001).

Universities have a de facto role in shaping society—to transmit a common culture and common standards of citizenship (Robbins 1963)—but also, conversely, they become the focus and refractors of the key social issues of the time. The key context in which the 21st century University operates is that of *sustainability* or *sustainable development*. Even to ignore this is to take a view. If universities become merely sources of tenure for academics and credits for students (Gibson 2001)—if they have become ways of gaming the system—then the world's path to sustainability (or extinction) will be, in Hobbes's well-known description, brutal and chaotic and the life of man, indeed, solitary, poor, nasty, brutish and short.

'Sustainability' is a contestable concept (like 'justice' which it subsumes) but it has come to be used so widely as to dilute its meaning.[1] It is also variously conflated with "green" issues such as double glazing and, most contentiously, limitations to car parking. Important as these are, they are often not seen as priorities for the management of research intensive universities, although in some cases they are part of the third stream community engagement. Where they are a high priority it is in the context of research funding for technology, so sustainability

[1] http://www.treehugger.com/culture/why-word-sustainable-becoming-unsustainable.html (accessed 31/3/16).

becomes the province of the engineers and chemists. Our view is quite the opposite. We believe that it is part of the intellectual challenge to academia to elucidate exactly what we might mean by a sustainable global society and how we might achieve this. In this view, sustainability becomes the overarching schema under which the role of the university is conceptualised. It embraces all disciplines: social sciences and economics, the arts and humanities as well science and technology. The successful university will manage effectively the interaction between science, technology, economics and society (Brennan and Lebeau 2004).

In some cases this, or something like it, has been part of the vision of senior management in our Universities with the appointment of academic leads in sustainability and strategies that go beyond wishful thinking and re-labelling of modules as "environmental science". In other cases any transition has begun at the grassroots level. It is only when these two approaches meet that significant progress becomes visible. This paper is about the challenges to such a closing of the gap at the University of Leicester.

The top-down approach focussed on the University estate and is described below in the section on environmental and social initiatives. The grassroots approach was initiated by Paul Warwick, now at Plymouth University, who established the ESD forum (education for sustainable development) for university staff. The key conclusion from these meetings was that there was a significant amount of sustainability activity at Leicester but that this was certainly uncoordinated and largely invisible. Several initiatives emerged from these meetings: a website that gives details of sustainability-related activities and events[2]; a series of events involving internal and external speakers; and a proposal for a module in sustainable development open to all undergraduate students. This has recently broadened out and presented in more detail in section below on ESD Forum and RCE (Regional Centre of Excellence).

2 Ideological Thinking Behind the Development of a University Wide Sustainable Development Strategy 2015–2020

This strategy sets out a vision, mission and key actions to drive this change. More work will be done on this once those with key responsibilities are in place.

Our vision is for the University of Leicester to be renowned for the quality of its graduates' understanding of complexity and the current and future global challenges they will face. It is about them having the ability to draw together economic, social, political and environmental dimensions when they are making decisions in their world of work. Likewise, our research has the potential to positively impact practices and procedures at every level of society as the University positions itself at the forefront of futures thinking and dilemma flipping.

[2]https://www2.le.ac.uk/offices/academic-practice/learning-and-teaching/esd (accessed 31/3/2016).

Yes, a more sustainable University is a more efficient one but the University of Leicester aims to be more than that. By underpinning its label as a Sustainable University it will be driving change through it research informed and informing practices. Thus, this sustainable University is not only a more efficient, effective, attractive and forward-thinking University, it is developing graduates and post-graduates who have the skills to work in 21st Century organisations who require them to be systems thinkers, dealing with and understanding complexity whilst engaging in real world thinking, dilemma flipping, visualisation and maker instinct (Johansen 2012).

Table 1 presents the key themes and priorities for the University of Leicester, Sustainable Plan for 2015–2020 (draft). This represents a framework focused on closing the gap between top-down, bottom-up approaches that engages all stakeholders.

Table 1 University strategic plan 2015–2020 (draft)

Key themes	Key priorities
Innovation, enterprise and partnership The Sustainable Environment Enterprise Development (SEED) Fund will encourage staff and students to come up with innovative solutions to sustainability challenges	**Excellence and interdisciplinary** Sustainability is one of the truest interdisciplinary areas, which we will encourage through our work with the ESD Forum, Sustainable Research Network and Innovation Hub
A student-centred approach and widening participation We offer many HEAR accredited activities including a student-led volunteering scheme and are responding to student feedback through the development of sustainability-related modules and online courses	**Synergy between research and learning** We use the University estate as both a classroom and laboratory to provide students with 'real world' project experience
Diversities, equalities and people who flourish Sustainability includes social justice and equality issues. Our engagement opportunities also enhance staff and student experience	**Discovery-enabling culture** Sustainability Hub within Professional Services led by a Head of CSR & Sustainability
Local impact and making a difference for our city and region CSR (corporate social responsibility) and sustainability both understand that the context, community and environment in which we operate are integral to our success but sustainability also considers the needs of the future generations. Thus, CSR looks backwards at performance while sustainability is more forward-facing. From reducing our carbon emissions, our Fairtrade status to our biodiversity work offering staff, students and local residents opportunities to get involved in and to learn more about their environment: we continue to make a positive impact	**Discovery-led research** Almost a third of impact case studies submitted to REF 2014 include some aspect of sustainability, which demonstrates that there we have a large amount of research making a positive global impact in sustainability

(continued)

Table 1 (continued)

Key themes	Key priorities
Global presence and a focus on internationalisation Globalisation is about interconnectedness of economic, environmental and social issues. Our academics and students need the knowledge and skills to overcome the world's complex challenges and, as an excellent educational institution, we are well placed to deliver this	**Discovery-led learning** We are creating new sustainability learning opportunities through the curriculum *Pathways* initiative We support student-led sustainability opportunities such as the Sustainable Student Working Group and our HEAR accredited activities
Welcoming and cohesive campus experience The University has 305 acres of green space, including a diverse range of habitats, and from pond clearing to food growing, we offer staff and students the opportunity to get involved and to learn about their environment as well as enjoying it	**Discovery-enabling environment** We embed environmental sustainability within Estates operations We support the Sustainable Procurement Policy We have a staff 'Positive Impact Network' We enhance biodiversity at the University and encourage staff, students and visitors to engage with our wildlife and natural environment We help to deliver the University's CSR strategy

3 Success Stories

In terms of success stories so far, we can see emerging partnerships and networks that are supporting the development for all stakeholders (though still limits in terms of participation). What needs to happen next is a more robust research model that allows us to learn from case studies so as to broaden engagement. Below we present some aspects of these success stories that include the Centre for Interdisciplinary Science and the Sustainable Futures Module, a Student Union Campaign and Hungry for Change Project, the ESD Forum and RCE, and Student Engagement.

4 Centre for Interdisciplinary Science

The module, developed by the Centre for interdisciplinary Science and called Sustainable Futures originally existed as optional first and second year 5 credit modules delivered by academics from throughout the university began as eight units each involving some pre-session preparation, a two-hour workshop and some follow-up activity, and each with an internal or external presenter. Topics ranged from archaeology to industrial chemistry, energy policy to international development.

Attitudes to these updated Sustainable Futures modules were evaluated by questionnaire after the first year of new content (2011/2012). 18 students responded to the questionnaire. All students surveyed agreed that the course improved their knowledge and understanding of sustainability. The majority (94 %) of students indicated that the course prompted them to consider different viewpoints and they had developed useful skills throughout the course, which would be valued by future employers and were suitable for a range of careers (78 %).

There was some popularity of the campus-based courses but reported issues with timetabling suggested there might be an appetite for an online version of these modules that would be accessible to a wider cohort. The Centre successfully applied via the University's Teaching Enhancement Fund for funding to adapt the campus delivered existing material into an online asynchronous course, available to all students at the University. The course was piloted in the 2014–2015 academic year (Raine et al. submitted). A total of 250 staff and students signed up for the course; 104 completed at least one multiple choice test and 49 successfully completed the essay assignment which ensured recognition on the students' Higher Education Achievement Report. Feedback from students completing the evaluation questionnaire was overwhelmingly positive. This was still non-credit bearing but contributed to the Higher Education Achievement Record (HEAR). A second run of the course in 2015–2016 attracted over 100 students. One of the units, on food security, had been developed as a student-staff partnership based, in part, on the practice-based Hungry for Change project run by the Students' Union.[3]

The Students' Union also ran a campaign to embed sustainability widely across existing taught modules which at least served to expose the areas where sustainability issues were discussed even if it did not itself lead to any curriculum development, although the Sabbatical Officers continue to campaign for this.

Whilst these modules were popular, the academics involved focused more on research into sustainability rather than "real-life" scenarios and applications. In response to this, and to create re-usable resources for teaching sustainability with Biological Science the Centre for Interdisciplinary Science and School of Biological Sciences successfully applied for £15,000 of funding from the HEA Bioscience Centre to develop resources for sustainability literacy teaching. This funding enabled the development of two new five credit modules and updating of the existing two modules resulting in a series of workshops each with accompanying online resources in the form of a pre-workshop lecture, a reading list and post—session multiple choice questions.

There were however significant difficulties with the programme mainly relating to scheduling, already mentioned. The Centre was also able to recruit a number of external experts to deliver workshops including a consultant from a local energy partnership and an Ecological Economist. As required by the funding, all the resources were made available as Open Education Resources and material from one of the workshops is now used by the University of Nottingham in their Sustainability and Engineering module.

The resulting modules were initially made available to students from Computer Science, Chemistry, Physics and Geography and in subsequent years as non-credit modules for students from programmes outside the College. These campus based

[3]https://www2.le.ac.uk/offices/estates/environment/getinvolved/hungryforchange (accessed 31/3/2016).

courses are well subscribed with over 100 students registering in total. Additionally the value of the modules was recognised externally by the Environmental Association for Universities and Colleges and shortlisted for their Courses Green Gown Award in 2012.

Nevertheless, we now have top-level management support for an optional subsidiary programme (a "minor" subject) amounting to 25 % of credits across all three years of a degree. This will bring together the Departments of Geography, Management, Politics and the Centre for Interdisciplinary Science. The introductory module will be based on the Sustainable Futures course. More importantly, it is helping to move away from discipline silos and starting to form a bridge between a bottom-up and top-down approach.

5 ESD Forum and RCE

The ESD Forum was introduced early in this chapter, but recent developments have meant an expansion of our networks through the development of our links with the East Midlands Regional Centre for Expertise in ESD. In addition to attending several meetings, we have also hosted the most recent meeting. This has widened and supported the RCE network and more specifically allowed us to form links with other ESD stakeholders. As the result of this we are currently exploring the possibility of collaborating with another Higher Education provider to deliver a Sustainability MOOC.

6 Environmental and Social Initiatives

The student-led 'Hungry for Change' project aims to change the way students think about what they're eating and why. They have turned over 3874 m^2 into growing space across the University with the aim of educating and empowering students to make their own informed food choices instead of relying on product labelling and presumed company ethics. With ethical and sustainable motivations, the project has engaged with hundreds of students by applying a practical approach to learning in a subject which we can all relate to and care about. There has been the creation of habitats in conservation efforts—Four Bug hotels around our sites. Along with tree planting across the campus, so that the University of Leicester now has almost 10,000 trees and is working with the Woodland Trust to increase this further. Current planting strategies focus on suitable species for pollinating, attracting bees and other crucial wildlife etc. In terms of Environmental Sustainability (*see scorecard in* Fig. 1) Initiatives include:

Fig. 1 Sustainability score card

- Reduced car parking to improve shared use areas
- Single occupancy cars travelling to our sites have been reduced by 12 % since 2010
  - 5 % more staff and 6 % more students now walk to campus
  - 20 % of cyclists use the underground bike park
  - 6 % more staff now use public transport
  - We have more than trebled the amount of bike storage on our sites
  - There's been a 51 % reduction in bike thefts since we introduced the 'D-lock swap' scheme 2 years ago
- Total carbon consumption has decreased by 9 %
  - We have increased our renewable electricity production by 19 %
- Energy consumption per m^2 being reduced by 27 % since 2004/2005 despite the University's significant growth:
  - We have 20 % more students
  - We have enhanced our student experience with two new sports centres with swimming pools and increased Halls of Residence (at a time when many universities are selling off such facilities)
  - We have responded to student feedback by keeping the library and the students union buildings open for longer

- A new (Sustainable) Procurement Strategy launched in 2015, along with new Procurement Policies (including for Sustainable Procurement) and a new set of Procurement (Financial) Regulations. These are supported by new procurement templates; guidance and training that prompt consideration and highlight the virtues of, sustainable procurement.

7 Student Engagement

Student engagement is a key focus for the development of the University of Leicester as a Sustainable University and the following key initiatives demonstrate *some* success in engaging the student population, however more work needs to be done:

- Sustainable Development Programme with 34 students registered in 2015/2016
- The Leicester Award saw 22 Second Year students complete the Sustainability option of the Award in 2015/2016 whilst 33 environmental volunteers achieved HEAR accreditation
- The Environmental Action Society (EAS) has over 100 members and manages our environmental volunteers as well as campaigning within and without the University on sustainability issues that matter to students.

8 Participation, Collaboration and Voice

Participation, collaboration and voice are all key elements underpinning sustainable development and the University of Leicester Strategic Plan (draft) aims for a 'student-centred approach and wider participation' that will nurture innovation and breed a culture of trust and integrity. We want this commitment to student experience and global responsibility to result in real change at the University of Leicester, but our plan has been to go beyond the usual formula that you might expect. Typically, there is a Student idea + Staff backing that equals = a nice success story, job done. At Leicester we are more concerned with the bigger picture of how we can instigate a long-term transition and encourage students to promote positive change in their future lives or careers. In 2012 our own survey backed up the consistent findings of the *HEA Sustainable Development Survey* by finding that students were less afraid of change than the bulk of society and seek success and the esteem of others. They are therefore perhaps more willing to take the lead on and embrace change. This is not surprising given that they are in a transition phase in their lives and University is an ideal time for them to begin to develop environmentally friendly/sustainable thinking habits (such as recycling and buying ethical food with less packaging, joined up thinking). Over 80 % of respondents felt that the University should be obliged to develop environmental and social issues and a further 62 % supported the idea of £5 of their tuition fees being ring-fenced for a fund for student-led environmental/sustainability schemes. Almost 60 % of students felt that the University should provide teaching and learning about environmental and social issues with 43 % feeling that it should be added to their current course and over 30 % being prepared to attend non-compulsory workshops. Therefore our students are increasingly demanding sustainability issues to be included in their course, regardless of subject studied.

Our student engagement programme can be summarised in three sections.

- PROJECTS: The Sustainable Enterprise and Environmental Development (SEED Fund)—a platform and resource for funding student projects and dissertations.
- EDUCATION: The Sustainable Development Programme as an educational course filled with workshops and voluntary opportunities.
- POLITICS & POLICY: The Student Sustainability Working Group (SSWG).

It is best understood as a new Sustainability Framework; we have exchanged their soap boxes for tangible tools, skills and power to turn their ideas into realistic business cases and project proposals. We offer a diverse range of opportunities from basic volunteering to political positions and value new ways of thinking by inviting people from all disciplines to be involved. Student engagement in sustainability is growing exponentially. Six years ago we started with a few keen volunteers litter picking and tree planting, now we have active student societies campaigning on all issues from divestment from fossil fuels, removal of paper cups from canteens, the food miles and packaging of foods sold on campus and even use of alternative sanitary products as well as running projects on biodiversity planning on campus, carbon offsetting modelling and many others.

Leicester students were prolific contributors to the HEA's 2015/2016 Sustainable Development Survey with almost 900 respondents. We were second only to Plymouth proving that focussing on Education for Sustainable Development is highly effective in engaging students with this agenda.

The participatory nature of education for sustainable development (ESD) has been driven by UNESCO through the United Nations Decade which is a global initiative with links to other United Nations (UN) initiatives, such as Education for All (EFA) and the Millennium Development Goals (MDGs). The decade of education for sustainable development has focused on developing transformative strategies that engage people in participative approaches to change. This helps shift education from the modern to the post-modern arena with a shift from constructivist to more transformative or even reformative (Giroux 2006; O'Sullivan 1999; Sterling 2002) approaches to learning. Through appropriate management practices that engage whole communities in a transformative change process this might offer hope for the 'vision' of the DESD to be realized. However complexities make this difficult in practice. The DESD did start to engage many universities in the United

Kingdom (and around the world) in a change process by providing unique opportunities to develop as sustainable communities of practice, whilst engaging them in aspects of whole systems approaches through participation and democratic decision-making (Shallcross et al. 2006; Tilbury 2004; Henderson and Tilbury 2004; UNESCO 2009).

There is still more work to be done. The Global Action Programme (UNESCO 2014) provides a roadmap that will help to facilitate and support the learning and training required within organisations (including Sustainable Universities) in closing the GAP to achieving the vision set out by the United Nations decade of education for sustainable development beyond Rio +20. It sets out priority areas and actions.

Situated learning, situated cognition and socio-cultural psychology relate to the work of Bruner (1996), Cole (1996), Engestrom et al. (1999), and Wertsch (1998) and are built on the work of Vygotsky (1978). Situated learning is concerned with identifying social engagement that provides social contexts for learning to take place, provides access to communities and learning in them, engages participants in legitimate peripheral participation and ensures learning from others is a two-way process (Lave and Wenger 1991). Situated learning provides opportunities for gaining confidence and identity; develops the roles, responsibilities and resources; and provides the development of analytical views on learning. Situated learning requires participation. It is with this in mind, that the Sustainable University needs to set up learning communities and social contexts that can allow this to take place.

Education for Sustainable Development (ESD) has been described as a participatory process requiring engagement with communication, decision-making processes and evaluation when searching for more innovative solutions to local problems through an on-going process of reflection (Reid and Nikel 2008). School/University/Community gardens are one way in which schools can engage in ESD to help foster participatory situated learning. At Leicester we have the example of Hungry for Change Project.

We know that situated or contextualized learning models allow practices to be shared (Bourdieu 1984; Bandura 1977; Karol 2007; Lave and Wenger 1991). The learning is facilitated through the engagement of people. In community gardens the newcomers (novice or apprentice) learn from the old-timers through legitimate peripheral participation (Lave and Wenger 1991). Situated learning is fundamental to whole systems approaches (Bandura 1977, 1986; Henderson and Tilbury 2004; Shallcross and Robinson 2007; Wenger 1998). Involvement in a community of practice in which participants engage in cooperative practices is an important aspect of situated learning. It is within the community of practice that participation, democratic decision-making and two-way learning takes place; which are all important aspects of sustainable development. Knowing how to participate in this type of social practice plays a crucial part in students' learning (Greeno et al. 1996) and is facilitated through 'learning by doing' (John Dewey).

Participation is an important but complex aspect of ESD steeped in notions of power, inequality, dynamic relationships, and interaction. Participation may be perceived as a means of consensus-making but what is important is the purpose of this consensus being understood, including how the consensus was reached and the power dynamics evident during that consensus-making process (Kapoor 2004). The imbalances of power that underlie participation can have profound influence on what decisions are made (Kapoor 2004). Participation can often be managed (Lewis and Naidoo 2004) and institutionalized. This leads to tokenism where the community engaged in this participative process is not empowered but becomes powerless where the status quo remains and oxymoron attitudes persist (Reed Johnson 2014). What this indicates is that the Sustainable University needs to be mindful of this in ensuring that these power dynamics are sympathetic to this type of consensus building and decision making.

Learning is fundamentally a social process, inherent in human nature, where there is a relationship between context and meaning. From this perspective learning is located in the processes of co-participation where '*learning is an integral and inseparable aspect of social practice*' (Lave and Wenger 1991: 31). It involves engaging in practice and thought to change who we are through processes of critical reflection on our participative practice (O'Sullivan 2001). Learning can also act as an inter-play between the local and global concepts (Wenger 1996). It is important at the University of Leicester that we draw on our success stories and ask the question,' 'to what extent is this happening?'

If learning is a situated practice, that allows participants to learn through legitimate peripheral participation, then it could be argued that universities engaging in ESD could also be referred to as 'communities of practice'. A community of practice is made up of three elements described as a *domain* of knowledge which defines a set of issues; a *community* of people who care about this knowledge and share their practices; and a *practice* that is shared by that community of people who are all developing to be effective in that domain (Wenger et al. 2002). Therefore, the University of Leicester needs to review the types of communities of practice it has and that these aspects are incorporated. The success stories presented above are made up of these three components, but an extension of these is required.

Current educational systems in many countries are believed to not always allow young people to engage in pedagogies of participation (Lotz-Sisitka 2004, O'Donoghue and Lotz-Sisitka 2006), democratic styles of learning and critical thinking (Sterling 2002). However, our success stories at the University of Leicester do indicate that these barriers are being broken down, but we do still need to explore the extent to which they are. We need to be asking the questions related to ways we view multiplicities in a 'one sole world' (Lotz-Sisitka 2010); how ecology (and ESD) is contributing to the political project of human emancipation where it is possible for everyone to live equitably and sustainably (as argued by Badiou 2007 and Morgensen and Schnack 2010), it may be possible to explore the participatory nature of these success stories presented in this paper and beyond to understand this further.

9 Challenges, Opportunities Emerging, and Lessons Learnt so Far

The challenges are fourfold—resources, senior management commitment and long term thinking, promotion of competences to include critical thinking, futures imagining, collaborative decision making.

The challenge is to embed sustainability into the heart of teaching and learning, and research. Graduates are becoming increasingly aware that their employability requires so much more than just a good degree. Meaningful projects and extracurricular activities are becoming an essential part of their degree programme.

Education for Sustainable Development requires a two-pronged approach:

- teaching students about sustainability issues (this can be formal or extra-curricular);
- Equipping graduates with the problem-solving skills necessary to deal with the sustainability challenges that the world faces now and in the future.

Therefore, ESD requires participatory teaching and learning methods that motivate and empower learners to change their behaviour and take action for sustainable development. ESD consequently promotes competencies like critical thinking, imagining future scenarios and making decisions in a collaborative way. Armed with the right knowledge and skills, Leicester graduates can be capable of contributing to a better world. The University of Leicester can be committed to sending students out into the world as global citizens, who are sustainability literate and have an appreciation of social and cultural diversity.

Multi-disciplinary teaching should be supported and encouraged wherever possible. Multi-disciplinary teaching staff should contribute to teaching resources and delivery.

We are working to develop the University itself as a 'laboratory' to help students (and staff) to engage with what is going on. Examining the organisation and how it's evolving, such as energy exchange etc. is an ideal case study for teaching that can provide students with the tools to move from concepts to implementation. There have been some examples of final year projects being carried out with Corporate Services staff such as Maths students calculating the carbon footprint of the University's procurement. This work is being formalised to enhance project opportunities for students and also provide added value for the University.

The aim of this new draft strategy (2015–2020) is to bring a new and coordinated approach across the University, building in multi-disciplinary approaches wherever possible.

There are challenges and this includes limited resources (financial or staff) and therefore we need to think creatively and be efficient. Another challenge is convincing senior management to think long term, but the sustainability team feel they are creating some small successes. This is most recently evidenced by opening of the Centre for Medicine, the largest Passivhaus building[4] in the UK. The University of Leicester has realised that vanity projects are not sustainable and instead has opted for setting industry standards by encompassing cost of life considerations into project designs, thus demonstrating a truly embedding of sustainability into University strategic operations.

10 Opportunities

The opportunities are great in that students do want to engage and there are agendas and communities of practice that can drive that engagement. These include the student marketplace for higher education and the growing interest in corporate social responsibility with the needs of the global citizen, along with the ESD Forum. There is a need for provocative narratives to stimulate agency, for conceptual and theoretical reflections. However, we point these out with trepidation because these are not the only drivers for change. What is really important is that young people entering the world of work understand how to deal with VUCA (Volatility, Uncertainty, Complexity and Ambiguity) in thoughtful ways. In dealing with complexities and decision making it is important young people are able to explore things through a 'sustainability lens', exploring things from all the angles of economic, environmental, political and social.

There are opportunities for engagement at the University of Leicester and students consistently demonstrate that they do care about sustainability and this convinces University management to listen. The university's new interest in Corporate Social Responsibility provides added impetus for sustainable development. There is a growing interest in the student experience agenda, through biophilia, mindfulness and community outreach which strengthen our business cases for top-down commitment. Sustainability is no longer the domain of the post 1992 Universities and with the opening up of the student marketplace it means that 'red bricks' such as Leeds and Nottingham have joined the revolution and are encouraging Leicester and other peers along the way.

An important role for the University is to develop 21st Century employability for its graduates, with a focus on sustainability that is agentic (*about, for and as sustainability*).

There are examples of where courses are developing problem solving skills, in formal teaching, for example 4th year Chemistry students have their lectures 'flipped' so they have open ended problems such as 'electric vehicles are green, discuss'. Students have to metro-size the problem, look at a series of papers from Green Chemistry and have a debate on the case studies. However, there are more opportunities, and particularly for graduates of the future to be able to dilemma flip (Institute for the Future) and scenario diagnosing.

[4]http://www.passivhaustrust.org.uk/what_is_passivhaus.php; http://www2.le.ac.uk/offices/estates/news/current-projects/ctrmed.

11 Lessons Learnt

The lessons learnt so far are that sustainability is far more bottom-up than top-down at the University of Leicester. Whilst senior engagement and support is essential, it is the embedding good practice into day to day operations which requires getting out there, really getting involved and listening to staff and students as key stakeholders.

We have also discovered that sustainability is everywhere! Often universities can get trapped in the environmental ruts of travel and waste, but at the University of Leicester we have found that a significant amount of our research has been related to the wider concept of sustainability and that every course curriculum can easily be adapted to include the essential problem-solving skills our graduates will require to solve the global challenges they will be presented with when they venture into the real world of work. Through this new remit it is hoped that this new vision will be coordinated across a wider Professional Services Directorate whose wider remit engages across both curriculum and research, and operations.

12 Success Criteria

In trying to understand our successes we have clarified some success criteria. These will require further refinement but we hope to indicate that sustainability has been embedded within the University when these milestones have been achieved (listed below). However, it is important that these are reviewed and updated regularly and that being a sustainable university is a living and breathing system and as with any living system, our indicators will evolve in evolutionary ways too.

- All staff take ownership of the targets set out in the Sustainability Plan and seek ways to incorporate sustainability in their own areas of work
- All strategic decisions made within the University consider the sustainability implications
- Our absolute carbon footprint is reduced
- Our campus facilitates staff, students and visitors to make sustainable choices
- All our students are aware of the University's commitment to sustainability
- All our students have access to formal and informal opportunities to learn more about sustainability issues
- Our curriculum helps students to develop the skills required to solve today's global challenges, including complex problem solving
- Our Sustainable Research Network is a truly innovative, interdisciplinary group who are able to compete for research funding
- We promote sustainable solutions locally and globally through our teaching, research and engagement work.

References

Bandura, A. (1977). *Social learning theory*. Englewood Cliffs, NJ: Prentice Hall.

Bourdieu, P. (1977). *Cultural reproduction and social reproduction*. Oxford: Oxford University Press.

Bourdieu, P. (1984). *Distinction: A social critique of the judgement of taste*. London: Routledge.

Brennan, J., King, R., & Lebeau, Y. (2004). *The role of universities in the transformation of societies*. Centre for Higher Education Research and Information, The Open University. Available at http://www.open.ac.uk/cheri/TRhome.htm

Bruner, J. S. (1996). *Culture of education*. Cambridge, MA: Harvard University Press.

Castells, M. (2001). Universities as dynamic systems of contradictory functions. In J. Muller, et al. (Eds.), *Challenges of globalisation. South African debates with Manuel Castells* (pp. 206–223). Cape Town: Maskew Miller Longman.

Cole, M. (1996). *Cultural psychology*. Cambridge, MA: Harvard University Press.

Gibson, C. (2001). *From inspiration to participation: A review of perspectives on youth civic engagement*. Berkeley, CA: The Grantmaker Forum on Community and National Service. Available at http://www.pacefunders.org/publications/pubs/Moving%20Youth%20report%20REV3.pdf. Accessed March 31, 2016.

Giroux, H. (2006). *The Giroux reader*. London: Paradigm Publishers. Global gateways: http://www.globalgateway.org.uk/

Greeno, J. G., Collins, A. M., & Resnick, L. B. (Eds.). (1996). *Cognitive and learning*. New York: Simon and Schuster Macmillan.

Henderson, K., & Tilbury, D. (2004). *Whole school approaches to sustainability: An international review of sustainable schools programs*. In ARIES (Eds.), Sydney: Macquarie University.

Johansen, B. (2012). *Leaders make the future: Ten new leadership skills for an uncertain world* (2nd ed.). San Fransico: Berrett-Koehler Publishers Inc.

Kapoor, I. (2004). Concluding remarks: The power of participation. Current issues in *Comparative Education*, *6*(2), 125–129. Teachers College, Columbia University.

Karol, J. (2007). *Bourdieu and education for sustainable development: Analysis of an interview*. In Paper presented AARE (pp. 1–18). http://www.aare.edu.au/06pap/kar06523.pdf

Karol, J., & Gale, T. (2004). *Bourdieu and sustainability: introducing 'environmental capital'*. In Paper presented at the AARE, Melbourne.

Lave, J., & Wenger, E. (1991, 2007). *Situated learning: Legitimate peripheral participation*. Cambridge: Cambridge University Press.

Lewis, S. G., & Naidoo. J. (2004). Whose theory of participation? School governance policy and practice in South Africa. Current issues in *Comparative Education, 6*(2), 100–112 (May 10, 2004), Teachers College, Columbia University.

Lotz-Sisitka, H. (2010). Changing social imaginaries, multiplicities and 'one sole world': Reading Scandinavian environmental and sustainability education research papers with Badiou and Taylor. *Environmental Education Research Journal, 16*(1), 133–142.

Macleod, J., & Yates, L. (2006). *Making modern lives: Subjectivity, schooling and social change*. New York: SUNY Press.

O'Sullivan, E. (1999, 2001). *Transforming learning: Educational vision for the 21st Century*. Toronto: University of Toronto Press.

Peden, M. I. (2008). Education for sustainable development: knowledge and environment in South African schooling. *Environmental Education Association of Southern Africa, 25*, 13–24.

Raine, D. J., Gretton, S., & Greenhill, D. (2016). *Sustainable futures online* (submitted to On the Horizon, Emerald Publishing).

Reed Johnson, J. A. (2014). *Education for sustainable development: A comparison of two eco-schools, one in England and the other in South Africa*. Lambert Academic Publishing.

Reid, A., Nikel, J. (2008).

Robinson, J., & Shallcross, J. (1998). Social change and education for sustainable living. *Curriculum Studies, 6*, 69–83.

Shallcross, T., Loubser, C., Roux, C., O'Donoghue, R., & Lupele, J. (2006). Promoting sustainable development through whole school approaches: An international, inter-cultural teacher education research and development project. *Journal of Education for Teaching, 32*, 283–301.

Shallcross, T., & Robinson, J. (2007). Is a decade of teacher education for sustainable development essential for survival? *Journal of Education for Teaching*, *33*(2), 137–147.

Sterling, S. (2002). *Sustainable education: Revisioning learning and change*. Devon: Green Books Ltd.

Engestrom, Y. Miettinen, R., & Panamaki, R. L. (Eds.) (1999, 2006). *Perspectives on activity theory*. Cambridge: Cambridge University Press.

The Robbins Report. (1963). *Higher education: Report of the committee appointed by the prime minister under the chairmanship of Lord Robbins* (p. 7). London: Her Majesty's Stationery Office. http://www.educationengland.org.uk/documents/robbins/robbins1963.html

Tilbury, D. (2004). Rising to the challenge: Education for sustainability in Australia. *Australian Journal of Environmental Education, 20*, 103–114.

UNESCO (2009) *DESD case studies report*: *Part 2*, UNESCO.

UNESCO (2014) *Roadmap: For implementing the global action programme on education for sustainable development*. United Nations Educational, Scientific and Cultural Organization, 7, place de Fontenoy, 75352 Paris 07 SP, France.

Wenger, E. (1998). *Communities of practice: Learning, meaning and identity*. Cambridge, UK, NY: Cambridge University Press.

Wenger, E., McDermott, R., & Snyder, W. (2002). *A guide to managing knowledge: Cultivating communities of practice*. Massachusetts, USA: Harvard Business School Press.

Wertsch, J. V. (1998). *Mind as action*. New York: Oxford University Press.

Environmental Awareness and Concern of the "carbon Cost" of Activities and Food Choice in Male University Gym Users, with Particular Reference to Protein Consumption

Kate E. Reed

Abstract

The aims of this study were to (a) determine the extent to which university gym-users understood the environmental impact of common activities & foods, particularly protein consumption, and (b) explore factors considered important when purchasing food, and determine whether knowledge was associated with behaviour. 43 males (18–24 years) completed a four part questionnaire. Responders were asked to (1) read a passage about the life an active male and consider which components contributed most to his 'carbon footprint', (2) rate the environmental impact of common foods (3) describe their own protein consumption habits, and (4) identify which factors they considered important when buying food. It was found that (1) Few responders considered diet as factor contributing to a carbon footprint, focusing on more 'visible' activities such as driving, (2) Most responders were unsure of environmental impact of foods, especially foods grown out of season and dairy produce, (3) 68 % of responders consumed protein supplements and (4) No responders stated that environmentally-friendly packaging or country of origin was 'very important'. In summary, there is low awareness of the environmental cost of activities and foods among university males. Ethical concern over food choice was somewhat higher in respondents with a higher environmental knowledge.

Keywords

Carbon footprint · Protein consumption · Food choice

K.E. Reed (✉)
Centre for Sports and Exercise, University of Essex, Essex CO4 3SQ, UK
e-mail: reedk@essex.ac.uk

W. Leal Filho (ed.), *Sustainable Development Research at Universities in the United Kingdom*, World Sustainability Series,
DOI 10.1007/978-3-319-47883-8_9

1 Introduction

High protein diets are promoted as methods to reduce body fat, promote muscle gain and, therefore, to improve sports performance. The recommended nutrient intake (RNI) for protein in the UK is 0.75 g of protein per kg body weight per day (g kg^{-1} day^{-1}). While a 70 kg male should, therefore, eat around 53 g of protein daily, the majority of UK adults, according to the National Diet and Nutrition Survey, consume around 50 % more protein than they require (MRC 2014). It is often stated that athletes have higher protein requirements than sedentary adults, with estimates of 1.2–1.7 g kg^{-1} day^{-1} (Rodriguez et al. 2009). There is little evidence to suggest that consuming excess protein (as much as double the RNI for protein) is damaging to health (Martin et al. 2005). Conversely there are also no particular benefits from very high protein diets aimed at promoting muscle mass, as there is a ceiling effect at around 20–25 g of protein in a single bolus. Meat and dairy sources of dietary protein have become inexpensive and readily available in most developed countries. It has also become simple to supplement protein-intake from food and consume atypically large quantities of protein via sports recovery drinks and shakes, offering the consumer as much as 40 g of protein (usually derived from the dairy source whey).

The livestock production associated with the meat and dairy industries results in large emissions of nitrogen, methane and other greenhouse gases. Crop-livestock production systems are the largest human cause of alterations in the nitrogen and phosphorus cycles, and reports suggest that consumption of farmed animal products should be curtailed in order to reduce anthropogenic greenhouse gas emission (Deckers 2010). It is estimated that around 51 % of all anthropogenic greenhouse gas emissions are from the farm animal sector (Goodland and Anhang 2009), and this value will continue to rise if our consumption of farmed animal products continues along its current trajectory. The dominant anthropogenic greenhouse gas is carbon dioxide, but other greenhouse gases, such as methane and nitrous oxide, are equally important in terms of environmental impact. The term carbon dioxide equivalent (CO_2e) will be used hereafter to cover the impact of all these emissions. CO_2e provides a single metric which can be used to calculate the environmental impact of multiple behaviours such as travel, consumption of goods and diet.

The food choices we make have a large impact on our individual contribution to sustainability. The decision to undertake a certain activity, or buy a certain product, has a direct impact on our environment, and several models exist to explain purchasing behaviour. Early attempts to explain buyer behaviour, (such as Howard and Sheth 1969), still form the basis of many contemporary buyer behaviour models. For example, the rational choice model contends that individuals consider the costs and benefits of a certain product, finally choosing the option that maximises their expected benefits. However, rational choices are only possible when there is adequate information available. Furthermore, it is apparent that consumers frequently by-pass the cognitive processing stage, reverting to habit and familiarity in many purchasing situations, with many everyday behaviours being carried out with little conscious deliberation. Likewise, the Theory of Planned Behaviour (Ajzen and Madden 1986) reasons that attitudes toward a behaviour, subjective norms and perceived control, all shape an individual's intentions and actions. It is beyond the scope of this paper to consider all models of consumer behaviour related to sustainable consumption, and readers are direct to an excellent and comprehensive review by Professor Tim Jackson (Jackson 2005).

In recent years, ethical and environmental considerations have come to play a greater role in the decision making processes governing purchasing food for many people. Tools such as the Food Choice Questionnaire (FCQ) (Steptoe et al. 1995) allow the researcher to examine the decision making process buyers undergo when purchasing food. The questionnaire assesses the relative importance of nine factors representing potentially important motives for food choice (*Health, Convenience, Sensory Appeal, Mood, Natural Content, Price, Weight Control, Familiarity and Ethical Concern*). Studies that have used the FCQ suggest the main reasons for buying a specific food product vary by gender, age and between countries but that that 'Ethical Concern' is the least important factor governing food choice (Januszewska et al. 2011; Prescott et al. 2002; Pollard et al. 1998).

Thus, there is little evidence to suggest the increasingly common practice of adopting a protein rich diet is beneficial but has potentially harmful environmental consequences. Such ethical concerns may affect certain food choices but informed, rational food choices require access to accurate information. On this basis the aims of the present study were as follows. First, to explore the knowledge-base of male university gym users, in relation to environmental impact (CO_2e) of common behaviours, including food choice. Second, we sought to determine the most frequently accessed sources of nutritional information used by this group. Finally, we aimed to identify specific factors influencing food choice. It was hypothesised that participants who had higher awareness of environmental impact would report a higher Ethical Concern in relation to food choice.

2 Methods

Subject selection: The target population for this study was male university gym users. Researchers positioned in the gym foyer approached males exiting the gym. Researchers were in position for a total of 12 h, over the course of 1 week, at various times of day. Respondents were informed that information was kept confidential, and provided informed consent. Ethical approval for the study was granted by the University of Essex Ethics Review Board.

Data were excluded if respondents had any dietary constraints (such as coeliac disease or lactose intolerance) or were varsity athletes. This final group were excluded as their diet is regulated by coaches and the team nutritionists.

The questionnaire was in 4 sections, designed to elicit knowledge and opinions related to sustainability.

Section (1) In this section respondents were asked to read a passage about the life a young active male ('John') and consider which 3 components contributed most to his 'carbon footprint'. The passage included information about John's travel habits (driving 50 km each week), diet (including details on meats, use of protein shakes and alcohol consumption) and selected recreational or lifestyle activities. All values given in the passage were realistic and an example from the passage is given below:

> 'In the evening he has a meal that includes a serving of chicken or lamb, with vegetables. He consumes 500 g of chicken and 500 g of lamb per week. This is followed by a second protein recovery drink...... He enjoys 2 bottle of German beer each night..... He watches TV for 2 h each evening....He washes and tumble dries his laundry once per week'

Participants were instructed not to consider the initial manufacture of the car or the household appliances. Participants listed the three activities they perceived to contribute most to carbon footprint. Those who correctly identified the three highest CO_2e activities were classified as 'very aware'; two correct activities, 'aware', one activity 'somewhat aware' and none—'unaware'.

Section (2) In this section respondents were asked to rate the CO_2e of 10 common foods and rank them from 'Highest' to 'Lowest'. The list included: fruits, vegetables, dairy produce, non-dairy beverages, meat and common snacks—all presented as a typical serving. Participants were instructed not to consider packaging in their ranking. No information was provided on other potential environmental considerations such as farming methods, processing or distribution. These were not mentioned so as to encourage participants give a spontaneous answer.

Section (3) In this section respondents were also asked to report the sources they use to obtain nutritional advice. Respondents could select from 4 categories: Friends, Gym staff/sports lecturers/coaches, Magazines/websites or Other. They were also asked to state if they took a protein supplement, and whether they had a target daily protein intake.

Section (4) This section used the Food Choice Questionnaire, to examine the relative importance of the nine motivational dimensions of food selection including Ethical Concern, Familiarity, Convenience, Health and Price. Respondents were asked to identify factors they considered important when buying food. Factors were rated from 1 (Not Important) to 4 (Very Important) on a Likert type scale.

The relationship between knowledge and food choice was examined. Respondents were classified as having High Awareness (very aware or aware) or Low Awareness (somewhat aware or unaware) using data from section 1. Ethical concern was determined from the FCQ. There was a potential maximum score of 12, and respondents with an Ethical Concern of 5 or less where classified as having 'Low Concern', those with 6 or above classified as 'High Concern'. Differences between the groups were examined using Chi Squared analysis, with alpha set at 0.05 (SPSS v 19).

There are limitations in the methodology. Estimates of CO_2e for a particular product vary enormously from country to country, and also according to specific components assumed in the calculation. Further, there is a small sample size, and the authors cannot ignore the potential for responder bias.

3 Results

A total of 62 gym users were asked to complete the questionnaire, with a response rate of 69 %. A total of 43 males (18–24 years) completed the questionnaire. No responders were vegan or vegetarian.

Section 1: Respondents were aware of several components of ‘John’s’ life that contributed to his carbon footprint. The three most cited components reported were; car driving, washing and tumble drying clothes, and using a disposable carrier bag each day. Few respondents commented at all on food and beverage choices. Only 4 respondents identified consumption of a relatively large amount of lamb (the meat with the highest CO_2e per kg), as a major contributor to carbon footprint. Only one respondent considered the weekly consumption of 320 g whey concentrate a relatively important factor. Figure 1, shows the components respondents considered to have the highest impact on the fictitious John’s carbon footprint. For example, 96 % of respondents selected ‘car driving’ somewhere in their top 3 CO_2e creating behaviours.

Figure 2 shows *approximate* kilograms of carbon dioxide equivalent. The estimates do not include manufacture of the car or the household goods, but do include manufacture and distribution of the food stuffs. Estimates of CO_2e where from a variety of sources (Berners-Lee 2010; Tesco 2012; Williamson 2014).

Only 1 respondent was classified as ‘very aware’, nine as ‘aware’ and the remaining 33 as ‘somewhat aware’

Section 2: Respondents ranked 10 common foods in order of CO_2e per serving. The foods in order of highest CO_2e, to lowest CO_2e were:

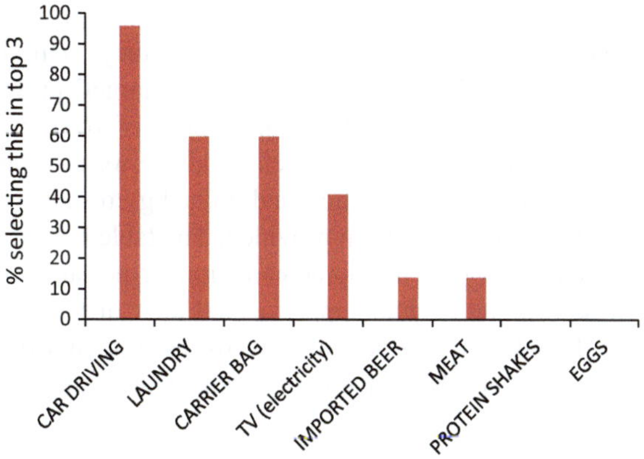

Fig. 1 Percent of respondents selecting each component in their top 3 CO_2e creating activities. No respondents selected ‘eggs’

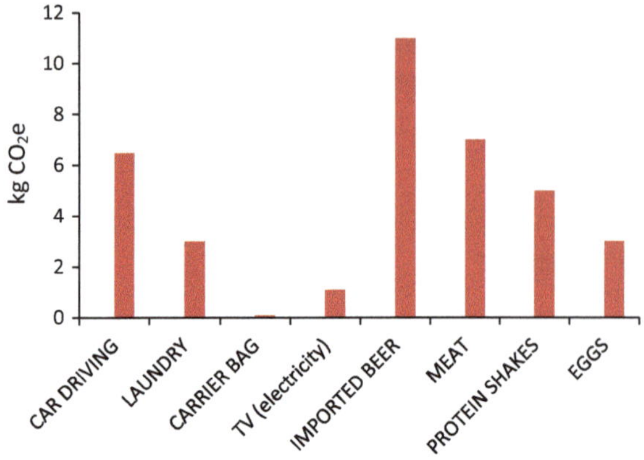

Fig. 2 Approximate CO_2e cost of each component over the course of the example week. (Values are kg CO_2e)

10 vine-ripened cherry tomatoes (grown in the UK in March), a beef burger (UK beef), a chicken breast, serving of tinned tuna, a pint of milk, a large café latte, an orange, a banana, a serving of crisps (30 g) and a cup of tea.

In general, respondents noted that the beef burger had a high carbon cost, but failed to note the high carbon cost of fruit grown out of season in the UK. Also, the banana and the orange were frequently ranked very high, despite them having a lower CO_2e than milk.

Only one respondent correctly identified the three foods with the highest CO_2e per serving, 13 identified two foods, 10 respondents identified 1 of the foods, and the remaining 19 failed to identify any of the three foods with the highest CO_2e.

Section 3: Despite many of the gym users having access to experts in Sports Nutrition, either as lecturers or the trained gym staff, the majority of responders turned to friends, and then websites for nutritional advice. There were a limited number of websites listed, and all but one were non peer-reviewed sites, hosted by companies that produce nutritional supplements, or are sponsored by supplement manufacturers.

Sixty eight percent of respondents consumed protein supplements, usually in the form of 'powdered' recovery drinks. Although most consumers of protein supplements were aware of the grams of protein provided by each drink, several were unaware of how much extra protein they were taking in. Just over half the respondents (51 %) reported having a 'target' protein consumption per day. This ranged from 100 to 300 g, with a mean of 163 g per day.

For analysis, sources of dietary advice were combined into two broad groups of friends/magazines/websites, and trained gym staff/coaches/lecturers to represent the reliability of information provided. The table below shows source of dietary advice according to protein supplementation or no supplementation. It can be seen that those responders who supplemented their diet with protein relied more frequently on friends, magazines and websites for nutritional information and advice. Respondents who did not supplement their diet with protein (although fewer in

Table 1 Source of dietary information according to category of supplementary protein intake. Data are percent in each category

Source of advice	Friends, magazines/websites	Staff/coaches/lecturers	Both
Supplemented with protein	66.7	16.3	17
No supplement	20.0	70.0	10.0

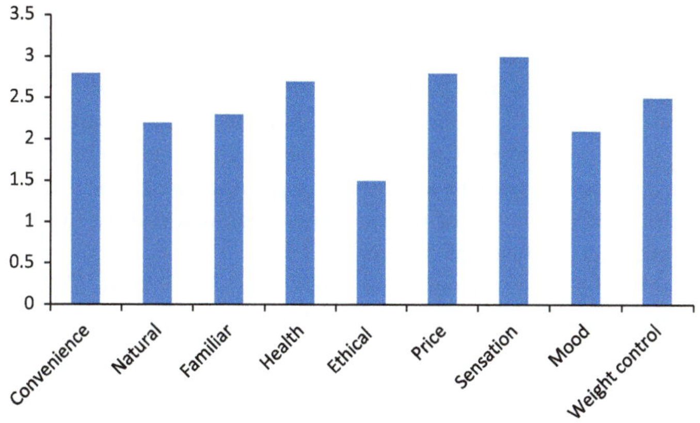

Fig. 3 Factors influencing food choice in gym users. Values are calculated from individual questions in the Food Choice Questionnaire, then averaged as some categories contain more questions. Scores could range from 1 to 4, from low to high importance

number) more typically obtained dietary advice from trained staff and coaches (Table 1).

Section 4: Fig. 3 shows a summary of results from the Food Choice Questionnaire. It can be seen that the biggest influence on food choice was 'Sensation', i.e. taste and texture of the food, closely followed by 'Price' and 'Convenience'. The category that had the lowest influence on food choice was 'Ethical Concern'.

Of those classified as having 'High Awareness' from section 1, 38 % were classified as having 'High Concern'. Of those classified as having 'Low Awareness', only 16 % had 'High Concern'. However, there was no statistical difference between groups in terms of likelihood of having High Concern ($\chi^2 = 2.41$, $p = 0.12$)

4 Discussion

The aims of this study were to explore the knowledge base of male university gym users in relation to environmental impact of common activities and foods, determine the common sources of information related to nutrition, and finally, to examine which factors influence food choice.

Knowledge of food and activities that had high relative environmental impact, in relation to greenhouse gas emission was fairly low. In recent years the UK government has undertaken a number of measures to reduce its greenhouse gas emission, and many of these have been widely publicised. Manufacturers of cars promote the economy of cars, reporting miles per gallon, and CO_2 emissions. Similarly, manufacturers of items such as washing machines, televisions, fridges and so on are now obliged to report the 'energy rating' of such products. These highly visible campaigns have clearly been successful in that each respondent in the study was aware of the environmental impact of activities such as driving and using a tumble drier. Further, many respondents included use of a carrier bag as an activity that would have a high carbon cost. At the time this study was undertaken (2015), it became law in England for large shops to make a charge for each new carrier bag a customer used. Carrier bags, although fraught with environmental issues, have a relatively low CO_2e (around 5 g per bag). It appears that respondents were aware that excessive use of plastic carrier bags was bad for the environment, but were somewhat confused as to why.

Interestingly, the understanding that foods can have a large CO_2e was rarely shown in this group. Lamb has the largest CO_2e of any meat, at around 17 kg CO_2e per kg lamb. Chicken is one of the meats with the lowest, at around 2–4 kg CO_2e per kg (depending on farming methods). Only one respondent considered the whey concentrate that was consumed. As a product of the dairy industry, whey has a large CO_2e at around 16–17 kg per kg of powder. It is the mostly commonly used protein in recovery shakes. Other common proteins used in powdered shakes are casein (derived from cow's milk) and soy. Soy protein has a CO_2e of around 2–4 kg per kg of protein (Schenck and Huizenga 2014). Soy milk has around half the CO_2e per litre (0.8 kg) than that of cow's milk (1.5 kg), contains a comparable amount of protein (30–35 g) and has been shown to support muscle protein synthesis (MPS) following resistance training (Tang et al. 2009). However, it has been shown than the rate of MPS is slightly slower in response to consuming soy protein post exercise, compared with whey protein (Wilkinson et al. 2007). For this reason, many post-exercise protein consumers prefer whey based supplements to those derived from soy protein.

The theory of planned behaviour contends that an individual is likely to embark on a course of action according, in part, to social influence. There is currently, a movement toward high protein consumption, both for weight (fat) loss and for muscle gain. Weight loss plans such as the Dukan Diet suggest eating only protein (plus a tablespoon of oat bran) for 10 days, followed by several *months* of only protein, in combination with non-starch vegetables, such as lettuce and broccoli. Websites dedicated to body building often promote protein consumption of up to 3.3 g kg^{-1} day^{-1} (suggesting a 70 kg man would need 231 g protein per day) (BodyBuilding.com 2015). Compare this with the UK government's RDI of around 0.75 g kg^{-1} day^{-1}. It seems feasible that if an individual's peers and those training with in gym environments are consuming such large quantities of protein, such practices may become the social norm. It is common place for protein shakes to contain 24 g of protein per serving, although some contain up to 40 g per serving.

The amount of protein per serving in recovery drinks is usually shown as % daily value on the packaging. A single 24 g serving of protein is around 48 % of an adults RDI, yet up to 5 servings per day are suggested by many manufacturers, as well as advice to consume a 'normal diet'. Following this advice would result in an individual consuming more than 250 % of the RDI for protein. One review examining high protein intake and renal damage stated that whilst consuming twice the RDI for protein did not cause permanent renal damage, acute changes in renal size or function were frequently reported (Martin et al. 2005). Importantly, many of the studies included in this review were short term (7 days) or examining protein consumptions that were only on the moderate side of high (1.2 g kg^{-1} day^{-1}). Studies examining the long-term (i.e. several years) health effects of protein consumption that is greater than twice the RDI, need to be conducted. In the current study, 2 out of 3 respondents consumed a protein supplement, with several aiming to consume 300 g of protein per day. This is easy to do, with the range of protein supplements available today. If this amount of protein were to be consumed in the form of food, it would equal around 1 kg of chicken, 1 kg of tuna, 3.75 kg of tofu, or 42 eggs.

In line with previous studies, there was little concern over environmental issues when purchasing food. Several researchers have used the FCQ to explore the major determinants of food buying behaviour. Only in Japan were Ethical Concerns rated as important food choice determinants (Prescott et al. 2002). In most European countries (Januszewska et al. 2011; Steptoe et al. 1995; Lindeman and Väänänen 2000), Asian countries and in New Zealand (Prescott et al. 2002), the most important factors were; Sensory Appeal, Price, Health and Convenience. The present study included only males. It has been reported that females place higher importance on Ethical Concern than males (Lindeman and Väänänen 2000), so inclusion of females in the current study many have altered results somewhat. However, given that the results of the FCQ in the present study mirror results from other studies, the inclusion of females would likely have only a small influence on findings. This is something that could be explored in future studies.

It was found that responders with a higher environmental awareness rated 'Ethical Concern' higher than responders classified as having lower environmental awareness, although there was no statistical difference. Respondents in this study were unable to reliably rank foods in terms of their CO_2e. Without the required knowledge, individuals are unable to make informed decisions. Information and knowledge are crucial for engagement in campaigns for change. However, the relationship between knowledge and action is not linear and behaviour change is dependent on a range of factors (Barr 2003). Too and Bajracharya (2015) explain the need for a combined approach if a 'Sustainable Campus' is to be achieved. Their model provides a holistic, 6-P Framework (psychological needs, personal motivation, public perception, price, physical facilities and policy) that is required. Although many universities in the UK are reducing their environmental impact through changes to their operational issues, this current study provides evidence that there is still room for improvement in relation to many of the P's in the framework. Specifically, the issues of public perception (social norms) and psychological needs (knowledge & values concerning the environment) need to be addressed. Importantly, given that 'Cost' and 'Convenience' are key factors affecting food choice in the current group, it is important that 'environmentally friendly foods' are readily available, at a price students can afford.

As state above, estimates of CO_2e vary, even for a single product. Thus, it was not the aim of this study to inform readers (or participants) specifically on the best and worst food choices in relation to greenhouse gas emission, but rather to gather an overall idea of interest and understanding. We acknowledge that relying of different sources to gain estimates of CO_2e is a potential limitation of this study. Similarly, the study sample which was young males is not representative of the university community as a whole, so results cannot be generalised too widely. However, this group is representative of gym users, who were the target population for this study.

5 Conclusions

In summary, it was found that knowledge of the carbon cost of most common foods & activities was poor, use of protein supplementation was high, particularly among respondent who relied on website for nutritional information. Further, in support of previous investigations, this study supports the fact that 'Ethical Concerns' in relation to food choice are not a priority for this group. Although not statistically significant, it did appear that respondents with a higher environmental awareness placed a higher priority on Ethical Concerns in relation to food choice. In order to make the campus that was the site of this study more sustainable, action needs to be taken to increase knowledge and alter social norms.

References

Ajzen, I., & Madden, T. (1986). Prediction of goal-directed behavior. Attitudes, intentions and perceived behavioral control. *Journal of Experimental Social Psychology, 22*(5), 453–475.

Barr, S. (2003). Strategies for sustainability. *Citizens and responsible environmental behaviour. Area, 35*(3), 227–240.

Berners-Lee, M. (2010). *How bad are bananas?* London, UK: Profile Books.

Body Building.com. (2015). How much protein should you consume? Available at: http://www.bodybuilding.com/fun/bbprotein.htm. Accessed Jan 13, 2016.

Deckers, J. (2010). Should the consumption of farmed animal products be restricted, and if so by how much? *Food Policy, 35*(6), 497–503.

Goodland, R., & Anhang, J. (2009). *Livestock and climate change*. Washington: Worldwatch Institute.

Jackson, T (2005). Motivating sustainable consumption. Report from Sustainable Development Research Network, Policy Studies Institute, London.

Januszewska, R., Pieniak, Z., & Verbeke, W. (2011). Food choice questionnaire revisited in four countries. Does it still measure the same? *Appetite, 57*(1), 94–98.

Lindeman, M., & Väänänen, M. (2000). Measurement of ethical food choice motives. *Appetite, 34* (1), 55–59.
Martin, W. F., Armstrong, L. E., & Rodriguez, N. R. (2005). Dietary protein intake and renal function. *Nutrition & Metabolism, 2*, 25.
MRC. (2014). *What do UK adults eat? A summary of key results from the UK National Diet and Nutrition Survey*, Available at: www.mrc-hnr.cam.ac.uk. Accessed Jan 2, 2016.
Pollard, T., Steptoe, A., & Wardle, J. (1998). Motives underlying healthy eating: using food choice questionnaire to explain variation in dietary intake. *Journal of Biosocial Science, 30*, 165–179.
Prescott, J., Young, O., OiNeil, L., Yau, N. J., & Stevens, R. (2002). Motives for food choice: a comparison of consumers from Japan, Taiwan, Malaysia and New Zealand. *Food Quality and Preference, 13*, 489–495.
Rodriguez, N., DiMarco, N., & Langley, S. (2009). Nutrition and athletic performance. *Medicine and Science in Sports and Exercise, 41*(3), 709–723.
Schenck, R., & Huizenga, D. (2014). In *Proceedings of the 9th International Conference on Life Cycle Assessment in the Agri-Food Sector*, San Francisco, USA, ACLCA Vashon, WA, USA, 8–10 Oct 2014.
Steptoe, A., Pollard, T. M., & Wardle, J. (1995). Development of a measure of the motives underlying the selection of food: the food choice questionnaire. *Appetite, 25*(3), 267–284.
Tang, J. E., Moore, D. R., Kujbida, G. W., Tarnopolsky, M. A., & Phillips, S. M. (2009). Ingestion of whey hydrolysate, casein, or soy protein isolate: Effects on mixed muscle protein synthesis at rest and following resistance exercise in young men. *Journal of Applied Physiology, 107*(3), 987–992.
Tesco. (2012). Product carbon footprint summary. Available at: https://www.tescoplc.com/assets/files/cms/Tesco_Product_Carbon_Footprints_Summary(1).pdf. Accessed Sept 8, 2015.
Too, L., & Bajracharya, B. (2015). Sustainable Campus: Engaging the community in sustainability. *International Journal of Sustainability in Higher Education, 16*(1), 57–71.
Wilkinson, S. B., Tarnopolsky, M. A., MacDonald, M. J., MacDonald, J. R., Armstrong, D., & Phillips, S. M. (2007). Consumption of fluid skim milk promotes greater muscle protein accretion after resistance exercise than does consumption of an isonitrogenous and isoenergetic soy-protein beverage. *American Journal of Clinical Nutrition, 85*(4), 1031–1040.
Williamson, D. (2014). Plant based eating; a viable solution for climate change and sustainability. Available at: http://www.alprofoundation.org/scientific-updates/sustainability/plant-based-eating-viable-solution-climate-change-sustainability/. Accessed Nov 20, 2015.

Author Biography

Dr Kate Reed has an undergraduate degree in Sports Studies from University of Chichester, and an MSc in Physiology (University of London). She then continued her post graduate studies at the University of British Columbia, Canada where she completed her PhD in Cardiovascular Health in Children. She currently works at the University of Essex, where her interests include sports nutrition and performance sport.

Governing Sustainability: Some Challenges Ahead

Rocío Valdivielso del Real

Abstract

The governance of sustainable development is a politically important issue but whose conceptualisation and implementation are testing the resilience of existing governance systems. The perceived failure to deliver effective change, moreover, is heightening pressures on governments and key stakeholders pushing for the greater prominence of sustainable development and sustainable governance on the policy agenda. Yet, opportunities to advance the notion of sustainability exist: reframing of the analysis of the issues related to sustainability; formulation of more effective sustainable development goals; and identification of novel ways to engage a greater number of actors in the sustainable development debate. Recent initiatives involving governments and private actors (i.e. business firms and civil society groups) constitute interesting developments in the study of the governance of sustainability at both at the national and transnational level. These initiatives are characterised by different forms—some being strictly private while others being collaborative efforts with the public sector (public-private partnership agreements, standard setting by multi-stakeholder initiatives). The chapter concludes with the idea that the crisis of sustainable development is primarily one of governance. Fundamental changes, in the actions of governments and in the life-choices of private citizens, will be required for a successful transition to a more sustainable world.

Keywords

Sustainability · Sustainable development · Sustainable governance · Policy outcomes · Challenges

R.V. del Real (✉)
Department Business and Management, Liverpool Business School,
Redmonds Building, Brownlow Hill, Liverpool L3 5UG, UK
e-mail: M.R.ValdivielsoDelReal@ljmu.ac.uk

W. Leal Filho (ed.), *Sustainable Development Research at Universities in the United Kingdom*, World Sustainability Series,
DOI 10.1007/978-3-319-47883-8_10

1 Introduction

In the context of high unemployment, the current focus on economic growth entails the risk of relegating important long-term questions of sustainability to the background. This is an issue since the process of generating economic growth, and material well-being, in the short-term is invariably linked to its reproduction over the long run. Linking different temporal perspectives in a complementary manner does not constitute an easy task. In fact, characterising and differentiating between important and urgent priorities is hard to establish (see Guillén and Ontiveros 2012: 105). It is at the intersection of these two points that the idea of sustainability lies.

From an economic point of view, the notion of sustainability refers with ensuring that meeting present needs does not come at the expense of doing so in the future. The notion also entails important social and political dimensions given the presence of intergenerational trade-offs. Thus, the concept of sustainability—the sustainable development of human societies—extends beyond matching the use of energy, natural resources, and the environment supporting life on the planet. It also encompasses all aspects of social, economic, and political life whereby present actions may place limitations on future actions (Guillén and Ontiveros 2012; Matthew 2012; Waughray 2011).

Yet, the complex interconnectedness that lies at the heart of the concept of sustainability does not mean that policy-makers face a situation of ungovernability characterised by an unattractive trade-offs of 'doing everything at once in the name of integrated approaches or doing nothing in the face of complexity' (UNEP 2007: 363). Instead, the ability of policy-makers to identify key interactive issues would paved the way for a more effective governance that, in turn, would translate in more effective responses at the national, regional and global levels with the aim to facilitate the transition towards a more sustainable society. Building on this view, governance approaches that are flexible, collaborative and learning-based are more likely to be in a better position to be responsive and adaptive to cope with key societal challenges such as social inequality, resource scarcity and demographic change. These societal challenges cut across policy sectors and extend beyond national borders, thereby placing a premium on the ability of policy-makers to adapt rapidly and learn from cross-national experiments in dealing with sustainability issues. Adaptive governance approaches are well placed to address complex interlinkages among these challenges as well as in managing uncertainty under contexts of rapid changes. For instance, the use of tools to manage these interlinkages—such as valuation techniques and integrated management approaches that link economic, social, and environmental issues—could constitute the seeds of the foundation for adaptive governance (see Stiftung 2015; UNEP 2007).

2 Are Sustainable Development and Sustainable Governance Elusive Concepts? A Review of the Literature

Sustainable development is a complex concept that involves different temporal and spatial scales and requires the involvement of multiple stakeholders (Martens 2006; Kemp and Martens 2007). Sustainable development invariably refers to a process of change whereby end goal might not be clearly outlined and is subject to modifications throughout the process. In particular, lack of consensus remains on what sustainable development entails and the scale and nature of needed reforms despite overall agreement on its importance by international agencies and conferences (Meadowcroft 2007; Van Zeijl-Rozema et al. 2008).

As a result, sustainable development does not constitute a single definition concept. Various positions and perspectives stand in competition to each other. Hueting and Reijnders (2004), for instance, argue that sustainability is defined by ecological boundaries that can be scientifically determined. McCool and Stankey (2004), on the other hand, highlight the societal dimension of the concept of sustainable development with highly normative implications. Furthermore, Gibson (2001) stresses that the involvement of societies invariably leads to a normative understanding of the concept of sustainable development.

Starting from a different angle, Brand and Karvonen (2007) argue that sustainability is also locally specific and, as a result, subject to local contextual interpretation that sit uneasily with the setting of common objectives and of universal goals. In the UK, for instance, policy-makers relate sustainable development to quality of life and well-being (DEFRA 2005; OECD 2010). In Bhutan, in contrast, sustainable development is guided by the philosophy of 'Gross National Happiness' based on equitable economic growth, ecological and cultural preservation and good governance (Rinzin et al. 2007). Finally, Williams and Millington (2004) introduce a sharp dichotomy between stronger sustainability, in which the importance of a change in values and behaviour is emphasised, and weaker sustainability, with its focus on technical solutions.

This brief overview informatively illustrates that there are different ways to conceptualise sustainable development. In academic circles, in particular, there have been frequent debates about whether sustainable development constitutes a philosophical-societal versus an economic concept. Although these academic discussions have generated important insights, they often assume that a rigorous operationalisation of the concept of sustainable development is the most important factor for the elaboration of specific policy prescriptions. These discussions have often missed the critical political point that this concept was not formulated as part of the technical vocabulary of social science, or as an operational rule that would allow policy outputs to be automatically read off from a list of situational inputs. Rather, as pointed out by Lafferty (1996, 2004) and Jacobs (1999), sustainable development constitutes a normative concept that, as for other liberal democratic ideas, is open to constant deliberation and re-interpretation.

Nevertheless, the concept of sustainable development has been associated with some core normative ideas with the 1987 report of the World Commission on Environment and Development: protection of the environment with an emphasis on the essential life support functions of the global ecosphere; promotion of human welfare; concern for the wellbeing of future generations; and broad public involvement in environment and development decision making. Moreover, it is widely recognised that discussions of the concept of sustainable has become more encompassing from its early focus on the environmental dimensions. More recent discussions privileges an understanding of sustainable development based on the integration of broader economic, environmental and social objectives (UNDESA 2001; UNDP 2014). The framing of the discussion is about achieving an appropriate balance between three 'pillars'—the environment, economy, and society. The concept of sustainable development is often presented as a pathway for the advancement of important societal goals (Holden and Linnerud 2007).

Another element of consensus—which has been forcefully expressed in the Report of the World Summit on Sustainable Development (United Nations 2002), the Millennium Ecosystem Assessment (2005), and Transforming our World: The 2030 Agenda for Sustainable Development (2015)—is that developments in the last three decades since the Brundtland Commission have been largely negative. Sustainable development, notwithstanding the different definitions, has become more difficult to achieve. This, in turn, has strengthened the case for the implementation of a new governance regime capable of putting society on a more sustainable track.

These demands take different forms: 'sustainable governance' (ECFESD 2000) and 'governance for sustainable development' (Ayre and Callway 2005; Nerwig et al. 2008). Others have called for 'reflexive governance of sustainable development' (Voss et al. 2006a, b). Grander aspirations have emphasised 'earth system governance' (Biermann 2007) and 'global environmental governance' (Speth and Hass 2006). Varieties in governance forms, nonetheless, convey the same assessment, namely that a new governance regime is needed if there is to be any realistic prospect of an orderly transition to sustainability.

The concept of governance is not new. The innovative element, however, is the increased emphasis on the changing responsibilities of public authorities, and the varied ways in which coordination generate outcomes in the area of sustainable development (Meadowcroft 2007). The more recent usage of the term has been shaped by debates about 'good governance' that emerged in international development circles in the late 1980s (World Bank 1991; DAC-OECD 1993). Despite variations in formulations by specific international bodies, 'governance' broadly refers to practices through which societies are governed. In turn, the concept of 'good governance' is associated with a diverse array of criteria: effectiveness and efficiency, the rule of law, participation, accountability, transparency, respect for human rights, the absence of corruption, toleration of difference, and gender equity (UNDP 1997; Plumptre and Graham 1999).

Within the field of political science, increased interest in 'governance' has been associated with attempts to understand changing patterns of state/societal interaction. An important framing is the link between 'governance' and new forms of socio/political interaction. Rhodes, for instance, defines 'governance' in terms of 'self-organising inter-organisational networks' that constitute 'an alternative to, not a hybrid of, markets and hierarchies' (Rhodes 1996: 659). Building from this specific framing, Jessop refers to 'governance' as a form of social co-ordination based on 'dialogic rationality', where goals are 'modified in and through ongoing negotiation and reflection' (Jessop 2000: 17).

In contrast, others have stressed the coordination function of governance. Kooiman (2000) presents 'social-political governance' as coordinating arrangements that facilitate the involvement of public and private actors in arenas aim at solving societal problems'. 'Diversity', 'complexity', and 'dynamics' are conceptualised as the outstanding features of modern society with the implication of the importance of developing more varied governance practices that would involve hybrids of three basic governing forms—'self-governance', 'co-governance' and 'hierarchical governance' (Kooiman 2003). Pierre and Peters (2000) advocate a similarly open textured notion of 'governance' that links institutional structures and interactive processes.

How do such broad debates about governance relate to sustainable development? The answer is that the notion of sustainability is 'a political concept replete with governance questions' (Farrell et al. 2005: 143). For instance, how is sustainability implemented into specific policies? Sustainability will not just occur in a pre-ordained manner. Sustainable development constitutes an internationally recognised objective which governments, and other organisations with governance responsibilities, ought to pursue (Meadowcroft 2007: 300). Sustainable development requires thoughtful deliberation before its implementation. These processes of deliberation and discussion constitute core constitutive features of governance that are not the same as governing—the latter referring activities which seek to 'guide, steer, control, or manage' societies (Kooiman 1993: 2). Governance, as illustrated by Ostrom (2005), emerges from the activities of diverse actors embedded in acceptable norms of behaviour.

Moreover, governance is not the equivalent to government. The concept of government centres on the institutions and actions of the state while the notion of governance enables the incorporation of non-state actors, such as business and non-governmental organisations (NGOs), into any analysis of social steering (Lemos and Agrawal 2006: 298). The notion of governance is insightful since it covers 'the whole range of institutions and relationships involved in the process of governing' (Pierre and Peters 2000: 1; see also Ostrom 2005). The use of the term governance, instead of government, highlights an important empirical feature, namely that policies are deliberated and implemented by a much wider array of public, private and voluntary organisations as compared to the more traditional governmental framework (Flinders 2002: 52).

One of the more prominent empirical element in this discussion is the growing use of new models of governance, such as voluntary agreements and market-based instruments, as well as systems of self-regulation, through which societal actors effectively steer themselves (Jordan and Schout 2006; Treib et al. 2007). In the international sphere, for instance, scholars point to the increasing involvement of multinational corporations, as illustrated by new forms of public-private and private-public co-operation, in interstate agreements as empirical evidence of governance (Levy and Newell 2004). This emphasis on global governance departs from traditional approaches in international relations viewed through the prism of sovereignty and statehood (Dingwerth and Pattberg 2006: 189-93; Biermann 2007).

However, and despite the positive features associated with an encompassing perspective, an excessively broad view of 'governance for sustainable development' carries the risk of resulting in different definitions that, in turn, are associated with different governance approaches. After all, many factors influence sustainable development: employment policy, fiscal management, the health care system, pension arrangements, housing policy, immigration law, the fight against crime, the management of natural resources (DEFRA 2005; Bertelsmann Stiftung 2015). Not surprisingly, these factors are invariably selected as indicator sets by governments in monitoring sustainable development.

Thus, the literature on governance and sustainability should become more amenable to empirical testing in order to assess the extent to which as shift from government to governance is indeed occurring—rather than remaining stuck in typology making (Jordan 2008; Kooiman 2003: 4–4; van Kersbergen and van Waarden 2004: 165). Moreover, the governance of sustainable development should incorporate both processes and outcomes (Adger and Jordan 2009: 29). If simply framed in terms of in the presence or absence of particular modes or instruments of governing, then governance risks becoming a static concept devoid of dynamism. Processes are important as they link in interactive manner interments of governance and outcomes (Pierre and Peter 2000: 22). Political legitimacy and public accountability are central to the process of governance and, as a result, influence sustainability outcomes (Rhodes 1997). Finally, studies of governance could become taxonomy exercise of classifying different instruments of policy-making at the expense of understanding how different processes of deliberation shape sustainability outcomes (Jordan 2008).

3 How Could We Achieve Sustainable Policy Outcomes? How Could a Long-Term Focus Inform Political Decision-Making?

The governance of sustainability is bound to be a long-term process for which political will is important. A number of recent studies have explored the issue of governance for sustainable development framed as a long-term perspective (OECD 2002a, b; Lafferty 2004; Kemp et al. 2005; Stiftung 2015). Two issues stand out. First, the governance of sustainable development constitutes an ambitious agenda for social change characterised by the large scale social transformation. To reconcile continued economic and social improvement with the preservation of global ecological systems and decoupling economic activity from environmental loading will be needed (UNEP 2007, 2012). In order to tackle this challenge in an effective manner, radical shifts in existing patterns of production and consumption are required (Toner 2006). This, in turn, is likely to profoundly impact on major socio-economic sectors such as energy, transport, agriculture, manufacturing and construction. These radical shifts have been presented as 'the next industrial revolution' (Hawken et al. 1999).

Second, the idea of governance for sustainable development as a long-term process embodies a specific 'steering logic' (Meadowcroft 2007). Sustainable development is not a spontaneous economic and social product, but instead involves the active involvement of governments and other actors. Ultimately, governance for sustainable development implies a process of 'societal self-steering' (see Adger and Jordan 2009; Kemp and Martens 2007) characterised by the involvement of society as a whole in the critical interrogation of existing practices in order to bring about change in a conscious effort. Thus it involves not only actions to implement policies, but also collective discussion in a deliberative democratic manner to define these policies. Choices about the kind of society we want to live in lie at the heart of governance for sustainable development.

In this context, it is important to note the power to influence outcomes is widely distributed. In representative democratic political systems, with privately owned productive assets and government regulation of market relations, no single group holds a monopoly on power (Meadowcroft 2007; van Zeijl-Rozema et al. 2008) Although the boundary between the economic and political spheres in modern democratic societies has changed over time (from nationalisation to privatisation for instance), it is difficult for a single actor to secure outcomes on their own. While the proportion of economic activity controlled by the largest firms has increased, the differentiation of their activities has also grown. On the government side, on the other hand, horizontal divisions have also become more complex: various ministries and departments, specialised in particular functions, reflect the diversity of tasks government undertakes. Therefore, there are a greater number and diversity of agents that are able to influence events.

Consequently, and returning to sustainable development, governance is embedded within a context of distributed centres of power. After all, problems of sustainability typically cut across functional administrative divisions, and nationally-based territorial jurisdictions, thereby increasing the number of relevant actors (see Adger and Jordan 2009; Meadowcroft 2007; van Zeijl-Rozema et al. 2008).

Diffused power should not be seen in a negative light (Grunwald 2007). The concentration of political and economic power, monopolised by a small group of economic and/or political elites, has often led to the proliferation of abuses in regard to sustainable development. Thus, the diffusion of power should not be conceptualised as a constraint for sustainable development, it could become an opportunity (Meadowcroft 2007). For instance, it could enable the formation of many channels for information to flow and provides (potentially) for multiple routes of intervention in order to encourage the turn towards sustainable development.

Obviously, the presence of diffused centres of power raises issues of coordination. The growth in the number of involved actors and the increase in institutionally-based veto points, contribute to complicate organised efforts for reform. This point is made prominently clear with the case of federal governmental system that has exhibited lower degrees of effectiveness in coordinating country wide engagement with sustainability (Lafferty and Meadowcroft 2000).

Nonetheless, the diffusion of power could serve as a beneficial constraint. That is, advancing an ambitious reform agenda in the context of power diffusion invariably requires a more interactive/reflective mode of governance. National governments, for instance, constitute just one (albeit one crucial) component of the overall process of governing sustainable development. Government actions could increase the likelihood that the governance process will evolve in the desired direction to promote a sustainable orientation towards development. However, no single actor could succeed on its own. The effectiveness of state policies is also contingent in the ability of governments to exploit interactions with non-state actors in order to acquire vital knowledge about different interests and perspectives on sustainable development. For instance, settings such as the International Conferences on Sustainable Development that bring together stakeholders from government, academia, international agencies, non-governmental organisations (NGOs), and grassroots organisers, constitute ideal forums to identify and share practical, evidence-based solutions towards the achievement of a more sustainable developmental path.

Interactions among government and non-state actors can facilitate the formulation and implementation of sustainability objectives within a long-term horizon. For instance, the Sustainable Governance Indicators (SGI), first published in spring 2009, and updated every two to three years, compare the ability of OECD member states to deal with economic, political and social challenges standing in the way of a more sustainable development. The updating of SGI provide national governments with reference points with respect to the assessment of progress, the formulation of goals, and the means to achieve them—thereby transforming what could be have been short-term objectives into a more long-term strategic vision of sustainable development.

Yet, another interesting initiative to achieve sustainable policy outcomes is the development of co-governance networks around specific issues, most notably those that draw together organisations from across the state/business/civil society divide. In some cases, governments may participate directly in these network-based management regimes. Alternatively, governments could limit their role as facilitator to such meetings and assume the role of monitor of deliberation and results—i.e. transitional management as a governance tool for sustainable development (Kemp et al. 2005; Voss et al. 2007; Pisano 2014).

The promotion of a vibrant 'public sphere' to facilitate discussion of social choices and critical reflection on the development path to sustainable development has the potential to shape discursive practices in both national and grassroots gatherings. The role of governments has facilitator for deliberation among a number of actors could encourage the consolidation of new ideas and practices, and renew political support for the continuing process of reform (Torgerson 1999). In settings where governments serve as facilitators, non-state actors could feel empowered to engage in shaping the content of governance policies in the public sphere (Lischinsky and Sjölander 2014). As a result, government and private actors, most notably, but not exclusively, private companies, can positively shape the transition toward a more sustainable development via their participation in networks that engage practically with collective problems in the context of diffused power.

The development of these interactive/reflexive networks that shape the development of governance mechanisms do not imply that governments cease to act at times in an unilateral manner. On the contrary, the undertaking of unilateral actions—in the form of legislative or institutional change—is often necessary to break the resistance of entrenched interests. The key issue is whether unilateral government actions that destabilise established ways of doing things will create pressures for further reform and adjustment along the lines of the involvement of a greater number of parties in a deliberative setting, or whether it will simply heighten feedback mechanisms that revert to the status quo.

Governments can unilaterally regulate the structural environment in which private actors operate by constraining, but not eliminating, strategic choices in the undertaking of adjustment paths. For instance, regulations that oblige industry to identify hazardous substances used at local facilities strengthen the bargaining power of environmental groups, via the release of public information, but do not determine outcomes.

In a similar vein, unilateral government actions could shape the structural environment in which private companies operate through the creation of new institutional actors. Financial and legal support by governments could facilitate the creation of new and autonomous actors who can then become active participants in the governance of sustainable development. The rise of independent regulatory agencies captures well this scenario characterised by the transfer of functions from the core of government to bodies that work at arm's length. The Sustainable Development Commission (SDC) in the UK, for instance, is an independent executive non-departmental public body that produce evidence-based policy advice and research on environmental issues by engaging with governmental and non-state actors. The SDC is interacting in a deliberative and participatory way with other actors in order to shape the content of policies on sustainable development. Thus, governments, by bringing together different groups of stakeholders to address issues of sustainable development, can encourage inter-organisational collaboration and the development of new patterns of interaction while, at the same time, encouraging shifts in the balance of power among these non-state actors (Higginson and Vredenburg 2010).

In a different way, unilateral government actions could serve to deliberately strengthen economic actors whose activities fit with desired social ends. A prominent example of this strategy is the overt support for the green business sector—i.e. renewable energy, organic farming. This type of unilateral government actions could secure direct environmental and economic gains (more green energy, growing export markets), but also strengthen interest groups and constituencies advocating specific, and often highly redistributive, changes such as the removal of subsidies to fossil fuels and reduction of pesticide usage.

4 Conclusion

This chapter highlighted the difficulties of finding a universally accepted notion of sustainability. Sustainable development is elusive because of the nature of global economic forces and the uneven distribution of political power. Important challenges, such as economic globalisation, social inequalities, and resource scarcity, cut across policy sectors and extend beyond national boundaries, thereby requiring policy-makers and private actors to deliberatively reflect about desired goals and to learn from the examples of others. Complexity and uncertainty about the future of sustainable development constitute difficult challenges to confront.

While unanimity on every single issue is impossible to achieve, there is a broad consensus on the importance of sustainable development. Ideally, governments should incorporate the long-term consequences of their policies. This involves implementing policies that, while maintaining/improving the quality of life for present generations, do not place an unfair burden on future generations. However, long-term thinking of this nature is often elusive as governments often act within the frame of the next electoral cycle in mind. The unequal allotment of participation opportunities and the wasteful exploitation of natural resources do not bode well for the sustainable development of countries as they entail negative implications for present and future generations.

Sustainable development cannot be achieved without governance transformation given the extent of the normative issues at stake. As a result, it is particularly important to elaborate strategies (a) to acquire more knowledge in critical areas, and (b) to take deliberatively-agreed decisions in the absence of full knowledge.

Finally, while there is no doubt that power is distributed widely in modern societies, the state's capacity to act remains crucially important for the future of sustainable development. The rise in prominence of new linkages among governance structures and new modes of public/private interaction might be well suited to cope with the collaborative and deliberative imperatives of the governance of sustainable development, but do not eliminate the importance of unilateral government actions in shaping the environment in which non-state actors operate. Thus, the ideology of minimal government not only cast the business community and the state as being at odds with each other, an important shortcoming given the importance of co-ordinating networks between governments and non-state actors; but fails to appreciate the positive consequences of the provision of beneficial constraints on the unregulated behaviour of private actors.

To sum up, governance for sustainability presents an enormous but unavoidable challenge. The current trend of unsustainable development is not a viable option. Instead, we need to establish governance structures and practices that facilitate the involvement of, and coordination among, of an encompassing group of actors on a vast complex of issues. This conception of governance captures the multiple and diverse strengths, preferences and capabilities, not just of governmental actors and traditional business interests, but of the full set of public, private and civil society players. The international system of independent states must overcome its anarchical nature via the introduction of new institutions for representation and decision making. The above discussion about diffused power fits well with the current multipolar structure in which several major global and regional powers will be the norm. The economic, socio-demographic, environmental, and political challenges that we are facing constitutes governance issues with implications for sustainable development.

References

Adger, W. N., & Jordan, A. (2009). Sustainability: Exploring the processes and outcomes of governance. In W. N. Adger & A. Jordan (Eds.), *Governing sustainability*. Cambridge: Cambridge University Press.

Ayre, G., & Callway, R. (Eds.) (2005). *Governance for sustainable development*. London: Earthscan.

Biermann, F. (2007). Earth system governance as a cross cutting theme of global change research. *Global Environmental Change, 17*, 326–337.

Brand, R., & Karvonen, A. (2007). The ecosystem of expertise: Complementary knowledges for sustainable development. *Sustainability: Science, Practice and Policy, 3*, 21–31.

DAC-OECD. (1993). *DAC orientations on participatory development and good governance*. Paris: OECD. http://www.agora-parl.org/node/1264. Last accessed January 10, 2016.

DEFRA. (2005). Securing the future: The UK government sustainable development strategy. DEFRA (pp. 1–186). https://www.gov.uk/government/uploads/system/uploads/attachment_data/file/69412/pb10589-securing-the-future-050307.pdf. Last accessed December 27, 2015.

Dingwerth, K., & Pattberg, P. (2006). Global governance as a perspective on world politics. *Global Governance, 12*, 185–203.

ECFESD. (2000). *Sustainable governance*. Brussels: European Consultative Forum on the Environment and Sustainable Development, European Commission.

Farrell, K. N., Kemp, R., Hinterberger, F., Rammel, C., & Ziegler, R. (2005). From 'for' to governance for sustainable development in Europe. *International Journal of Sustainable Development, 8*, 127–150.

Flinders, M. (2002). Governance in Whitehall. *Public Administration, 80*, 51–80.

Gibson, R. B. (2001). Specification of sustainability-based environmental assessment decision criteria and implications for determining 'significance' in environmental assessment. Report, 46. Canadian Environmental Assessment Agency Research and Development Programme.

Grunwald, A. (2007). Working towards sustainable development in the face of uncertainty and incomplete knowledge. *Journal of Environmental Policy and Planning, 9*(3–4), 245–262.

Guillén, M. F., & Ontiveros, E. (2012). *Global turning points: Understanding the challenges for business in the 21st century*. Cambridge: Cambridge University Press.

Hawken, P., Lovins, A., & Lovins, A. (1999). *Natural capitalism: Creating the next industrial revolution*. Brown and Company, New York: Little.

Higginson, N., & Vredenburg, H. (2010). Collaborating for sustainability: Strategic knowledge networks, natural resource management and regional development. *International Journal of Sustainable Economy, 2*(3), 334–351.

Holden, R., & Linnerud, K. (2007). The sustainable development area: Satisfying basic needs and safeguarding ecological sustainability. *Sustainable Development, 15*, 174–187.

Hueting, R., & Reijnders, L. (2004). Broad sustainability contra sustainability: The proper construction of sustainability indicators. *Ecological Economics, 50*, 249–260.

Jacobs, M. (1999). Sustainable development as a contested concept. In A. Dobson (Ed.), *Fairness and futurity*. Oxford: Oxford University Press.

Jessop, B. (2000). Governance failure. In G. Stoker (Ed.), *The new politics of british local governance* (pp. 11–32). Basingstoke: Macmillan.

Jordan, A. J. (2008). The governance of sustainable development: Taking stock and looking forwards. *Environment and Planning C: Government and Policy, 26*, 17–33.

Jordan, A. J., & Schout, A. (2006). *The coordination of the European Union*. Oxford: Oxford University Press.

Kemp, R., & Martens, P. (2007). Sustainable development: How to manage something that is subjective and can never be achieved? *Sustainability: SciencePractice and Policy, 3*, 1–10.

Kemp, R., Parto, S., & Gibson, R. (2005). Governance for sustainable development: Moving from theory to practice. *International Journal of Sustainable Development, 8*, 12–30.

Kooiman, J. (1993). *Modern governance*. London: Sage.

Kooiman, J. (2000). Societal governance: Levels, modes and orders of social-political interaction. In J. Pierre (Ed.), *Debating governance: Authority, steering and democracy*. Cambridge: Cambridge University Press.

Kooiman, J. (2003). *Governing and governance*. London: Sage.

Lafferty, W. (1996). The politics of sustainable development: Global norms for national implementation. *Environmental Politics, 5*, 185–208.

Lafferty, W. (Ed.) (2004). *Governance for sustainable development*. Cheltenham, UK: Edward Elgar.

Lafferty, W., & Meadowcroft, J. (Eds.) (2000) *Implementing sustainable development: Strategies and initiatives in high consumption societies*. Oxford: Oxford University Press.

Lemos, M. C., & Agraval, A. (2006). Environmental governance. *Annual Review of Environmental Resources, 31*, 297–325.

Levy, D., & Newell, P. (Eds.) (2004) *The business of global environmental governance*. Cambridge, MA: MIT Press.

Lischinsky, A., & Sjölander, A. E. (2014). Talking green in the public sphere press releases, corporate voices and the environment. *Nordicom Review, 3*, 125–139.

Martens, P. (2006). Sustainability: science or fiction? *Sustainability: Science, Practice and Policy, 2*, 1–5.

Matthew, R. (2012). Demography, climate change, and conflict. In J. A. Goldstone, E. P. Kaufmann, & M. D. Toft (Eds.), *Political demography: How population changes are reshaping international security and national politics* (pp. 133–146). Boulder, CO: Paradigm Publishers.

McCool, S. F., & Stankey, G. H. (2004). Indicators of sustainability: Challenges and opportunities at the interface of science and policy. *Environmental Management, 33*, 294–305.

Meadowcroft, J. (2007). Who is in charge here? Governance for sustainable development in a complex world. *Journal of Environmental Policy and Planning, 9*(3–4), 299–314.

Millennium Ecosystem Assessment (MA). (2005). *Ecosystems and human well-being: Synthesis*. Washington DC: Island Press.

Newig, J., Voss, J. P., & Monstadt, J. (Eds.). (2008). *Governance for sustainable development: Steering in contexts of ambivalence, uncertainty and Power*. London: Routledge.

OECD. (2002a). *Governance for sustainable development: Five OECD case studies*. Paris: OECD.

OECD. (2002b). *Improving policy coherence and integration for sustainable development: A checklist*. Paris: OECD.

OECD. (2010). *United Kingdom: Policies for a sustainable recovery*. Paris: OECD.

Ostrom, E. (2005). *Understanding institutional diversity*. Princeton, NJ: Princeton University Press.

Pierre, J., & Peters, G. (2000). *Governance politics and the state*. Basingstoke: Macmillian.

Pisano, U. (2014). *Transition management as a governance tool for sustainable Development, European sustainable development network (ESDN) case study N° 17*. Vienna: ESDN.

Plumptre, T., & Graham, J. (1999). *Governance and good governance: International and aboriginal perspectives*. Ottawa: Institute on Governance.

Rhodes, R. (1996). The new governance: Governing without government. *Political Studies, 44*, 652–666.

Rhodes, R. (1997). *Understanding governance*. Milton Keynes, UK: Open University Press.

Rinzin, C., Vermeulen, W., & Glasbergen, P. (2007). Public perceptions of Bhutan's approach to sustainable development in practice. *Sustainable Development, 15*, 52–68.

Speth, J., & Haas, P. (2006). *Global environmental governance*. Washington, DC: Island Press.

Stiftung, B. (2015). *Policy performance and governance capacities in the OECD and EU: Sustainable governance indicators 2015*. Gütersloh: BertelsmannStiftung Foundation.

Toner, G. (2006). *Sustainable production: Building Canadian capacity*. Vancouver: UBC Press.

Torgerson, D. (1999). *The promise of green politics*. Durham, NC: Duke University Press.

Treib, O., Bähr, H., & Falkner, G. (2007). Modes of governance: Towards a conceptual clarification. *Journal of European Public Policy, 14*, 1–20.

UNDESA. (2001). Guidance in Preparing a national sustainable development strategy: Managing sustainable development in the new millennium. Report DESA/DSD/PC2/BP13.

UNDP. (1997). *Governance and sustainable human development, United Nations development programme*. New York, Oxford: Oxford University Press.

UNDP. (2014). *Governance for sustainable development: Integrating governance in the post-2015 development framework, United nations development programme, discussion paper*. New York: United Nations Development Programme.

UNEP. (2007). *Interlinkages: governance for sustainability' in global environmental outlook: GEO-4 environment for development, United Nations environmental programme*. Malta: Malta by Progress Press Ltd.

UNEP. (2012). *21 Issues for the 21st century. Results of the UNEP foresight process on emerging environmental issues*. Nairobi, Kenya: United Nations Environmental Programme.

United Nations. (2002). *Report of the world summit, on sustainable development. A/CONF. 199/20 and A/CONF.199/20/corr.1*. New York: United Nations. http://www.un-documents.net/aconf199-20.pdf. Last accessed January 29, 2016.

United Nations. (2015). *Transforming our world: The 2030 agenda for sustainable development A/RES/70/1*. New York: United Nations. http://www.unric.org/html/english/library/backgrounders/sdgs_eng.pdf. Last accessed January 15, 2016.

Van Kersbergen, K., & van Waarden, F. (2004). Governance as a bridge between disciplines. *European Journal of Political Research, 43*, 143–171.

Van Zeijl-Rozema, A., Cövers, R., Kemp, R., & Martens, P. (2008). Governance for sustainable development: A framework. *Sustainable Development, 16*, 410–421.

Voss, J-P., Bauknecht, D., & Kemp, R. (2006a). Preface. In J. P. Voss, D. Bauknecht, & R. Kemp (Eds.), *Reflexive governance for sustainable development* (pp. xiv–xvii). Cheltenham, UK: Edward Elgar.

Voss J-P., Bauknecht, D., & Kemp, R. (Eds.) (2006b). *Reflexive governance for sustainable development.* Cheltenham, UK: Edward Elgar.
Voss, J. P., Newig, J., Kastens, J., Monstadt, J., & Nölting, B. (2007). Steering for sustainable development: A typology of problems and strategies with respect to ambivalence, uncertainty and distributed power. *Journal of Environmental Policy and Planning, 9*(3–4), 193–212.
Waughray, D. (Ed.) (2011) *Water security: The water-food-energy-climatenexus.* Washington, DC: Island Press/World Economic Forum.
WCED. (1987). *Our common future.* Oxford: World Commission on Environment and Development, Oxford University Press.
Williams, C. C., & Millington, A. C. (2004). The diverse and contested meaning of sustainable development. *The Geographical Journal, 170*, 99–104.
World Bank. (1991). *Managing development: The governance dimension.* Washington: World Bank.

Author Biography

Dr. Rocío Valdivielso del Real is currently a Senior Lecturer in Management/Business at Liverpool Business School where she is module leader in 'Globalisation and Business' and 'Strategy for Sustainability'. She is also a Teaching Associate at Warwick Business School, a Research Associate at the London Centre for Corporate Governance and Business Ethics, Birkbeck College, University of London, and a Research Associate at ABM Analytics London.

Her research interest in comparative political economy, corporate governance, sustainable energy policy in Europe, re-regulation and governance have been central in the successful completion of a second PhD, in Management/International Business, at the University of London (November 2009). The PhD highlighted the interactions highlighted the interactions between institutional differences (Varieties of Capitalism) across EU member states with an incorporation of the role of policy-makers (state activism) in order to account for the different trajectories followed by EU energy companies in the context of liberalisation policies. She also holds a M.Sc. in Regional and Urban Planning Studies from the London School of Economics, a MRes in Management/Organizational Behaviour at Birkbeck College, and a PhD in Political Science from Complutense University (Madrid).

Dr. Rocío Valdivielso del Real also served as a Teaching Assistant in Managerial Economics and Research Methods for Management and Business at Birkbeck College; Lecturer in Industrial Sociology (European University of Madrid); Assistant Professor in Public Policy and Public Administration (Carlos III University of Madrid); Lecturer in Comparative Political Systems at UNED (Spanish Learning Distance University), and Visiting Academic and Part-Time Lecturer at the Centre of Globalisation and Regionalisation (CSGR), University of Warwick.

Rocío has a developed record as project manager in both academia and private consultancy gained from a broad background of successful applied research in organisational analysis—as well as in the economics and politics of the liberalisation of network utilities.

Higher Education Support to Small Medium Enterprises: A Local Experience in Energy Efficiency

Richard Allarton

Abstract

Industry accounts for 29 % of UK energy use, with 86 % derived from non-renewable sources, placing energy efficiency in this sector as a fundamental to sustainable development. Given that some 99 % of UK industrial companies are Small and Medium Enterprises (SME), who are least able to devote specific resources to energy analysis, supportive initiatives in this area have the potential for significant savings and success. This paper provides example and advice on methods of support from Higher Education (HE) to improve the sustainability of industrial SMEs. This is exampled through the experiences of a 24-month project called "SUSTAIN Lincolnshire" with a focus on SME energy efficiency. A critical analysis, starting from the existing literature, will centre on the problems of co-ordinating and encouraging a large number of SME to become pro-active in this area. This uses a project life-cycle approach, discussing the importance of clearly defined requirements, SME engagement, lessons learned and further work beyond the project. Current initiatives in Higher Education/Industry cooperation make this paper particularly timely and its critical analysis will provide HE institutions with guidance and advice when developing similar projects. The paper identifies resources and techniques whilst highlighting the difficulties in developing higher-level strategies to the hard-pressed SME communities. It shows the importance of persistence in this area for the initiating bodies and the benefits of building on outcomes in a coherent manner. This paper has does not analyse the most appropriate areas for project deployment or the format in which they should be provided. It does, however, provide insights into the exploitation of local implementations.

R. Allarton (✉)
School of Engineering, University of Lincoln, Lincoln, UK
e-mail: richallarton@lincoln.ac.uk

W. Leal Filho (ed.), *Sustainable Development Research at Universities in the United Kingdom*, World Sustainability Series,
DOI 10.1007/978-3-319-47883-8_11

Keywords
Higher education · SMEs · Energy efficiency · Lincoln · Local level

1 Introduction

SMEs form over 90 % of UK industries (Quayle 2003), but generally they lag behind larger companies in their organisational and operational efficiency savings. In simple terms, they do not have the resources to dedicate to efficiency identification and implementation. Concepts such as Lean or Agile Engineering often require specialist knowledge that is not available in smaller firms. Implementation of such strategies in a large company, such as Toyota, are achievable because it can be an enforced change within the company bounds. These changes can be extended to the supply chains through mandate, and can also be enforced between departments within a large company, ensuring the efficiency of both inter-action and intra-action is maximised. Such knowledge is available within HE institutions and its provision to SMEs could significantly improve their competitiveness in supply chain roles, both nationally and internationally.

In his paper reviewing Business-University collaboration (Wilson (Emer. Prof. University of Hertfordshire 2012)), Wilson identifies the weakness of the supportive links between HE and SMEs, recommending specific and targeted government intervention to strengthen this activity. However, Wilson's support of the Knowledge Transfer Programme (KTP) and in particular the innovation voucher scheme does not recognise the financial barriers that prevent SMEs from engaging in KTPs, or the KTP's failure to provide the lower level support that can have significant impact on SME operation. This paper seeks to address this by demonstrating how resources and techniques that can practically support SMEs can be deployed, highlighting the difficulties of implementing higher-level strategies, such as Industrial Symbiosis (IS), into the existing communities of hard-pressed SMEs. It shows the importance of persistence in this area for the initiating bodies and the benefit of building on outcomes in a coherent manner. The critical analysis in this paper will allow HE institutions to take advantage of the lessons learned when developing similar projects.

Knowledge transfer from HE to industry should be attainable. The organisational and operational efficiency techniques used by larger companies are also readily taught in HE Business and Engineering Schools across the UK. However, graduates from these programmes are under-represented in SMEs and where present, their educationally delivered skills are not effectively utilised (Sear et al. 2012). Additionally, knowledge transfer should be available through Knowledge Transfer Partnerships (KTP), research consultancy, student projects and summer placements, etc., but again, SMEs tend to be under-represented in these activities, often displaying an apprehension in engagement. With over 3000 SMEs within Lincolnshire, less than twenty have collaborated effectively with the University of Lincoln's School of Engineering (School) since its inception some six years ago. The overarching aim of this project was therefore to provide impetus to HE/SME collaboration within the local engineering domain.

1.1 Background

The County of Lincolnshire is the second largest (by area) in England and is predominantly agricultural/light manufacturing. Mechanisation of farming at the start of the 20th century focused significantly on Lincolnshire with several major engineering companies, such as Fosters (who built the first tank), Richard Hornsby & Sons and Rustons developing to meet this industrial need. This received further impetus in the second World War, in support of the many airfields that were built and operated in the county. Significant engineering activity still takes place, with a significant potential for supply chain support to the developing North Sea wind energy industry.

From this background, the University of Lincoln and Lincolnshire County Council initiated a 24-month project to support the local Engineering SME community. This paper is a deep reflection on the 2 years of the SUSTAIN Lincolnshire project, and presents key insights into:

- Organisational strategies for industrial SME support.
- Practical support of individual companies.
- Practical support for company collaboration.
- Encouraging a sustainable approach to energy.

2 Problem Bounds

In seeking to support energy efficiency policies for SMEs in the locality, a series of viewpoints are relevant, shown in Table 1. Firstly, there are efficiency savings that can be made within individual companies. These intra-company company initiatives may be obvious and relatively simple, but are often not initiated. Secondly, there are efficiency savings from the rationalisation of inter-company activity. To be of interest, these should go beyond simple group interaction to the synergy of teams, using the concepts of Industrial Symbiosis. Finally, the more altruistic viewpoint from Circular Economy concepts should be espoused.

Table 1 Efficiency perspectives

Viewpoint	Focus areas		
Intra-company barriers	Economic constraints	Behavioural constraints	Organisational constraints
Inter-company barriers	Achievement of synergy	Locality	
Circular economy	Recycle and reuse		

2.1 Intra-company—SME Barriers

In their paper on SME efficiency, Trianni and Cagno identify the requirements within European Directive 2009/28/EC (European Parliament 2009) for 20 % GHG reductions, 20 % of renewable energy and a 20 % increase in efficiency (Trianni and Cagno 2012). These can be seen as drivers for the SUSTAIN Lincolnshire project. Significantly, Trianni and Cagno also identify the barriers that prevent SMEs from achieving the levels of energy efficiency already attained by their larger counterparts. These are classified as either Economic, Behavioural or Organisational.

Economic barriers may typically include:

- Unforeseen costs due to lack of understanding or inappropriateness of efficiency application,
- Missed opportunities due to lack of appreciation,
- Staff bias and perception,
- Customer bias and perception,
- costs of application and analysis,
- cost of implementation and production disruptions,
- lack of funding and inability to absorb long term investment,
- Risk aversion (Trianni and Cagno 2012).

Inconsistency of terms and semantics may be due to differences within business contexts and cultures typically exist among different enterprises dynamically participating in supply chains (Ye et al. 2008). In addition, supply chain partners own heterogeneous applications and legacy systems, developed independently with different knowledge modelling schemata. This is typified by the same term being used to denote different concepts, and different terms represent the same entity or concept (European Parliament 2009).

Behavioural barriers may typically include:

- decisions made without appropriate investigation or rigour,
- information format influences decision making,
- individuals in the decision making process are adverse to change (Trianni and Cagno 2012).

SMEs have often grown organically with personnel styles playing a large role in their interaction rather than any industrial strategy (Singh et al. 2008). This has been somewhat uncontrolled, though in an evolutionary manner, the more effective co-operations have favoured the survival of participants. It is accepted that the interaction in supply chains is often hindered by inconsistent terms and semantics applied by participants to the descriptions of their knowledge (Desouza and Awazu 2006).

Behavioural barriers may typically include:

- energy management not seen as a priority,
- culturally not significant (Trianni and Cagno 2012).

By characterising these barriers, Trianni and Cagno provide templates against which solutions may be developed. Whilst Trianni and Cagno develop these barrier types in some detail, it is relevant only to recognise the general classifications for the purposes of this project, given its time constraints.

2.2 Inter-company—Industrial Symbiosis

Industrial Symbiosis is considered as the interaction between companies for mutual benefit. It is often characterised through the example of the Kalundborg industrial region, situated on the Danish island of Zealand. Chertow captures the concepts well through this example, describing its key facets that include the focus on (1) flow of materials and energy between collaborators and (2) the regionality of activity (Chertow 2000).

Symbiosis suggests a long term interaction for mutual benefit and as such, offers little more than a descriptor for supply-chain relationships that have always existed. Whether embedded or complimentary, synergy can be considered as additional output above simple summing, that occurs when companies interact (Evans 1996). It is the synergistic aspect of IS that provides the opportunity to leverage further competitive advantage from company interactions. Chertow suggests that it is geographical proximity that allows synergy to occur, though highlights the importance of the physical exchange of materials, energy, water, and/or by-products. Whilst definitely capable of reducing costs, assuming transportation exists, its reduction is not truly synergistic.

Whilst the reduction in transport cost can have a significant impact in production efficiency, the re-use of by-products/waste can also be significant. Waste disposal is a costly activity in both terms of financial penalty and ecological effect (Gandy 2014). In addition, the inefficient use of resources is becoming a significant issue as we use the finite set of world resources (See Circular Economy below).

Many examples of IS have been developed, including United States based examples espoused by Chertow (2000), as well as developments in Australia and China (Roberts 2004; Zhu et al. 2008; Geng et al. 2009). Although a key factor in

Chertow' s assessment of IS, its synergistic value, is not always included in these further examples.

In the United Kingdom, a number of initiatives have been tried, primarily focused on the Kalundborg model. These include the Humberside Industrial Symbiosis Programme (HISP) and similar activities in the West Midlands and Merseyside, under a National Industrial Symbiosis Programme (NISP). It is recognised that industrial symbiosis is in its infancy in the UK and that no metrics or success criteria were defined to measure the success of individual programmes (Mirata 2004). This is significant as benefits arise from supply chain interactions, but if industrial symbiosis is to regarded as different and its benefits argued, the characteristics that make it unique need to be defined, as do the metrics by which its value are measured. Within the SUSTAIN Lincolnshire project, a series of success criteria were defined, but only those which could be attributed to locality or shown to be synergistic would be seen as supporting industrial symbiosis.

2.3 Global Perspective—Circular Economy

In 2010, The Ellen McArthur Foundation was set up to promote the concepts of a Circular Economy. The Circular Economy recognises the linear product lifecycle, from design through to disposal, resulted in the depletion of finite resources. It therefore promoted the concept of a circular lifecycle that saw the reuse of a product or its components at the end of their useful life. This concept is described as 'cradle to cradle' (rather than 'cradle to grave') and intended to retain the earth's resources by replacing the concept of sustainability with recycling (Macarthur 2013).

Still in its infancy, the Circular Economy can be seen as an ideal, which has spawned a number of initiatives, focused on recycling. Such activities promote efficiency and align with IS in more general energy saving application.

3 Methodology

The main aim of the project was to provide sustainability support to local SMEs from within the HE environment. To achieve this, the University and County council agreed a set of requirements which focused on energy efficiency.

A set of auditable deliverables (see Appendix 1) were agreed to achieve the following high level aims:

- increase the awareness of resource efficiency techniques and benefits;
- improve the resource efficiency of participating businesses;
- reduce the carbon footprint of Lincolnshire businesses;
- reduce waste from county supply chains;
- increase the number of businesses accredited to ISO14001 or equivalent
- demonstrate resource efficient or renewable technologies;

- create a climate for business collaborations;
- support government, regional and local strategies and commitments to reduce CO_2 emissions (Lincolnshire County Council 2012).

The Circular Economy (Macarthur 2013) was identified as a guiding principle to achieve sustainable SME growth, strengthen the industry sector, reinforce partnerships and inform policy. An initial funding of was allocated as: £22k University staff support, £44k engineering consultancy, £142k capital funding for support and demonstration equipment purchases, allowing for one University member and one consultant to be dedicated to the project in part-time roles.

Given the Council's generic remit, the project was limited to the needs of the County's industries. The limitation on resources (particularly staffing) bounded the project in its exposure and exploitation. At best, it would be a pilot within the County and not touch the majority of its 3000 SMEs. Industrial Symbiosis and Circular Economy principles would not be implemented in full, but would be used as guiding principles. This paper offers a deep reflection of the activities in the belief that there are learning points to assist in duplication and extension of this template in other geographical and industrial areas.

These objectives would be achieved through a series of presentations on efficiency topics to increase understanding and change the mind-set of participants. The School would provide underpinning knowledge through academic delivery on related topics such as Industrial Symbiosis, the Circular Economy and their specific current research. Subject specific knowledge would also be provided in presentations by Exemplars such as larger companies or other SMEs. These would provide practical examples and experiences from proven efficiency activities. Further presentations from specialists such as Technology Centres and other Universities would complete the presentational delivery.

In addition, energy audits would be provided to assess and advise individual companies on increasing their resource efficiency and reduce the carbon footprint. Engagement of companies accredited to provide ISO14001 training would add further, practical support to the engaged SMEs.

Mind-set change and longer term engagement would only be sustained if the SMEs felt part of the project, rather than as subjects or observers. Individual research activities would therefore be initiated and supported through the purchase of specialist equipment and grant support assistance. Engagement was seen as key to the success of the project and would be achieved through the engagement of specialist consultants with expertise in marketing and communication.

Finally, in order for the project to be successful, it would require clear and directed coordination. The County Council would provide funding and auditing of requirements, whilst the day to day running would be the responsibility of the School.

A set of key stakeholders were identified to achieve this, as summarised in Table 2. In addition to the University and the local Council, it included an industrial consultancy firm, larger enterprises from within the county, an advertising company and specialist support.

Table 2 Key stakeholders in SUSTAIN Lincolnshire

Stakeholders		University of Lincoln	Lincolnshire county council	Oakwell management services	Peterborough environment city trust	Lava	SME	Larger companies
Status		Collaborative partner	Collaborative partner	Consultancy	Audit COMPANY	PR	SME	Exemplar
Number of people in the organisation		1300	4000	3	10	3		2000
Number of people involved		3	2	2	2	3		2
Role		Co-ordination and expertise	Project audit and administration	Co-ordination and expertise	Environmental auditing	Web	Actors	Expertise
Working hours effort		900	500	1000	500	100		
WP	Engage with SMR	X	X	X		X	X	
	SMR collaborations	X	X	X		X	X	
	Audits				X		X	
	Examples	X		X			X	X

Oakwell Management Services (OMS) were engaged, as the industrial consultant, to establish contacts with local companies. In addition, they provided much of the day-to-day running of the project, facilitating the events and collaborations and reporting on activities.

Peterborough Environment City Trust (PECT) is an independent charity founded by the Peterborough City Council. PECT was set up 1993 to support environmental projects of local and national significance and works with over 200 different partners (PECT 2016b). Their role in the Project was to deliver specialist resource efficiency support to engaged businesses. Key activities included energy audits of the SMEs and delivery of presentations on energy efficiency techniques. PECT were able to provide advice on the viability of renewable and low carbon technology use, within an overarching strategy to implement an environmental management system to a recognised standard such as ISO14001 or Investors in the Environment. Their activities are funded by donations from users.

Over 100 SMEs were engaged in the project attending quarterly meetings and engaging in energy audits and research collaborations. Networking, stimulated by presentations allowed SMEs to form mutually beneficial links as part of the Project.

A number of larger companies, including Siemens Industrial Turbo-machinery, were engaged to present on their use and experience of a number of energy initiatives.

3.1 Project Lifecycle

These high level aims were achieved through a 3 phase lifecycle, which looked to develop, exploit and prolong SME support.

Initial planning grouped the SMEs into three clusters, defined by company manufacturing output. The companies were taken from the consultant's pre-existing contacts. Whilst this could be seen as being limiting, it provided an effective way of engaging a large number of companies rapidly. An initial website was set up on the Council Portal (Lincolnshire County Council, n.d.) to attract further membership, but its effect was minimal when compared to the consultant's activities. This initial cohort formed the majority of the clustering groups for the project lifecycle. The three clusters of homogenous technologies were Plastics manufacturing, Electrical/Electronics manufacturing and General Engineering. The initial groupings were developed from the rationale that common interests and activities would promote the dissemination of best practice. This stemmed from the concepts of Trianni and Cagno (Trianni and Cagno 2012) and the value of disseminating proven best practices, rather than developing bespoke efficiency solutions.

The project ran as a series of University delivered events, focused around cluster activities and took the format of:

- an appraisal of current activities,
- an academic presentation by the School of Engineering,

- a presentation by an Exemplar,
- a visit, tour or demonstration.

The first meeting included a presentation on Industrial Symbiosis by the University, using Kalundborg (Chertow 2000) as the example. The contradiction between the homogeneous grouping of the clusters and the heterogeneous nature of symbiotic industry was apparent to the organisers, but justified through the intent to promote best practice. This contradiction would dog the project for its lifecycle.

A website (Lincolnshire County Council, n.d.) was set up on the Council Portal to advertise activities. This was not fully considered in the initial planning and its use was not well defined. All files generated at the meetings were stored in a cloud database, available to the participants. A further series of initiatives to use the web-space were considered, but it was not effectively populated until the engagement of a dedicated marketing company.

Funding was available for specialist equipment purchase, which would have been beyond the budget of individual SMEs. Such equipment would need to meet certain criteria, including generic use beyond initial purchase and value outside of the initial purchasing group. The equipment should also stimulate research and development rather than be for everyday use. However, equipment used, such as flow metering equipment, thermal and high-speed cameras were all available as University resources.

An Arburg All-Rounder 270S Injection Moulding Machine (Arburg, n.d.), obtained by the County Council was made available for the Plastics cluster use and was re-located to the School of Engineering, providing a central location and research support. The machine was specialist, and would only meet the research needs of companies who owned such equipment already, but its availability made this a worthwhile inclusion.

Events were delivered every 3 months in different formats, including all clusters together, individual cluster meetings and repeat to individual clusters (same day). No real grouping format was preferred and from more than 100 SMEs attracted, group sizes at any event would typically be approximately 20–30. This was considered acceptable; as hard pressed SMEs were only likely to attend events of direct relevance. The delivery timescale meant that the possibility of disengagement with the project as a whole was high and it became apparent that there was a much smaller cohort of regular supporters.

Presentations by the University were provided to stimulate and provoke company policy and focus, with the intention of bringing efficiency and symbiosis to the fore. Academic Presentations on industrial symbiosis informed SMEs about current thinking on the subject. In addition, an Interface Questionnaire (Appendix 3) was issued to all attending SMEs, which sought to elicit the inputs and outputs across each company's interface. This identified wastes, suitable for re-use and focused SME's on their fellow companies' resource requirements. In this way, synergy could be significantly increased in the locality. Presentations on recycle and reuse were delivered to SMEs to inform them of the concepts and the work done by the Ellen MacArthur Foundation on the Circular Economy. Information from the

Rethinking Progress Conference and links to the foundation website were provided to stimulate interest and a resource pro forma (Appendix 3) was issued with the aim of identifying recycle and re-use opportunities. In addition, School research topics in energy efficiency, such as energy harvesting and laser use in manufacturing were included. These were bolstered by industrial deliveries on topics such as 3-D printing. A presentation was also given on interfacing of companies in a symbiotic manner. This included a practical, ongoing activity to illicit information on material and waste to initiate a form of industrial symbiosis within the Lincolnshire area. This could be seen as a key aim of the project, but lack of impetus and support meant that it achieved little traction.

A presentation on interfacing of companies in a symbiotic manner initiated an initiative to understand cluster raw materials use and waste production. A simple form was devised to capture material flows in and out of the SMEs that could be used to identify symbiotic relationships and reduce SME waste. The intention was that his would form the basis of an industrial symbiosis within the Lincolnshire area and be a key output of the project, but lack of impetus and support meant that it achieved little traction.

Exemplar delivery was intended to provide advice and expertise from the larger manufacturers within the area (but also SMEs where an appropriate skill existed). It was recognised that larger enterprises are able to devote personnel to specific efficiency activities (e.g. Siemens, Lincoln, Business Improvement Team), that is not possible in SMEs (Desouza and Awazu 2006). This emphasis allows larger businesses to become significantly more efficient in their activities relative to SME equivalents and this can be seen, as an anomaly that this project sought to address. Exemplars were sought for each cluster and their presentations proved largely successful.

However, best practice from SMEs themselves added valuable contributions to the discussion. In particular, the use of voltage regulation mechanisms (Trust 2011) was claimed by one company to have achieved an 8 % saving in electrical costs. Another described the use of accurate flow meters to measure and control waste effluent disposal, again with significant savings.

A specialist Exemplar was PECT (PECT, n.d.), who are a city council (Peterborough, Cambridgeshire) initiative focused on environmental issues. Within their project portfolio is Investors in Environment (iiE), a not for profit accreditation scheme providing the business sector with advice and auditing on energy saving measures. Through this scheme, PECT were able to provide presentations on simple energy initiatives as well as carry out energy auditing activities with individual SMEs. Acting autonomously in these activities, the University was able to utilise their specialist capabilities as a significant aspect of SUSTAIN Lincolnshire support.

Use of the industrial consultant provided a rapidly grown participant base, which was able to expand further organically. However greater exposure of the project beyond its participants and the development of an effective website was beyond the scope of the consultant and the School. The introduction of a dedicated marketing company allowed this aspect to develop more effectively whilst freeing other participants to concentrate more fully on their core responsibilities. A Facebook page (SUSTAIN Lincolnshire) and mail shots were added to the dedicated web pages.

Whilst initially considered somewhat specialist, the Arburg 270S Injection Moulder spawned a series of initiatives with companies who did not have such equipment. The idea of developing products using injection-moulded components, which, if successful, could be sub-contracted to specialist fabricators, was attractive to a number of SMEs. However, traction with individual initiatives was difficult to maintain as the complex support of these activities (die manufacture, training, etc.) stretched the resource capabilities of the project.

Funding and support was withdrawn from the project at the end of its two-year period. It had been intended that the clusters would have become self-sufficient, though there was little cohesiveness, particularly without an effective website tool and the underpinning understanding of its use. The greater aims of symbiotic interaction between companies and across clusters was largely unfulfilled and not only greater time, but greater intensity would have been required to make this a reality. However, the outstanding projects, particularly using the Arburg 270S were simply terminated. It remained with the School of Engineering to provide continued collaboration to those SMEs who wished to remain engaged.

4 Results

The measurable commitments to the project were fully met, as outlined in Table 3.

This included support to gain grants worth €420,106 and leveraged funding of €770,182.

PECT were able to offer free attendance to their ISO14001 Design and Implementation Workshop (worth €1100) in March of the second year. This was attended by 8 SMEs (PECT, n.d.).

Table 3 Engineering deliverables

Item	Deliverable	Required	Achieved
1	Assist SMEs through collaboration	11	16
2	Engage SMEs in interaction with Knowledge Base	18	29
3	Hold demonstration/knowledge sharing events	5	16
4	Develop research proposals with SMEs	9	19
5	Assist SMEs with gaining grants/knowledge transfer funding	3	12
6	Provide SMEs with specialist equipment	6	9
7	Evidence new jobs created		13
8	Evidence improved SME performance		14

Examples of PECT activity included (PECT, n.d.):

Deepings Building & Plumbing Supplies, and sister company Eco Building Products has worked with SUSTAIN Lincolnshire to improve their resource efficiency and install a range of renewable technologies saving over £17,000 a year.

P&R Plant Hire achieved ISO14001, helping them to maintain a competitive edge when tendering and retain a key contract with the Environment Agency. They have also recycled over 300 tonnes of material from the sites they have worked on.

Parrot Zoo has developed a new environmentally friendly visitor management centre, including ground source heat pump and wind turbine. This will attract new customers whilst significantly reducing running costs.

However, within the remit of the project there were a number of higher level outcomes including the following.

4.1 Positive

Non-sector engagement occurred during the project, with non-cluster companies taking advantage of the cluster support infrastructure. This included a smoke machine manufacturer, who engaged in some significant research in a novel method of liquid atomisation. Two separate University start-up design companies used the Arburg 270S and manufacturing advice to develop new products.

Oakwell Management Services engagement was instrumental in developing a significant SME cohort within an acceptable project timescale. Their intimate knowledge of the field and close relationships with the SME's and the School provided impetus to relationship.

Arburg 270S Plastic Injection Moulding Machine was an expensive purchase that was only commissioned well into the project lifecycle, with little practical output achieved within the project lifecycle. However, it acted as a stimulus for a number of SMEs, who would not have engaged without the exposure. In addition, it also attracted exemplars who were able to provide advice and support to non-experts considering injection moulding as a manufacturing solution. As a statement piece, it stimulated discussion within the other clusters on the types of support equipment pertinent to their fields. Within the University, it inspired a number of manufacturing projects, both from School and from Art and Product Design students within the School of Architecture. Its presence also provided practical demonstrations for the School's material teaching. From this perspective, the Arburg 270S was a key element of the overall project, being associated with most of the engineering activities and in use more often that other specialist equipment purchased for other strands of the SUSTAIN Lincolnshire programme.

Peterborough Environmental City Trust provided a direct and tangible addition to the project that was of immediate benefit to participants. The auditing activities were low level, using recognised concepts for energy saving, but they provided an impetus and a schedule for SMEs to achieve direct efficiency gains.

Lava Public Relations provided appropriate skills for further exposure and coordination of web based activities. This proved significant in the engagement and invigoration of participants.

4.2 Non-optimal

The Requirements Specification was deficient in a number of areas. A more detailed development by the prime initiators (University and County Council) could have improved the project flow. In particular:

- 'Demonstrating new technologies' was interesting but of little practical value. The research oriented University was naturally focused on such activities, but the SMEs' focus on immediate, practical solutions meant that they were quick to disengage from the more theoretical aspects of the project deliveries. This is supported by Trianni and Cagno (2012), who show that the best returns are achieved through the implementation of mature technologies.
- 'Development of the green supply chain' was vague. If the intent was to develop savings and synergies through Industrial Symbiosis, then this was unlikely to be achievable within the project lifecycle. Its inclusion blurred the focus of operation.
- The rationale for SME Clustering was unclear with 2 potential strategies of diverse aims. Clustering of similar activities was chosen, allowing the potential for communities to form and learn from each other. However, a second strategy of clustering for symbiosis would have met high-level aims more closely.
- The setup of the specialist Arburg injection moulder, inevitably took time, slowing the momentum of activity.
- There was no defined termination strategy, resulting in lost value from truncated activities.

4.3 Negative

Symbiosis did not become relevant for many SMEs despite the academic deliveries on the subject. Engagement of companies was difficult and would need to be far more pro-active and structured to ensure implementation, such as that employed by industry through Lean Manufacturing Rapid Improvement Teams (Feld 2000).

Exemplar engagement proved difficult without clear benefit to the exemplar.

The learning points to be taken forward to other projects were not collated or disseminated.

Cloud sharing of data was ineffective as not all members were cognisant of the process or necessarily inclined to activate the initial invitation. The data therefore became dormant to most participants.

5 Conclusions and Recommendations

This paper presents an in depth analysis of a local government sponsored energy efficiency intervention by Higher Education in the industrial field. This type of linking is not uncommon within local activities, though little analysis is publicly available on their high-level achievements. This paper has not sought to analyse the most appropriate areas where such projects should be deployed, or provide a pattern for future implementation. It does, however, provide insights into the best exploitation of this form of project and makes recommendations for future implementations. It is valuable because this form of project could have far-reaching implications for UK industry in both economic and environmental terms.

The following conclusions and recommendations were drawn from the fundamentals of the project:

- Symbiosis naturally exists between companies, but it is the synergy of interaction that will maximise efficiency gains. Low levels of synergy can occur fortuitously, but significant gains are achieved only through planning synergistic interaction.
- The concept of symbiosis cannot be achieved effectively by simply grouping companies together. Appropriate symbiosis strategies need to be developed at least at county level (or beyond), and then implemented at a local level when full visibility of all company input/outs has been achieved.
- The clustering of similar companies was not effective as synergy is not promoted and potential conflicts of interest in sharing best practice were likely.

The following generic conclusions and recommendations were drawn from the project:

- Selling the concept to the SMEs was vital and should have been a significant part of the tasking, delivered as a separate, initiating phase.
- Use of PECT was fortuitous, but of great value. A direct interactor of similar style will provide immediate improvements, adding impetus and marketing value to project as well as demonstrable achievement.
- Use of the Oakwell Management Services was significant to this running of the project. However, this use of this industrial consultancy has to be placed in the context of the development of the School of Engineering (in existence for less than 5 years). Defined outcomes from such services may still make their use cost effective to more mature organisations, but not fundamental, as it was in this case.
- Failure to develop an integrated website suite was significant. Such a website could have provided space for publicity and marketing, but also the storage of data that would be of value to the participants, such as address details for interaction. In addition, it could have included an interactive tool to provide effective interfacing of companies in a symbiotic manner.

- The failure of the cloud storage to act as an effective communication medium and provide group cohesion was predictable, given its limited abilities and passive nature. It was, however, surprising to note that cloud storage concepts were not always embraced within SME's, or other actors in the project.
- The use of circular economy and green supply chain concepts was undeveloped. Concepts such as criticality of resource were not addressed and would have proven far too complex for this level of project (Graedel and Reck 2015)
- Industrial symbiosis implies an efficiency gain, but not necessarily in energy use. Its generic nature means that it is a useful concept to apply, but perhaps not the optimum. Similarly, the Circular Economy concepts, focus on re-use, and aim to preserve resources rather than improve energy efficiency. These two aims may be mutually supportive, but the positive association is not guaranteed. Trianni and Cagno [5] suggest that it is the known techniques, simply implemented that provide the best return on efficiency investment and this is the case in this project. The easy efficiency gains achieved through the PECT auditing provided the most tangible and easily quantified savings within the project.
- The inclusion of circular economy and industrial symbiosis required a mind-set change amongst the participants, particularly the SMEs. The time required to achieve this is a significant overhead to the project and requires a subset audience who are willing to invest time and resources in its implementation.

The following conclusions and recommendations were drawn from the management of the project:

A specific skills-set is required to project manage at this level and it is questionable whether University academics have either the skills or time to carry out this activity effectively. Specialist support in this aspect may prove worthwhile, dependant on the capabilities of the University. The following should be implemented:

- Use prior project experience (such as this) to shape new projects.
- Develop comprehensive requirements, to provide a sharper focus on the aims and to reference during project execution. The use of concepts, if not the formal tools, as suggested by Kaindl et al. is beneficial (Kaindl et al. 2002). This could include a simple Checklist such as that provided at Appendix 2.
- Definition of success metrics are required to validate the project. The lower level deliverables defined.
- Appropriate termination strategy.
- New initiatives should not be commenced within the project unless there is a realistic chance of their completion within the project lifecycle.
- Carry out a project debriefing to understand the strengths and weakness of the project's management.

5.1 Further Development

The Interface Questionnaire (Appendix 3) was issued with some response, but the value of the information gleaned and the expansion of the database of information was largely unexploited. Its expansion and potential development as an on-line tool could provide significant savings to those engaged.

The 'best practice' exemplar deliveries and PECT auditing activities have great potential for automation through the concept of Linked Data (Bizer et al. 2009) to provide local and national support for sustainable industrial development.

Appendix 1

SUSTAIN Lincolnshire auditable targets

- Assist 11 businesses through collaboration
- Engage 18 new businesses in interactions with KB
- Inaugural meeting of sector groups
- Hold 5 demonstration/knowledge sharing events
- Evidence clustering activity
- Develop research proposals with 2 existing businesses
- Develop research proposals with 7 new businesses
- Assist 3 businesses with applying for R&D grants/KTPs
- Evidence investment in technology/leverage of funds
- Initiate 6 student projects with SME businesses
- 6 businesses make use of Sustain equipment
- Evidence new jobs created
- Evidence businesses improving performance
- Evidence GVA as a result of businesses improving performance
- Attend Steering Group meetings
- Report on progress to UL
- 5 case studies as requested.

Appendix 2

Checklist for Government Initiated Industrial Support Projects

Derived from experiences in SUSTAIN Lincolnshire Project

1. Use experts and stakeholders to define high-level aims and concepts.
2. Bound problem aims and expectations.
3. Assess viability of project aims.
4. Ensure appropriate resources available.

5. Sell to the SMEs. Without them on-board, the project is worthless, but these are hard-pressed people with many different priorities. This should be a separate initiating phase to the project.
6. Communicate project aims and individual targets to stakeholders.
7. Engage consultant with experience/contacts in target area to rapidly build audience.
8. Assess viability of target audience.
9. Engage media expert to market and facilitate communication.
10. Monitor progress, not only of initially set targets, but of holistic progression to ensure developments are maximised.
11. Terminate project coherently with all aspects bought to a close to ensure no wasted effort on incomplete threads.
12. Debrief to learn lessons for future projects.

Appendix 3

Industrial symbiosis - flows across the company interface						
Company name			Contact name			
Inputs			Outputs			
Description	Significance or volume/month	Seasonal?	Description	Significance or volume/month	Seasonal?	Destination (product/by-product/waste)

Description: what passes across the interface. **Significance or Volume/month**: How much either qualitatively or quantitatively. **Seasonal**: Is this a steady flow across the interface or is it at certain times of the year. **Destination**: is it going out as product, by-product or waste
Note Inputs and outputs may be heat, water, data as well as more traditional materials
Please complete and return to Richard Allarton at University of Lincoln, LN6 7TS. An electronic version is available on request for return to **richallarton@lincoln.ac.uk**

References

Arburg, Hydraulic ALLROUNDERs. Available at: http://www.arburg.com/fileadmin/redaktion/mediathek/prospekte/ARBURG_hydraulic_ALLROUNDERs_680472_en_GB.pdf

Bizer, C., Heath, T., & Berners-Lee, T. (2009). Linked data-the story so far. *International journal on Semantic Web and Information Systems, Southampton, UK, 5*(3), 1–22.

Chertow, M. R. (2000). Industrial symbiosis: Literature and taxonomy. *Annual review of energy and the environment, Connecticut, USA, 25*, 313–337.

Desouza, K. C., & Awazu, Y. (2006). Knowledge management at SMEs: Five peculiarities. *Journal of Knowledge Management, Emerald, Bingley, UK, 10*(1), 32–43.

Parliament, European. (2009). Directive 2009/28/EC of the European Parliament and of the Council of 23 April 2009. *Official Journal of the European Union, Brussels, Belgium, 140*(16), 16–62.

Evans, P. (1996). Government action, social capital and development: Reviewing the evidence on synergy. *World Development, London, UK, 24*(6), 1119–1132.

Feld, W. M. (2000). *Lean manufacturing: Tools, techniques, and how to use them*. Boca Raton, FL, USA: CRC Press.

Gandy, M. (2014). *Recycling and the politics of urban waste*. London, UK: Routledge.

Geng, Y., et al. (2009). Assessment of the national eco-industrial park standard for promoting industrial symbiosis in China. *Journal of Industrial Ecology, 13*(1), 15–26 (Yale, Mass., USA).

Graedel, T. E., & Reck, B. K. (2015). Six years of criticality assessments: What have we learned so far? *Journal of Industrial Ecology* (Yale, Mass., USA).

Kaindl, H., et al. (2002). Requirements engineering and technology transfer: Obstacles, incentives and improvement agenda. *Requirements Engineering*, *7*(3), 113–123 (Springer, New York, USA).

Lincolnshire County Council, Sustain Lincolnshire. Available at: http://www.epic-lincolnshire.org/sustain/. Accessed Sept 16, 2015. Lincoln, UK.

Lincolnshire County Council. (2012). SUSTAIN Lincolnshire phase 2 innovation call for activity October 2012. Lincoln, UK.

Macarthur, E. (2013). *Towards the circular economy: Opportunities for the consumer goods sector* (2nd ed.). Cowes, UK: Ellen MacArthur Foundation.

Mirata, M. (2004). Experiences from early stages of a national industrial symbiosis programme in the UK: determinants and coordination challenges. *Journal of Cleaner Production*, *12*(8–10), 967–983 (Elsevier. Amsterdam, Netherlands). Accessed Jan 25, 2014.

PECT, PECT investor in environment. Available at: http://www.pect.org.uk/working-with-us/business. Accessed June 1, 2016a.

PECT, Peterborough Environment City Trust. Available at: http://www.pect.org.uk/. Accessed Sept 16, 2016b.

Quayle, M. (2003). A study of supply chain management practice in UK industrial SMEs. *Supply Chain Management: An International Journal*, *8*(1), 79–86 (Emerald. Bingley, UK).

Roberts, B. H. (2004). The application of industrial ecology principles and planning guidelines for the development of eco-industrial parks: An Australian case study. *Journal of Cleaner Production*, *12*(8–10), 997–1010 (Elsevier. Amsterdam, Netherlands).

Sear, L. et al. (2012). Graduate t to SMEs recruitment final report prepared for: Ian Drumm, innovation and skills ond, Department for Business. Available at: www.sfedi.co.uk. Accessed June 20, 2016.

Singh, R. K., Garg, S. K. & Deshmukh, S. G. (2008). Strategy development by SMEs for competitiveness: a review. *Benchmarking: An International Journal*, *15*(5), 525–547 (Emerald. Bingley, UK).

Trianni, A., & Cagno, E. (2012). Dealing with barriers to energy efficiency and SMEs: Some empirical evidences. *Energy*, *37*(1), 494–504 (Elsevier. Amsterdam, Netherlands).

Trust, C. (2011). Voltage management. *Carbon Trust*, p. 24. Available at: http://www.carbontrust.com/media/77191/ctg045.pdf. Accessed June 16, 2016.

Wilson, T. (February, 2012). A review of business—university collaboration. In *The Royal Academy* (pp. 1–890). London, UK (Emer. Prof. University of Hertfordshire, H., 2012).
Ye, Y., et al. (2008). Ontology-based semantic models for supply chain management. *International Journal of Advanced Manufacturing Technology, 37*(11–12), 1250–1260.
Zhu, Q., et al. (2008). Industrial symbiosis in China: A case study of the Guitang group. *Journal of Industrial Ecology, 11*(1), 31–42 (Wiley. New Jersey, USA). Available at: http://doi.wiley.com/10.1162/jiec.2007.929. Accessed Sept 14, 2015.

Individual Upcycling in the UK: Insights for Scaling up Towards Sustainable Development

Kyungeun Sung, Tim Cooper and Sarah Kettley

Abstract

Community-level innovation or action for sustainability is an important strand for sustainable development. As such, researchers investigated grassroots innovations, community-driven development or bottom-up approach. Many studies have focused on expert-led poverty alleviation projects, market-led social enterprises, or activists-led social movements for sustainable development. Relatively little attention has been paid to rather spontaneous, unorganised, citizen's collective actions. This paper, therefore, aims to analyse one such example in the UK from the perspective of Design for Sustainable Behaviour; and to suggest how behavioural insights could feed into the development of strategies for scaling up collective actions towards sustainability. The selected action (or behaviour) is individual upcycling—creation or modification of any product from used materials for a product of higher quality or value than the original. Interviews with 23 British residents with practical upcycling experiences were analysed to identify some characteristics in individual upcycling behaviour. The results expand current understanding of individual upcycling in terms of the variance in behaviour, behavioural context and potential group differences based on demographic attributes. The paper further links the analytic insights to the ideas of scaling-up.

K. Sung (✉) · T. Cooper · S. Kettley
Nottingham Trent University, Burton Street, Nottingham NG1 4BU, UK
e-mail: kyungeun.sung2013@my.ntu.ac.uk

T. Cooper
e-mail: t.h.cooper@ntu.ac.uk

S. Kettley
e-mail: sarah.kettley@ntu.ac.uk

W. Leal Filho (ed.), *Sustainable Development Research at Universities in the United Kingdom*, World Sustainability Series,
DOI 10.1007/978-3-319-47883-8_12

Keywords
Scaling-up · Sustainable design · Sustainable development · Upcycling

1 Introduction

Community-level innovation or action for sustainability is an important strand for sustainable development. As such, researchers investigated grassroots innovations (e.g. Davies and Mullin 2011; Horwitch and Mulloth 2010; Longhurst and Seyfang 2011; Middlemiss and Parrish 2010; Monaghan 2009; Scott-Cato and Hillier 2010; Seyfang et al. 2014; Walker 2011), community-driven development (e.g. Alkire et al. 2001; Binswanger and Nguyen 2005; Bowen 2005; Dongier et al. 2003; Gillespie 2004; Grootaert 2003; Krishna 2003; Mathie and Cunningham 2003) or bottom-up approach (e.g. Akpomuvie 2010; Danish 1995; El Asmar et al. 2012; Fraser et al. 2006; Rayner 2010; Smith 2008).

The research in grassroots innovation (GI) may largely fall into two categories: the first focuses on social enterprises and social movements for greener economy in developed countries (mostly in the UK and USA), and the second focuses on poverty alleviation and capacity building of the poor in developing countries (mostly in India). The former included the studies in GI by innovative network of activists and organisations for sustainable development (Seyfang and Smith 2006, 2007) through, for example, low-carbon housing (Seyfang 2008, 2010) or an organic food producer cooperative (Bekin et al. 2007) in the UK and USA, GI in community renewable energy in the UK (Walker et al. 2007; Walker 2011), Cleantech activities—collaborative activities involving diverse social entrepreneurs, grassroots movements, firms, public policy action and hubs of innovation—in the USA (Horwitch and Mulloth 2010), social enterprises in Ireland (e.g. Davies and Mullin 2011), and Transition Towns in the UK (Longhurst 2012; Scott-Cato and Hillier 2010). The latter included the studies in: GI for solving local problems including poverty through the Honey Bee network—loose platform to share knowledge, innovations and sustainable practices in India and 75 other countries (Gupta 1995, 2000; Gupta et al. 2003a); GI by ecopreneurs (ecological entrepreneurs) in alternative agriculture in India (Pastakia 1998); capacity building aspect of GI by the people with limited power, resources and ability (Middlemiss and Parrish 2010); preliminary framework for scaling up and commercialisation of GI in India (De Keersmaecker et al. 2012); and GI in Information Technology for the bottom of the economic pyramid (Heeks 2012).

The research in community-driven development (CDD) dealt with the cases in developing countries to identify (1) obstacles to scaling-up of CDD (Binswanger and Aiyar 2003), and factors and guidelines (or model) for successful scaling-up (Binswanger and Nguyen 2005; Bowen 2005; Gillespie 2004; Lavery et al. 2005; Mansuri and Rao 2004; Sharma 2004); (2) role for the state and state-community synergies in CDD (Gupta et al. 2003b, 2004); (3) relationship with empowerment and social capital (Grootaert 2003; Krishna 2003); (4) problems and challenges in CDD (Bebbington et al. 2004; Platteau and Gaspart 2003), or to suggest asset-based community development rather than needs-based approaches to development (Mathie and Cunningham 2003, 2005).

The research in bottom-up approach was also primarily development studies in developing countries such as the case of self-help strategy for rural development in Nigeria (Akpomuvie 2010), decentralised energy planning in India (Hiremath et al. 2010), sustainable urban development in Lebanon (El Asmar et al. 2012), subsistence marketplaces in India to suggest bottom-up orientation to business policy development (Viswanathan et al. 2012), and the desertification convention focusing on local developmental issues and the marginalised people (Danish 1995).

It appears that many studies have focused on expert-led poverty alleviation projects, market-led social enterprises, or activists-led social movements for sustainable development. Relatively little attention has been paid to more spontaneous, unorganised, citizen's collective actions. The aims of the paper, therefore, are twofold. The first is to analyse one such example in the UK from the perspective of Design for Sustainable Behaviour. And the second is to show how behavioural insights could feed into the development of strategies for scaling up citizen's collective actions towards sustainability.

2 Setting the Scene: Scaling-up Individual Upcycling

2.1 Project Background

The UK is legally obliged to reduce its greenhouse gas emissions by at least 80 % from 1990 levels by 2050 (UK Government 2015). As part of the Government's commitment to achieving this reduction, the Research Council's UK Energy Programme established six End Use Energy Demand centres (EUED 2015). Centre for Industrial Energy, Materials and Products (CIE-MAP) is one of six centres and focuses on materials and embodied energy reduction in the UK. The first author's PhD is part of CIE-MAP and it intends to explore the emerging household behaviour of individual upcycling in the UK as an important opportunity at the household level and beyond for sustainable production and consumption by reducing carbon emissions related to materials and energy.

2.2 Individual Upcycling as Environmentally Significant Behaviour

Environmentally significant behaviour is, according to Stern (2000), the behaviour that "changes the availability of materials or energy from the environment, or alters the structure and dynamics of ecosystems or the biosphere itself" (Stern 2000,

p. 408). It is often used interchangeably with pro-environmental behaviour, green consumer behaviour, environmentally responsible behaviour, environmentally friendly behaviour, ecological behaviour, sustainable behaviour or sustainable lifestyle. Despite differences in terminology, the common denominator is the idea that individual behaviour can collectively impact positively on the environment.

Individual upcycling, the creation or modification of any product from used materials, components, or products in an attempt to generate a product of higher quality or value than the compositional elements (Sung et al. 2014) by individuals, is another example of environmentally significant behaviour. The term, upcycling, was recently coined and can be traced back to the interview with Riner Pilz (Kay 1994). Pilz, in the context of architecture and interior design, said, "Recycling, I call it down-cycling. They smash bricks, they smash everything. What we need is upcycling, where old products are given more value, not less." (Kay 1994, p. 14). The more widely understood meaning of upcycling in academia, however, comes from MacDonough and Braungart (2002). They see upcycling as the process that maintains or upgrades materials' value and/or quality in their second life and beyond in a closed-loop industrial cycle. This paper uses the perspective from Pilz and adds 'individual' in front of 'upcycling' in order to reflect the emerging, contemporary individual activities of upycling—with over 50 % (64 out of 120 books) of the published books on upcycling since 1999 categorised as craft and hobbies (Sung 2015)—as well as to distinguish it from 'industrial upcycling' which more often refers to improved recycling rather than product recreation.

Individual upcycling may be assumed to be a more sustainable way of making, crafting or personalising products for individuals than doing so with virgin materials only. When scaled up to a meaningful level with appropriate interventions, it could, in theory, significantly reduce the need for new products as well as municipal solid waste. Reduced need for new products would lessen the amount of materials and industrial energy used in production with new materials, and therefore contribute to reductions in greenhouse gas emissions (Ali et al. 2013; Goldsmith 2009; Szaky 2014). In addition, the decreased amount of municipal solid waste may obviate the need for additional landfill spaces. In addition, individual upcycling has the potential to extend product lifetime by improving the user-product relationship through, for instance, self-expression and memories of upcycling (Sung et al. 2015). The benefit of individual upcycling is not only limited to positive environmental benefits. It can save money for individuals—fulfilling needs with fewer financial resources (Frank 2013; Lang 2013)—and, in theory, lead to new jobs in small- or medium-sized enterprises (e.g. Sarah Turner in Sung and Cooper 2015). It can, furthermore, provide participants with socio-cultural and psychological benefits such as learning, empowering, a sense of community and relaxing (Sung et al. 2014).

2.3 Scaling-up of Individual Upcycling

Many anecdotal evidences suggest that the overall number of people who upcycle used materials, components or products has recently increased in developed countries, including the UK, possibly as a response to the contemporary Maker[1] Movement (Anderson 2012; Lang 2013), physical resources (e.g. Maker Faire, Hackspace/Makerspace) and digital resources (e.g. Instructables, Etsy). Despite this growing interest, individual upcycling is evidently still a marginal activity. Considering the potential benefits of individual upcycling environmentally, as well as economically and socio-culturally, one of the pertinent questions from the perspective of sustainable design may concern scaling-up (Ceschin 2012; van den Bosch 2010). How can this emerging, yet still marginal activity, be scaled up into a mainstream everyday activity in households (and possibly also in industries) to make a bigger impact on the environment and society?

3 Methods

A qualitative approach was selected to explore the behaviour of individual upcycling. Semi-structured interviews with 23 British residents with practical upcycling experiences were conducted. The data was collected between April and July 2014. Interviews were varied in time, but typically lasted around 30–90 min.

3.1 Interview Schedule

Semi-structured interviews were conducted, chosen for their flexibility which allows for probing when necessary, while keeping the pre-determined interview schedule (Barriball and While 1994). The interview questions included the behaviour variance (how often to upcycle; with what materials—how or where to get them, and why to choose particular materials; and what to do with end products) and context of the behaviour (when, where, with whom to upcycle). At the beginning of the interview session, basic demographic information was collected.

3.2 Sampling of Participants

Hackspace/Makerspace was considered as an appropriate starting point for recruiting people with practical upcycling experiences. Hackspaces provide local residents with a membership including access to tools, materials and expertise. Hackspaces have increased in number since 2009 and are now available in more than 90 different locations in the UK (Nesta 2015; UK Hackspace Foundation 2015). Ten workshops

[1]The term, 'Maker', could apply to potentially everyone in the sense that "we are all makers. We are born makers: just watch a child's fascination with blocks, Lego, etc. It's not just about workshops, garages and man caves. If you love to cook, you are a kitchen Maker and your stove is your workbench. If you love to plant, you are a garden Maker. Knitting and sewing, scrap-booking, beading, and cross-stitching—all Making." (Anderson, 2012, p.13).

in ten different cities of nine different regions in England were selected. The selecting criteria were accessibility (i.e. whether or not the community has a Google group or other online forums) and activeness (based on the number of postings) of the members. A recruiting advertisement was posted on Google groups or forums of the ten workshops, with the only inclusion criterion being previous experience of practicing upcycling. Thirteen participants directly answered the advertisement and another ten were identified by snowball sampling.

The total of 23 face-to-face interviews were conducted. Participants were from nine different cities and aged between 24 and 66 years old. 17 (74 %) were British and 6 (26 %) were non-British. 15 (65 %) were male and 8 (35 %) were female. 12 (52 %) worked in science and engineering, 7 (30 %) in art and design, and 4 (17 %) in other areas (health service, business and management) or were unemployed.

3.3 Analysis

All interviews were transcribed and analysed by the interviewer. The transcripts were anonymised and entered into QSR NVivo 10 software. A thematic analysis (Braun and Clarke 2006) was conducted, with each transcript examined line by line and categorised into five behavioural variances (how often, what materials, how and where to get materials, why particular materials, and what to do with end products) and three behaviour contexts (when, where, with whom to upcycle). Within these eight categories, grounded codes were identified and constantly revised to fine-tune the coherent collective themes.

3.4 Limitations of the Work

The results presented in this paper may not be generalisable to the overall UK populations or UK Makers or upcyclers on the basis of the sampling method applied and a limited sample used. In addition, this study focused on particular aspects of the behaviour including the frequency of the behaviour, selection and attainment of the used materials, use of the upcycled products, and limited contextual information. Potentially interesting other aspects such as drivers and facilitators for and barriers to the behaviour, or skills level and tools involved in the behaviour are not within the scope of this study.

4 Results

This results section describes the variance in individual upcycling and behavioural context of the participants. Full quotes are included in the appendix.

4.1 Behaviour Variance

Seven participants mentioned that the frequency of upcycling varies and depends on the project—participants often called their upcycling 'a project'. One male participant (M01) stated, for example, electronics takes longer time whereas woodwork is relatively quick and easy. They appeared to have one project at one time period, finish it and move on to the next project. Four participants mentioned that they have been upcycling things almost every day—"maybe 1.5 hours a day" (M13); "probably four days a week" (M09); "4 to 5 days a week, 4 to 6 hours a day" (M12)—or all the time (M02). Two participants mentioned that their upcycling "tends to spread out through a very long period of time" (F05) sometimes in such a way that upcycling "is interwoven into [their] lives" (F03). One participant stated that upcycling frequency depends on her job situation: she (F05) said "If I have no contracts, then I have been here [Hackspace] up to 5 days a week […] If I do have work maybe once or twice a month." Upcycling frequency from other five participants varied from once a week (M06) to once a month (F01), once every three months (M11), twice a year (M04) or once a year (M07).

Eight participants mentioned that they utilise wood and furniture—e.g. used furniture (F02; M03); old pallets and used plywood (F06); bits of wood (M11); wooden pegs (M10). Five participants said that they use anything "[they] come across" (F02), "lying around" (M11), or "in [their] hands" (M13). Another five participants said they use metal—e.g. nuts and bolts (F04); "metal and wires and stuff with copper" (F07); aluminium (F08). Four participants stated that they use electronics. Three participants stated fabrics—e.g. T-shirts (F02); different kind of fabrics (F03; F05; M09). Three participants said they use package—e.g. containers, boxes, shelves (M06); paper cardboard (M07); general packaging (F03). Three participants (M03; M09; M11) mentioned that they use anything required for their particular project. More miscellaneous materials included "waste from glass industry" (M08), "watches and jewellery" (F04) and plastics (M12).

Seven participants answered that they get used materials (including used components and products) from online shops or networks—e.g. ebay (F02; M02; M06), gumtree (F02), freecycle or freegle (F02; M03; M06; M07). Six participants mentioned that they get used materials from anywhere everywhere: one participant said "from all kinds of places […] I look out for stuff that are on the street […] I've got a lot of stuff from neighbours leaving things out […] I am looking at skips and those places where the buildings are renovated." Another six participants stated that they find used materials from skips. Four participants mentioned that they have utilised their own unused items: "my own consumables" (M06); "excess on stuff that I may have bought for another purpose"; or broken items—"the child swing is actually something we had in our garden, but it had fallen apart. And I used the steel poles" (M04). Four participants said that they go to charity shops and other local shops to buy or get used materials. Three participants mentioned that the used materials were given by other people; another three stated that they go to car boot sales. Building site was mentioned by two participants. Other miscellaneous places included recycling centre (F06), local factory (M08) and Hackspace (M02).

The most frequently mentioned selection criterion for particular used materials was project requirement by nine participants (i.e. applicable to what is needed for the project). Five participants said that they consider potential value: for instance, one participant said "I see things: compressor and electronics. And I see if it's repairable. […] I see value, if I can clean it or if it is recoverable." (M13). Four participants mentioned about financial saving perspective. Another four participants stated that they consider (relatively) high quality: e.g. solid wood rather than cheap MDF (F02); no rotten or moulded wood (M13); colour and texture of the fabric (F03); clean and in good condition (M04). Another four participants said they do not have any criteria. They may start with materials (crafting based on available materials) and not the other way around (designing first and gathering materials accordingly)(M02) or do not mind trying varied range of materials (M10). Three participants said that the materials chosen were something they liked—e.g. "what catches my eyes" (M03) or "pretty things, smallest things" (F07). Two participants pointed out that used materials need to be easy to handle—e.g. easy to saw, stick, paint, turn into anything (F02) or easy to cut and fix without much tools (M06). Other miscellaneous answers included: depending on the person who wants it when upcycling for someone else (F03); depending on what I have (F08); unrecyclable materials (F08); and relatively unused materials (M04).

The most frequently answered use of the end products after upcycling was 'use for home or myself' by 15 participants. Amongst these 15, two participants added that the end products are not good enough to give to someone else (M06; M13). Eight participants said that they give the upcycled products to family, friends or acquaintances—e.g. when they no longer want it (F02); when they think the product is relevant for someone (F05); for my daughter (F03); or as a birthday present (F08). Among these eight, seven mentioned that they occasionally give it away, and one participant (M04) said he usually does so. Seven participants stated that they considered the option of selling to others. Three of them have actually sold some upcycled products through craft shows or fairs (F01; F04), internet market places such as Folksy or Etsy (F01; F04) or a physical shop (M10). Despite their consideration for commercialisation, some upcycled products have not been sold because they "have not put an effort to investigate how feasible it is" (M07); they "faced some legal issues […] and safety issues" when using broken and discarded electronics parts (M08); or they have not found a market for it (M10). Three participants mentioned that some of their upcycled objects were used for exhibitions (M08; M11) or Maker Faire (F03). Two participants (M11; M13) mentioned that sometimes it is not so much for the output at the end but just for fun. One participant (M01) used upcycling electronics as part of his degree project.

4.2 Behavioural Context

Half of participants (11 out of 23) stated that they upcycle anytime that suits them: the timing may depend on the job situation (F05; M03; M09; M10), amount of free time and distractions (M06), or working space—e.g. mostly during summer since the participant upcycles at the patio or in the garden (F06). Three participants (F03; M03; M13) stated that they have upcycled all the time. Two participants (F03; F08) mentioned that they sometimes responded to particular events. Other miscellaneous answers included when they find the material they have been looking for (M03); when they feel like upcycling (F04; M02); and when there is need (M03).

Regarding the place for upcycling, seven participants stated that they upcycle at home, without specifying any particular spaces. Six participants reported that they use their shed or garage. Five participants mentioned particular rooms at home: living room (F04); office room (M03); workshop room (M09); dining room (F03); and bedroom (M10). Two participants said that they use patio. Six participants said that they use local Hackspace or Makerspace mostly for tools (M01; M07) and space (F05). Three participants stated that they have their individual or shared studio or workshop outside home.

When the participants were asked about with whom they upcycle, most participants (17 out of 23) answered that it is just by themselves because they could not find people with similar interest (M01; M06); they could be more productive on their own (M02); they tried collaboration and it did not work (F05); or they do not want to be interrupted nor told what to do (M10). Six participants mentioned about local experts for consultation (M03), mutual help (M02), or collaboration (F08; M05; M11; M12). Three participants stated that their partner is a collaborator (F01; F04) or a companion—not necessarily working on the same project (F06). Two participants mentioned about other family members—father for consultation (M03) or daughter for collaboration (M04). Two participants (M08; F03) said that they worked together with expert friends; another two (M03; M07) said they got help from the people in online communities; another two (M03; F03) stated that it depends on the project.

4.3 Summary of the Results

The frequency of upcycling from the participants varied from 'all the time' to 'once a year', sometimes depending on the project or job situation. The upcycling practitioners seemed to vary from enthusiastic hobbyists (or environmentalists) to more pragmatists (or pragmatic Makers upcycling only when it is necessary). Participants' frequently used materials appeared to be wood and furniture, followed by metal, electronics, fabric and package. The most popular place to get used materials was online shops (e.g. ebay, gumtree) and online networks (e.g. freecycle, freegle), followed by skips, charity shops and car boot sales. General material selection criteria included potential value, financial saving, (relatively) high quality, easiness to handle and un-recyclability. The use of end products was mostly for oneself, followed by gifts to family or friends, and selling. Even though one third of the participants considered the option of selling, the people who have actually sold anything were only three for the issues of safety (especially electronics), and business feasibility and viability. More than half of participants expressed that they upcycle anytime that suits them (as a hobby) or all the time (as a lifestyle). Their upcycling place is mainly home (either rooms or shed, garage, patio) but some go to Hackspace/Makerspace mainly for tools and bigger space. Most participants engaged in upcycling just by themselves because of the difficulty in finding similar-interest people, previous bad collaboration experience, expected increased productivity, or preferences towards no interruption and instruction.

The differences from demographic characteristics are not conclusive. Nevertheless, there were some meaningful observations. For instance, male participants appeared to more frequently utilise online shops and networks, and skips than female participants. 50-and-over participants did not mention charity shops, car boot sales, building site, Hackspace, local factory or recycling centre as their used material source. 50-and-over participants did not state about potential value, financial saving or easiness to handle as used material selection criteria. Mostly under-30 male participants reported that they have been to Hackspace/Makerspace for upcycling.

5 Discussions: Implications for Successful Scaling-up

Scaling up of individual upcycling may include making enthusiastic upcyclers become an entrepreneur (e.g. Sarah Turner in Sung and Cooper 2015), enabling more pragmatic Makers (not necessarily utilising used materials for every making project) to practice upcycling more frequently, and attracting non-Makers to engage in making and upcycling. How do we make such changes or transitions? The results may suggest the following possibilities. Considering the frequently used materials, it might be helpful if any attempt to improve materials provision first targets wood, electronics, fabric and package. Based on the popular places to get materials, if each local authority runs a unified used material centre (instead of many scattered places) of which collection service is in line with the existing waste collection system, and provides online service on which users can search for materials (similar to freegle/freecycle but more organised, top-down service), it might facilitate both enthusiastic and pragmatic Makers to become more frequently engaged in upcycling. Taking into account the common material selection criteria, any online service for material provision might be more helpful for decision making if it provides users with the estimated potential value in the material with estimated money saving (comparing to new materials) and quality rate for each item assessed by experts. Reflecting on the aspirations from some people for selling their upcycled products without any commercialisation experience, some specialised services by relevant actors (e.g. local authority, academic institutions, design council, social enterprises) may lower the barriers for those enthusiastic upcyclers to become entrepreneurs. Such services may include: (1) business feasibility test based on, for instance, financial analysis, technical analysis, and risk analysis; (2) technical safety test (especially for electronics); and (3) suitable niche market identification. Based on the predominant use of home for upcycling and reasons why people use Hackspace, if people can hire or rent tools to work at home, or use the tools and space in

Hackspace/Makerspace for a day or shorter for some pence instead of paying significant amount of money for long-term membership, it might be more convenient for existing Makers, and potentially attract more non-Makers to try out making and upcycling. Considering the difficulty in finding similar-interest people or good collaborators, it might be helpful if a community event (e.g. mini Maker Faire or Hackspace-initiated event) on a regular basis plays an active role in enabling people to find their hobby friends, companions, collaborators or even potential business partners.

The scaling up of individual upcycling can go beyond hobbies and niche enterprises. In-house designers in big multinational corporations, for example, can also learn from individual upcycling. They just need to look for the products worth mass-production, and the production technique worth scaling up in terms of cost-effectiveness and sustainability in a large scale. They can also design more effective and efficient systems and services to take back products and packages for upcycling as extended producer responsibility. Furthermore, considering individual upcycling as one of niches in the multi levels of sustainability transitions (Geels 2011), the mechanism of 'broadening'—getting different niches linked together (e.g. linking individual upcycling with repair, reuse, and other types of sustainable Do It Yourself activities)—could also lead to a niche-cluster which can eventually grow into a niche-regime (van den Bosch 2010). Since such a niche-regime exists at a higher scale level within the multi-level perspective (de Haan and Rotmans 2011) it could be more stable and influential to challenge the power of the regime (van den Bosch 2010).

6 Conclusions

Recognising the relatively little attention paid to more spontaneous and unorganised citizen's collective actions as part of community-level innovations for sustainable development, this paper introduced one such example of individual upcycling in the UK. The interview study with 23 British residents with practical upcycling experiences expanded the current understanding of individual upcycling in terms of the variance in behaviour, behavioural context and potential group differences based on demographic attributes. The paper further discussed how the results (behavioural insights) can be utilised to develop strategies for scaling up individual upcycling, and how the scaling up can go beyond hobbies and niche enterprises.

The future of individual upcycling looks promising with the developing, global 'circular economy' debate (Ellen MacArthur Foundation 2015)—in which repair, reuse and refurbish are important inner circles—and recent design movement for sustainable change including design activism (Fuad-Luke 2013) and design for social innovation (Manzini 2015). However, whether or not individual upcycling scales up enough to gain critical mass would depend on how every stakeholder—industry, government, NGOs, citizens, etc.—in society acts and reacts to it. It is, therefore, our hope that the proponents of circular economy, design activism, design for social innovation, and other relevant concepts and thoughts are inspired and informed by this article, and contribute to sustainable scaling up of individual upcycling in the UK and beyond.

Acknowledgments This work was funded by Nottingham Trent University with support from the RCUK Energy Programme's funding for the Centre for Industrial Energy, Materials and Products, grant reference EP/N022645/1. The participation in the Symposium on Sustainable Development Research at Universities in the United Kingdom was funded by the Design Research Society.

Appendix: Participant Answer Quotations

Theme	Age	Participant answers
How often or how long		
Depends on the project (7)	Under 30 (2Fs & 2Ms)	F01: "depends on what craft projects I am doing. Because I've been doing craft, knitting, crochet, sewing, and it depends on what project is, sometimes I upcycle things, and sometimes I buy things new." F02: "depends on I've got in really. If I find something that I like, I work on it until it's finished, and it would be so over lunch or any evenings, or if I just need a break. Sometimes I have more than one thing at the same time, and sometimes I don't do anything for a couple of weeks and I pick up something new. I won't pick up anything I don't like. So if I don't see anything, I won't, I won't do." M01: "in terms of the raspberry pi thing, electronics, that's taking a long, long time, a year or so. ehh. in terms of things to do with woodwork, it's very quick coz it's easy. It's hard to make mistakes, if you are with something like electronics. It is very easy to make a little mistake on things don't work. So, it's completely different, electronics, takes long time, woodwork, takes not very long at all. With the patio, and the path, umm, the only reason I take a long time is because it's a lot of labour, a lot of work to do it, to lift the big stones and everything. It takes lots of efforts." M03: "It's one of those things I generally get into it and finish it and then not do anything for a while and then pick up something else and finish it and then move on. So it's in fits and starts rather than every weekend type of thing."
	30–49 (1F & 1M)	F04: "Probably a couple of times a week, for an hour or so. I tend to have a specific project that I do that I spend longer on over a shorter space of time. If I have got a project, I will spend hours a day for a week or so. But if I haven't got a project on the go, I might not touch for days on it." M05: "It really varies. […] it varies so much. sometimes… it just really varies. Well, the majority of my work is admin and correspond to projects. The actual making has been about

(continued)

(continued)

Theme	Age	Participant answers
		maybe 10 % of my work? And I sort of work more or less 40 h a week. […] when I am really intense in making, I was working for 2 weeks non-stop making. When I got my drawing machine, I just demonstrated machine, non-stop for a period of days."
	50 and over (1F)	F08: "depends on what I am actually making. I realized when I was preparing my display for the next week library exhibition, I haven't got very many of the wider bangles and I thought I must make some more because I know that they are already sold really because people have shown their interest in them. But I only have limited amount of time. So I started yesterday to cut up some more of wires […] I tend to work over 2–3 weeks depends on what time I have in between doing the other things that I am doing."
All the time (4)	Under 30 (1M)	M13: "it was definitely 2–3 days a week, when I was studying. I was always working on something. Whether it's upcycling or based on something new. Now I am working, I have less time to actually work on projects, and also strangely with less money? I have to pay council tax and transport and bills now, they are expensive. So I don't have money, which is frustrating. […] Maybe 1.5 h a day… maybe about 4.5 h a week?"
	30–49 (2Ms)	M02: "It's kind of… I don't know, all the time? So, I go to the Hackspace, at least one probably two nights a week. and probably one day in a weekend, every month. So, it's something like that. So one day, maybe, two or three days a week for a certain amount of time." M09: "you can count 80 % of every evening during the week. […] They will be probably 4 days a week. Sometimes, at least one of the two days of the weekends."
	50 and over (1M)	M12: "4–5 days a week? 4–6 h a day. So maybe I say 30 h a week, something like that."
Spread over a long time period (2)	Under 30 (1F)	F05: "not often a lot. They tend to be spread out through a very long period of time. It's mostly a hobby. So, it's kind of… I do want to make something, then I start looking out for materials I can use for, and I usually gather those over a few months, and then kind of do the work in fits and starts. If I've got a day for a weekend, I spend a day working on it. Or, spend a couple of hours in the evening, but it's usually spread out… like not particularly organized… I am just bit like… carrying there. It's pretty much always ongoing but not that frequent."
	30–49	NA
	50 and over (1F)	F03: "i probably spend four full days actually taping and measuring and testing. If I just sat down to do it, it probably takes about a week. The dress, for example, a month, but it was because I do something and I sit back and think 'what do i do next? Should I do this or that?' i have to find another fabric. So that took me about six months. But it's not every day. It's interwoven into my life."

(continued)

(continued)

Theme	Age	Participant answers
Depends on the job situation (1)	Under 30	NA
	30–49 (1F)	F05: "it depends on how my work is going. If I have no contracts, then I have been here up to 5 days a week for probably… up to 9 h each day. If I do have work maybe once or twice a month for 5 to 7 h each. It's, you know, unless you can be paid for, it's hard to justify, what am I doing? Playing with like rubbish?"
	50 and over	NA
Once a week to once a year (5)	Under 30 (1F & 3Ms)	M06—once a week: "once a week? 2 h a day? It is relaxing." F01—once a month: "usually one project every month? But at the moment, probably about three projects a week." M11—once every three months: "I'd say at least one every three months. I would say that. Over the year, a couple of every other six months. Not massive amount but depends on what scale we are talking. They are perhaps bigger projects… but some are bits and bobs. You repurpose something and it might only be a small item like a screw or something like that. […] we've done things little things but we don't really think about it." M07—once a year: "looking at this list, it's about one a year. These are the best kind of examples I am thinking of. One or two more, given that the first one was in 2007. It seems to be about one a year."
	30–49	NA
	50 and over (1M)	M04—twice a year: "well, probably not that often. I mean I would think probably I do about 2 things this year, perhaps."
With what materials, components or products		
Wood and furniture (8)	Under 30 (2Fs & 2Ms)	F02: "pick up furniture in a car boot, even broken bits to make up new things like a coffee table that I showed you at the end of the pictures, it's actually two chairs. […] just wood stuff. […] I think, I tend to use wood because it's easily paintable." F06: "I use mostly woods, so recycling old pallets and used plywood." M03: "So when I did my jigsaw table I told you about. What it was that when we moved in, I've been meaning to get a coffee table for a while, and I was at the village fete, helping my parents out there they are having a barbeque every year and I had a big box to try to find interesting things, and there was a table, I brought it home and the space… the table was a bit big so I cut the top into four jigsaw pieces and added new lags, figured out how to make joints and I built four tables three of which are matched but one of which is slightly shorter so I am doing it again now. […] mostly wood work. […] Probably that's the core of what I am doing. The table, workbench, basically woodworking projects. [… things about woodwork] M11: "and all sorts of bits of wood"
	30–49 (1F)	F05: "Sometimes wood."
	50 and over (3Ms)	M04: "it's mostly wood." M10: "wood. Wooden pegs or matches. Sometimes bits of materials if I need to use it to cover things." M12: "I normally use wood."

(continued)

(continued)

Theme	Age	Participant answers
Anything I come across (5)	Under 30 (1F & 2Ms)	F02: "I use anything I come across to make stuff." M11: "and anything lying around really." M13: "just anything in my hands really. I live near garages, so they have tires and stuff. And I've taken one of those and cut it down, and get some pipe pips, and screw them up, and I use them to put it on new shoes. That's good. […]"
	30–49 (1F & 1M)	F04: "I did have a plan to make a dragon. It's big like 5 to 10 foot long dragon. But I have not found a thing that will spark it to generation. I've got an umbrella, an old dead umbrella nobody can use it anymore, I've got an old hack's wrenches to use for claws, things like that, but I haven't got the one thing that will be the start but everything else will grow around. But I have all the stuff in my loft to be waiting for that time. When all these things I've picked up from the floor can come to flourish in a new life as a dragon." M05: "all sorts of really. I sort of find myself looking at products thinking how I can make something out of it. So I use everything from bin bags, plastic bottles, state agency signs, train tickets, measuring tape, playing cards, ya, all sorts of different things."
	50 and over	NA
Metal (5)	Under 30	NA
	30–49 (2Fs & 1M)	F04: "nuts and bolts, or bits of metal" F07: "metal and wires and stuff with copper, so anything that comes along" M09: "metal, plastics, fabrics, anything that I need to use."
	50 and over (1F & 1M)	F08: "I use aluminium and upcycle some of previous art work." M12: "I sometimes use metal and plastics and electronics."
Electronics (4)	Under 30 (1M)	M11: "electronics mainly, I would say"
	30–49 (1F)	F04: "I use printed circuit board"
	50 and over (1F & 1M)	F08: "At the moment, I've been making jewellery for the upcoming exhibition and that is from recycled data cabling, so computer cabling and electric wiring I have a big store of that, which I acquire from the company who gave it to me about 2 years ago when I was working on my MA and I was wanting cabling to integrate it into my art." M12: "I sometimes use metal and plastics and electronics. […] I always save my electronics bits because my background is electronics engineer."
Fabric (3)	Under 30 (1F)	F02: "cushions out of t-shirts […] cushions and blankets and that kind of stuff."
	30–49 (1F & 1M)	F05: "mainly fabric." M09: "metal, plastics, fabrics, anything that I need to use."
	50 and over (1F)	F03: "I work with fabric, different kind of fabric, I like a range of texture."

(continued)

(continued)

Theme	Age	Participant answers
Package (3)	Under 30 (2Ms)	M06: "I mostly use storage things […] just like the containers, boxes, shelves." M07: "I use a lot of paper cardboard, glue, bamboo, stuff like that."
	30–49	NA
	50 and over (1F)	F03: "I saved packaging and used that."
Anything required for the project (3)	Under 30 (2Ms)	M03: "depends on the project, really." M11: "and anything lying around really. It depends on what needs to be done."
	30–49 (1M)	M09: "metal, plastics, fabrics, anything that I need to use. […] there's really no limitation to materials."
	50 and over	NA
Plastics, glass, watches and jewellery (3)	Under 30 (1M)	M08—waste from glass industry: "this one is cork, chemistry beaker and glass. This is like a waste from the glass industry because the bottom is not very even. They can't sell it so these become waste. And they are from one of the biggest glass manufacturers in the world. They have like tonnes of bottles they can't sell. They normally will melt and cast them again, but they nicely offered us, gave us some bottles with defects."
	30–49 (1F)	F04—watches and jewellery: "I use watches, and bits of old jewellery and things like that."
	50 and over (1M)	M12—plastics: "I sometimes use metal and plastics and electronics."
How or where to get the materials, materials, components or products		
Online shops and networks (7)	Under 30 (1F & 3Ms)	F02: "sometimes online, ebay or gumtree.com. pick up things from there… a lot of people… or freecycle. It's another one. A lot of people who don't want to sell it, just want to get rid of it, put it on freecycle." M03: "so, most of the things I've done recently, parts came from freecycle, freegle,… freegle is… it used to be called freecycle which is much better name and is a sort of yahoo group. It's an international loose group of organizations, so you join up for the local yahoo group and people post "I have such and such for free." "I want such and such for free" what they want is usually hilarious. "I want a car!" "I want a computer!" ya, giving away things they don't want anymore… that's where I found the door for the workbench or at the fete… I don't go there very often, but when I am helping my parents, I can find [something]." M06: "I bought i-Mac G4 for 10 lb from ebay. […] the piano was from freecycle for free." M07: "I've got some stuff from freecycle."
	30–49 (2Ms)	M02: "Second hand things from ebay as well sometimes." M09: "usually I use internet. There's a shop online."
	50 and over (1M)	M12: "electronics, I normally get from ebay or from Farnell [Electronic components online shop]."

(continued)

(continued)

Theme	Age	Participant answers
Anywhere everywhere (6)	Under 30 (2Fs & 1M)	F01: "just collecting things." F06: "from all kinds of places. […] I look out for stuff that are on the street like lots of people put stuff outside the houses, for people to collect. So I've got a lot of stuff from there, like neighbours leaving things out and I am walking passed, and picking up things useful. I am looking at skips and those places where the buildings are renovated." M13: "just like wherever someone has thrown away really. I've never really found a lot near home actually because it's suburban area. But here in city, if you go to back alleys, then people just throw everything away."
	30–49 (2Fs & 1M)	F04: "wherever I go, I am always keeping my eyes on the floor, because there's all sort of things you find, people have just lost or disregarded. I found (showing things to the interviewer) that old nuts and bolts and that piece of plastic thing, just on the ground while I was walking two days ago. And I just pick things up and collect them." F05: "either it's stuff lying around maybe my housemate, she just bought and doesn't want it anymore, or like… I don't know… various places like… I don't think I ever pull something out of the garbage can but I would if I saw something that was good enough…" M02: "Anywhere everywhere really."
	50 and over	NA
Skips (6)	Under 30 (1F & 1M)	F06: "I am looking at skips" M13: "And there's a dump out of the street. I found even a humidifier. I mean a good one. You know, big one with a refrigeration and heat pumps."
	30–49 (3Ms)	M02: "I also used to work at the university and the things like… things been thrown away, what they consider it as waste from the project, I would go through the skips and find things there. And they are brilliant. Some stuff are amazing. So yes, the materials that are being… before moving out of the space, the materials are all over." M05: "some of them are freely available for the… for example, I found state agency signs from the skips, sort of bins around the town, where just people discard them." M09: "those are from the street, from the bins, from the skips […] I know it sounds weird, but I look inside the bins and especially when there is construction, I look inside the skips, trying to see if there is any material I can reuse. and sometimes those materials look useful and I just take them."
	50 and over (1M)	M04: "I get them from skips usually. I quite often also collect things from people who get rid of them."

(continued)

(continued)

Theme	Age	Participant answers
Broken or unused items (4)	Under 30 (3Ms)	M06: "they are just rubbish. They are just free. I just use… my own consumables." M07: "it's probably just excess on stuff that I may have bought for another purpose." M11: "something I have already."
	30–49	NA
	50 and over (1M)	M04: "some of the things, the child swing is actually something we had it in our garden, but it had fallen apart. And I used the steel poles for that."
Charity shops and other local shops (4)	Under 30 (1F & 1M)	F01—charity shops: "charity shops" M03—charity shops: "Charity shops occasionally."
	30–49 (1F & 1M)	F05—local shops: "sometimes the offcut from the fabric shops. The cutoff edge of the curtain or something. And they say… you know they try to sell it, but nobody buys then they throw away. So before that happens I will buy it for a pound." M05—local shops: "Other ones like… lamp shades out of plastic bottles and the local café, one of the waitresses very kindly stored them for me and gave me a big plastic bag full of them."
	50 and over	NA
Given by people (3)	Under 30 (2Ms)	M07: "if someone has something that they are obviously not using and they don't want, and I see some potential in it then I will ask them if they want to get rid of it." M11: "other stuff… where I get originally is being donated by somebody"
	30–49	NA
	50 and over (1F)	F08: "some are given to me."
Car boot sales (3)	Under 30 (2Fs)	F01: "car boot sales […] mainly, we go to car boot sales" F02: "sometimes in the car boot sales"
	30–49 (1M)	M02: "so, big source of my materials are carboot sales. […] But probably the carboot sales are the main supplier of second hand parts. Carboot sales and hackspace."
	50 and over	NA
Building site (2)	Under 30 (1F & 1M)	F06: "those places where the buildings are renovated." M03: "And… in future, not yet, I plan to be going around and sort of looking at building sites and asking "do you need that wooden pallets?" but I don't really have a space to do that at the moment."
	30–49	NA
	50 and over	NA

(continued)

(continued)

Theme	Age	Participant answers
Hackspace, local factory, recycling centre (3)	Under 30 (1F & 1M)	F06—recycling centre: "There's a place called 'Brighton wood recycling centre' that sells used woods for cheap. And I get any particular sizes of woods I need from there. And they are really good prices." M08—local factory: "this one is cork, chemistry beaker and glass. This is like a waste from the glass industry because the bottom is not very even. They can't sell it so these become waste. And they are from one of the biggest glass manufacturers in the world. They have like tonnes of bottles they can't sell. They normally will melt and cast them again, but they nicely offered us, gave us some bottles with defects. [...] we contacted the glass manufacturer. And the cork, we contacted the biggest cork manufacturer. We've got some of the parts in this space [shared, community workshop garage], so we've got basic parts here."
	30– 49 (1M)	M02: "Hackspace, donations to hackspace. [...] But probably the carboot sales are the main supplier of second hand parts. Carboot sales and hackspace."
	50 and over	NA
Why to choose particular materials, materials, components or products		
Project requirement (8)	Under 30 (1F & 3Ms)	F06: "I try to be quite strict about not just picking up stuff that I don't need for the particular projects. Because everything needed for my projects is on the hallway of my flat, and my partner is annoyed by keeping stuff like that. So I picked up something recently from the neighbour's house, and that was like old flooring, because I wanted to turn that into a workbench. And it's like quite thin and quite strong." M03: "What specific thing that I am looking after. The tables, I was looking for a coffee table and I found a coffee table and I thought I could do and it turned out to be big so I adapted it. The workbench I was waiting for three months to find the right bit of material on the freecycle." M07: "it's almost entirely functional. I don't usually tend to think much about how things look. It's more what fits the structure and it has the kind of mechanical properties. I guess the bamboo lamp, I chose it because it's attractive material I like. But ultimately, it's just something I had it in my hands and it fitted the purpose I wanted. With more effort I could make it better but, my materials choice is usually driven by functions. Most of the things I am making are functional items. They are not kind of artistic or decorative. It's mostly something... I want to make something that does something." M11: "It depends on what kind of thing it is. For example, we've got that old CRT monitor that we want to make it into our arcade machine. So we pretty much built the whole thing around that. So, we did have to get the material, wood and so on, to fabricate around it. So, that's the kind of thing we are talking."
	30–49 (1F & 1M)	F05: "it has to be applicable for the goal. So I look at the goal, what do I need?"

(continued)

(continued)

Theme	Age	Participant answers
		M09: "not particularly. The criteria will be like depending on what I want to build. […] and depending on what the material will be used for. […] it's purely case-by-case based."
	50 and over (1F & 1M)	F08: "obviously a lot depends on what I have in stock and what I am aiming to do with the materials at any one time." M12 "when you are choosing electronic components, you choose the components that are appropriate for the job. Each electronic component has got its own set of characteristics so if you are designing something you sort of are roaming through your box of bits and find the most appropriate things. You know things fit with any range values that will handle the amount of power, will handle voltage, things like that. […] it really comes down to when you are reusing the piece of wood. What the wood is going to be useful for, whether it requires the soft wood or hard wood. […] So, depending on what you are doing, whether it is a shelf that you want it to look nice, or whether a shelf in a garage you don't care what it looks like, or a shelf in home, you obviously want it to look nice. If it's the shelf in your workshop, you don't care as long as it's strong, and stands the weight of the tools you are putting on it. It really depends on what you are making, what it is going to be useful for, and whether you are using it more than once. Sometimes, you just make a jig to mill out a particular shape, so you just cut this jig out, clean it all up, wrought it out, use it and throw it away or put it on the fire?"
Potential value (5)	Under 30 (1F & 2Ms)	F01: "Or I sometimes see the potential in things that it might be in a bit stated or a bit of over repaired. (2:53) I can see the potential in it. And yeah, go for it." M11: "sometimes… you might see something and that would give you an idea. So, perhaps something that is somewhat inspiring? That might be the criteria. […] Or something that might have function." M13: "Value is other thing. I knew a refrigerator, I lifted it up to see if it's heavy, because you can tell the type of unit, if it's heavy, then there's refrigerator, that kind of system costs you some hundred pounds to buy. That's a lot of money. […] I see things, compressor, electronics, and I see if it's repairable, is it something that I can pick up and repair? Is it something I can pick up and strip outside off? is it smashed into pieces yet? […] more often someone has already smashed it to take some parts in it. […] so I see, quality, value, and can I clean it? Is it recoverable? I guess. […]"
	30–49 (2Fs)	F05: "And I don't take crap. Sometimes I will go to my way a little bit. I will use lower quality stuff than I normally would if I can fix the quality gap by working on it more?" F07: "I found these (metallic button-shape screws) from an old lamp, and I found them very beautiful, I keep them, and I will use them some way. So anything like that, this is how I see and mentally visualise use examples, then I take it and use it for that purpose."
	50 and over	NA

(continued)

(continued)

Theme	Age	Participant answers
Financial saving (4)	Under 30 (1F & 2Ms)	F01: "we go to car boot sales simply because we're moving a house, we don't have a lot of money for furniture, so we are just going to car boot sales and picking things up for a pound and just making things our own really." M03: "Mostly I tend to find something cheap or free or second hand and build on to it with new materials." M11: "it's usually price-led. So, I am not going to, if I am upcycling, whole point is to make something, repurpose it, I don't want to cost too much, so it would be price-led really."
	30–49 (1M)	M09: "sometimes you've got something in your head, and you don't want to go to the shops, or don't have money for it, then you are looking into a bin for particular items. And that can be a piece of MDF, or old copper tubes, or whatever you want to build something with. It can be like an electronic component. You get an old computer and just extract a fan for a project, something like that."
	50 and over	NA
High quality (4)	Under 30 (1F & 1M)	F02: "I prefer to use things that are solid wood. So I don't tend to pick up anything that's sort of veneered, so anything that's made of cheap board or MDF (medium-density fibreboard) that's got like pretended on the top, so I use it, I just prefer to use anything that's solid which means I can sand it down and paper tape it? Properly. That's it, really." M13: "obviously I check the quality. […] I take it for quality. If the wood is rotten, mould, then I can't clean up. I can dehumidify it."
	30–49	NA
	50 and over (1F & 1M)	F03: "I had a very expensive dress which was given to me by a friend. It didn't fit me. It was velvet. The fabric was nice but the design was bad. [..] the decisions are to do with colour, texture, or the person who wants that." M04: "it has got to be clean, it's got to be reasonable size, and in good condition."
No criteria (4)	Under 30	NA
	30–49 (1F & 1M)	F04—collect things not knowing where to use: "I would never know when I am going to need it or what it might be useful for. So, I collect them with the intention of one day making something with them. But I know what I want to make when I know what I want to make. I will have all the bits then hopefully." M02—start with materials, not the other way around: "I think more of my projects are defined by the materials I have rather than choosing materials for the projects. So solar power charger for example, I was given maybe 15 small solar panels and I needed to use them. I have them, and I don't want them to go to the waste. So I was thinking what can I do, what I can make that I use these and also what other people would like. So I did it as a kits so that other people can make as well at the

(continued)

(continued)

Theme	Age	Participant answers
		hackspace. So more of my process has got this thing: what can I make out of it rather than the other way round."
	50 and over (1F & 1M)	F03—given to me: "I had a very expensive dress which was given to me by a friend." M10—just trying to do varied range of things: "not really. I just try to do varied range of things from a rocking chair to a little bench, little garden benches, and tables. I do quite a range of things. […]"
Something I like (3)	Under 30 (1F & 1M)	F01: "It's just something that I like. Because I am a creative person. I kind of see things I like." M03: "mostly I kind of look at things on a… I guess I find things on an individual basis. I look at something… it's what catches my eyes"
	30–49 (1F)	F07: "I just like pretty things. smallest things really."
	50 and over	NA
Easy to handle (2)	Under 30 (1F & 1M)	F02: "I tend to use wood because it's easily paintable. And I can make it into something completely different. Easy to saw, easy to stick, easy to turn into anything that I want really." M06: "something that I can use without much tools. Something like cardboard or plastic because it's so easy to cut and so easy to fix."
	30–49	NA
	50 and over	NA
Miscellaneous (3)	Under 30	NA
	30–49	NA
	50 and over (2Fs & 1M)	F03—depends on the person who wants it: "the decisions are to do with colour, texture, or the person who wants that." F08—depends on what I have: "obviously a lot depends on what I have in stock" F08—unrecyclable: "I guess I pick up the things which are generally not recyclable with the exception of aluminium and copper which clearly are." M04—relatively unused: "for most purposes, it has to be relatively unused. What you find is, builders, they buy big piece of sheet of plywood and then they cut the big bit off and then the rest of it might be 2 feet wide, could be quite long, but they actually can't use that because it's too small to make any use out of. So they throw it to the skip. As long as it's not covered with rubbish, then I would take it, if I can find it."
What to do with end products		
Use for home or myself (15)	Under 30 (3Fs & 5Ms)	F01: "we save them for in our house. So, at the moment, we kind of upcycle them we leave them in our garage and will go into our home." F02: "a lot are in my house. So I usually pick up stuff that we tend to need" F06: "all pretty much functional stuff. Coz I made like furniture, and storage and stuff like that. So yes, kind of stuff that I can put it in my flat for particular purpose."

(continued)

(continued)

Theme	Age	Participant answers
		M01: "okay, so, the trellising is the... it goes on the fence [...] And the patio, obviously is for dining in the summer." M06: "they are not good enough to give to someone. They are primarily for my own use." M07: "these are all for my own usage." M11: "and our arcade machine for example that is sort of attraction for this space here. Sort of a piece of central piece. Equally, if you did that kind of thing at home, then just for fun, I suppose. Just entertainment, I think for other times. A friend of mine, he recently got an old stereo system like 1950s one, and took all the bits out of it and redid that with using Raspberry pi and now it's streaming radio system. So, it's that kind of thing. At the end, you might not necessarily have a purpose for it, but it looks good and it has sort of feature you would want to see in your home or vehicle or whatever." M13: "everything I do is... because of attention deficit, I have a real trouble in finishing things, so I always felt like my stuff is not really good enough to give to someone else. [...] I never really thought about giving it to someone really."
	30–49 (2Fs & 2Ms)	F04: "the printed circuit board stuff, I kept some, two of the biggest projects for us." F05: "most of my projects are for myself so far." M02: "things have been done either for myself or for people I know." M09: "I usually use them inside of my house, I use them inside of my workshop, I use them in my day to day life. If I don't use it, then give it to someone or put it back into a bin? Or kick it on the side and try to use some of the parts of it."
	50 and over (1F & 2Ms)	F03: "sometimes it's practical stuff. I built a fence from the old fence wood materials from neighbours." M04: "I use it myself, most often I use it myself." M12: "sometimes I make pieces of furniture for myself. I have at home some lovely Parana pine and it's very scarce these days. If you look at this wood, it's big, thick, long and very heavy, and very strong. And I've been carrying this about for 20 years, because I won't throw it away. It's too good. And I am going to make some cabinets for my Hi-Fi for music and television and things like that. So, I hope to get into that project in the next year or two now. So I save this wood, very precious, it's very nice wood. It was a bed that I made years ago because I wanted a bed to fit in a particular room, particular corner of the room. It didn't have to be that high. It needed a big drawer in it. So I used huge planks of wood and that's it. So I've got that and I keep all sorts of scrap wood. You know this wardrobe here, this tool box, it was a big wardrobe, it was all rickety and falling apart, so it was like either take it apart and re-glue it all or scrip it down or buy a nice new one. And I used all wood in it and made these tool boxes."

(continued)

(continued)

Theme	Age	Participant answers
Give to family or friends (8)	Under 30 (1F & 1M)	F02: "anything that I decide I don't want then send it to family or friends and I do that a lot." M13: "I do jobs for other people. I repair computers and laptops. I recycle stuff for that, like sound card."
	30–49 (2Fs & 1M)	F04: "I've given some pieces to friends. We made a piece for a couple when they got married." F05: "I would give something to somebody if I thought it is relevant for them." M02: "things have been done either for myself or for people I know."
	50 and over (2Fs & 1M)	F03: "prom dress was for my daughter." F08: "the items that I have been making with the data cabling are bangles and bracelets and I am going to sell them for the first time. And I've shown, and given a couple to people as a birthday present." M04: "I usually give it away."
Sell to others (7)	Under 30 (1F & 2Ms)	F01: "So I made a lot of cushions with them and little bunting, and bags, handbags, little purses and stuff. And those things I actually do sell. Do quite a lot of craft shows around Christmas time. Christmas decoration and things. and I've also got a folksy account. Folksy is like a craft-based website where you can sell your own. It's same as etsy. But it's just based in the UK. And I sell all my craft bit on there as well." M07: "I have been wondering about making things for the purpose of selling them because you know, a couple of times, people have seen the things I've made, and then they said, oh, you could sell these on Etsy or whatever. And really that's just I haven't put an effort to investigate how feasible it is." M08: "at the beginning of the project, our aim was to start production here, but we faced some legal issues, like selling the products, because we can't trust the parts, it's like products go to the recycling centre, such as broken kettles, and broken toasters, and we disassembled them, we can't trust them. And we can't say where it comes from. So if we sell it to somebody, like the kettle, then we are the responsible for all the safety issues because we can't say that these parts are from so and so manufacturers. This is a big problem for this project."
	30–49 (2Fs)	F04: "I do sell my upcycled jewellery. […] I use etsy, and I go to certain craft fairs around about." F05: "I do have an etsy shop. But I haven't sold stuff there yet."
	50 and over (1F & 1M)	F08: "the items that I have been making with the data cabling are bangles and bracelets and I am going to sell them for the first time." M10: "I only sold the bowls at the moment. I am not selling the rocky chairs at the moment. If I can find a market for it in this place, I will try do it."

(continued)

(continued)

Theme	Age	Participant answers
For a faire, exhibition or performance (3)	Under 30 (2Ms)	M08: "we haven't sold anything. We did many exhibitions with this kind of project but we didn't sell them. And we decided to stop the project for our financial issues." M11: "our arcade machine for example that is sort of attraction for this space here. Sort of a piece of central piece."
	30–49	NA
	50 and over (1F)	F03: "Black box was for the maker faire in Newcastle. […] I did performance once, […] and I built a big projection set, and at the end of that making, […] it was the combination using old carpets that I saved and recycled some pieces for other things, […] perfection doesn't happen."
Just for fun (2)	Under 30 (2Ms)	M11: "and our arcade machine for example that is sort of attraction for this space here. Sort of a piece of central piece. Equally, if you did that kind of thing at home, then just for fun, I suppose. Just entertainment, I think for other times. A friend of mine, he recently got an old stereo system like 1950s one, and took all the bits out of it and redid that with using Raspberry pi and now it's streaming radio system. So, it's that kind of thing. At the end, you might not necessarily have a purpose for it, but it looks good and it has sort of feature you would want to see in your home or vehicle or whatever." M13: "ya, I usually do things for sort of pleasure in doing it. […] I do it as a hobby, for fun, I do it as I like it. I mean, I did IT support, and I've always been fixing things and making things. And that was all about fixing things and problem solving. And that's all about engineering, it's a problem solving."
	30–49	NA
	50 and over	NA
Part of degree project (1)	Under 30 (1M)	M01: "The raspberry pi project, that's electronics project, umm, it's a prototype for some running some computer software that I am trying to write as part of my degree."
	30–49	NA
	50 and over	NA
When to upcycle		
Anytime that suits me (11)	Under 30 (1F & 4Ms)	F06: "I work full time. So I do it at weekends. Generally during the summer. I don't have much space to work in my flat, so I do it outside the garden, more like at the patio. So I don't get a lot done during the winter really." M03: "usually at the weekends I guess because I work full time. That really depends on what I am doing and when I am doing. It depends on what I doing at the time being." M06: "if I feel like I have the chunk of time, a block of time, and there is no distractions, then I can dedicate my efforts into tinkering, into doing something. If I know that I have something other more important, then I am not even starting, because if I have a little time before I open the computer,

(continued)

(continued)

Theme	Age	Participant answers
		before I open the browser, before I open the previous notes, it's just taking too much time. So I know that I need to have an allocated chunk of time with no distraction. Only then I can start working. […] it's on weekends." M07: "whenever I have free time, really. So, historically, it's mostly weekday evening because weekends are, well I have time to go and socialize. Weekday evening often I find something I do to keep myself entertained, I make things for the fun of making things as much as for what it is I end up with. It's more productive than watching TV it seems." M11: "just when I get spare time."
	30–49 (2Fs & 2Ms)	F05: "it's when I can just find around my work." F07: "during week days." M05: "I work when I am able. I am sort of juggling a family life and so forth." M09: "I usually work in the evening. Mostly after work or weekends."
	50 and over (2Ms)	M04: "any day that suits me really. Coz I am retired, so I can work during the week. Usually I am more busy with the family at weekends. So it's the other way around for me. It's usually weekdays when I am working on things. […] probably more in the afternoon." M10: "when I can, coz I am working in the shop. So it totally depends on when I can get to do something. Probably about a couple of days a week to do it. In the evening after the work."
All the time (3)	Under 30 (2Ms)	M03: "It's something that I've always kind of been used to doing I guess, from my parents kind of told me to be using recycle as much as possible, and I kind of always have." M13: "I am thinking about it every day."
	30–49	NA
	50 and over (1F)	F03: "I realised that I've done it thorough all my life. So it is not specific thing I set time to do, it's I need something, want something, have urge to make something, my first thing is… do I have anything here I can use around me? Building whatever it is. So, it's kind of, I guess I do it all the time. Even when I am cooking, I take something that's left over, and turn it into something else. It's kind of interesting thing to think about. Some of reasons are economic, and some of them are just challenges that I make this happen, without having to go to store and buy things."
When responding to a particular event (2)	Under 30	NA
	30–49	NA
	50 and over (2Fs)	F03: "One was in response to a particular event, by curiosity club team, so we've done something outside, and response to that was how can we make it to the maker faire. So that was the challenge. So that's where we started internal dialogues and dialogues with other people to talk about what can I do for it. and then I made a box, the box was constructed out of plastic I saved from work, I just thought this would be useful,

(continued)

(continued)

Theme	Age	Participant answers
		and plucked it, and it's been sitting around for years, I don't even know what it is, I thought it was packing materials. I put them together with a tape to make it look pretty." F08: "it is a variety of time because I am working and with various people communications work and obviously I am gathering materials at any time from the people I am contact with and that can be through work, through business, or through being on seashore and finding drifting woods and like that foraging and actually doing the work itself it depends on what events are coming up such as exhibitions."
When I am triggered, I feel like it, or there is need (5)	Under 30 (1M)	M03—when triggered by the wanted materials: "some cases it's what I have been meaning to do for ages and I managed to get the materials […] I think it's mostly finding the materials. It's… I don't often go out and buy new bits and pieces. But when I find something that I want to do something, "ohh, have that!" so, that has been the driver for the last a couple."
	30–49 (1F & 2Ms)	F04—when I feel like it: "whenever I feel in the mood for it. It's definitely hobby rather than anything else. So it's just when I got time, and I feel like it. […] because it's quite random when I do it. I don't have any particular time that I spend on it." M02—when I feel like it: "But it's difficult because I have my jobs and also use the Hackspace based on my job. So, I am self-employed. So I don't have anyone telling me exactly what to do. So, when I am interested in something, I will just do it. and it doesn't matter; there's no set time when I have to do these." M03—when there is need: "Some cases, it's need"
	50 and over (1M)	M12—when I feel like it: "I am retired, so I do it anytime. Sometimes 2-3 o'clock in the morning if there isn't any loud work or noise involved with machines or things like that."
Where to upcycle		
At home (not specified) (7)	Under 30 (4Ms)	M06: "I have all the equipment. I am able to make all times. And I have a workspace. And everything is within my reach." M07: "I've been making things since before Makespace existed. So mostly I just do it at home with kind of whatever tools I have. […] until the Makespace existed, it was my home." M11: "either at home" M13: "it's usually at home."
	30–49 (1F & 1M)	F05: "for a long time, I used to do that at home, when I had a flat that I was sharing" M05: "at home"
	50 and over (1M)	M12: "inside my house."
Hackspace or Makerspace (6)	Under 30 (4Ms)	M01: "Almost always, either in Hackspace, here, because of all the tools are here […] when… for the past about a year, so I probably come here maybe twice a month? Something like that? Not very often. I live about an hour's drive from here so commute to come here so I am not here every day like a lot of people."

(continued)

(continued)

Theme	Age	Participant answers
		M03: "I do some stuff down in the Hackspace." M07: "So, until the Makespace existed, it was my home. And it still is to an extent, I still do kind of simple stuff that won't generate much mess at home. But when I am working on something more involved I tend to be here especially just because of the access to tools, the stuff like lathe and cutters, they are incredibly useful." M11: "in Hackspace."
	30–49 (1F)	F05: "So now, I use Hackspace, because now I have a room instead of a flat. So the less room I have, the more I work at the Hackspace."
	50 and over (1M)	M12: "here at Hackspace."
Bedroom, dining room, living room, office room or workshop room at home (5)	Under 30	NA
	30–49 (1F & 2Ms)	F04—living room: "my living room." M03—office room: "my office when I do stuff indoors. And the office used to be a bedroom and I keep all my tools in the clothes cupboard because the shed is less secure." M09—workshop room: "in my house. It's inside of my house. I got a little room that can be used as a bedroom and I converted it into a workshop."
	50 and over (1F & 1M)	F03—dining room: "Sometimes I do it in my dining room." M10—bedroom: "I do have in my bedroom."
Shed or garage (6)	Under 30 (2Ms)	M01: "or at home in my garage, coz it has got a work bench, so either that." M03: "I have a little shed in the back garden."
	30–49	NA
	50 and over (2Fs & 2Ms)	F03: "I have a shed. This shed itself is part of upcycling. I built the shed behind my house. There were loads of doors, old-fashioned sliding doors, and I bought them, my shed is made out of, big portion of it is made out of these doors. And I have sewing machine, table, and bed. I have equipment so that I can play muc and video or play videos or DVDs." F08: "I do have a studio in the garden in my home. I store all my materials there." M04: "I have a workshop because our house has a built-in garage, but we don't keep the car in the garage, we just use it as a workshop." M12: "in my garage"
Studio or workshop (outside home) (3)	Under 30 (1M)	M08: "yes. This is our workshop in London and we have another in Paris. My associate lives in Paris and I live in London. So, he comes here or I go to Paris or we work through Skype."
	30–49 (1F & 1M)	F07: "yes, my sometimes workshop." M05: "at my studio […] I really like it when I am doing it in my studio where I have my tools and sort of things."
	50 and over	NA

(continued)

(continued)

Theme	Age	Participant answers
Patio (2)	Under 30 (1F)	F06: "I don't have much space to work in my flat, so I do it outside the garden, more like at the patio. So I don't get a lot done during the winter really."
	30–49	NA
	50 and over (1F)	F03: "Sometimes I do it in my patio, under sunlight, it's bright and warm."
With whom to upcycle		
Just myself (17)	Under 30 (6Ms)	M01: "always by myself, coz it's sort of… none of my family or friends are kind of interested in that kind of thing." M03: "mostly, I've been doing things on my own. In terms of actually physically doing things, I am doing it on my own." M06: "I am doing this primarily on my own. I wish I could have someone close to me who is sharing similar interest but it's very difficult to find people with similar interest. Of course, I am on the London Hackspace mailing list and Cambridge Makespace mailing list. But these people are… they are busy, they have their own schedules, and it's very difficult to coordinate and find a suitable date." M07: "most of the actual making is by myself. And a number of the projects I have made have been inspired by stuff other people have done. And I got a lot of ideas from reading stuff I found on instructables and various blogs." M11: "has been traditionally more by myself. And now I get involved in this Hackspace, I do a lot more with other people, groups of people, getting involved in doing things together." M13: "it's usually me. I'd love to work with someone who is more experienced in electronics. […] I've always been a hacker, I've always been putting things together. So I've always been looking on the websites or forums. Google you go when you have an idea or questions, electronics or programming or whatever it is."
	30–49 (3Fs & 3Ms)	F04: "nobody else. Just me and him [husband] pretty much." F05: "by myself. I try to involve people but it doesn't usually work." F07: "I work on my own." M02: "Generally, I am probably most productive when I am on my own." M05: "traditionally, on my own" M09: "yes, just by myself."
	50 and over (2Fs & 3Ms)	F03: "I often do it myself." F08: "it tends to be just me." M04: "mostly by myself." M10: "I do all by myself which I like. Nobody can interfere with me or tell me what to do." M12: "if I am home, I work on my own, if I am here at Hackspace, I work with whoever else is around. It depends what the project is and how many people are required."

(continued)

(continued)

Theme	Age	Participant answers
Local experts (6)	Under 30 (2Ms)	M03: "There are a couple of things that I consulted with people about or ask questions about. But mostly I do, I work by myself. […] people at the Hackspace and people on the internet. Mostly." M11: "has been traditionally more by myself. And now I get involved in this Hackspace, I do a lot more with other people, groups of people, getting involved in doing things together."
	30–49 (2Ms)	M02: "Generally, I am probably most productive when I am on my own. But it's… I learn a lot from other people, and I ask people for help on things, and people give me help. Sometimes I help other people, but I can't give you one… generally I work on my own. But I use a lot of other people's resources." M05: "but the project done more recently I had my assistants. So I work with fabric artists for a couple of years to do sort of fabric projects, and also […] end of last year, I brought some assistants making a big Christmas structure. […] I get volunteers to help build things. It's like passing on skills, getting people involved in making process, which I enjoy."
	50 and over (1F & 1M)	F08: "Occasionally if I need to do something much more constructional which I don't have the ability to do myself then I will seek out local artist or craft person to be in partnership with me. And something I'd like to do more of is enhance upcycling business." M12: "if I am home, I work on my own, if I am here at Hackspace, I work with whoever else is around. It depends what the project is and how many people are required."
A partner (3)	Under 30 (2Fs)	F01: "with my boyfriend." F06: "my partner works on projects as well. We don't do a project together but we both work on anything at the same time really."
	30–49 (1F)	F04: "nobody else. Just me and him [husband] pretty much. […] if we are making one of the big PCB (Printed Circuit Board) project, and obviously I can show you the pictures, and send them on to you if you'd like. He is very good at spotting what will look right in different places on the panel, but it was my original idea and I have a lot of input into, I collect all the stuff, I get the ideas on what I want to make, he helps me make them and make them as nice as possible. I love it, it's nice to do things together."
	50 and over	NA
Other family members (2)	Under 30 (1M)	M03—father: "people at the Hackspace and people on the internet. Mostly. Occasionally my dad [that I am asking for consultation]. He has done a lot of things himself."
	30–49	NA
	50 and over (1M)	M04—daughter: "Yes. Occasionally, I do things with my daughter. I've got one daughter who lives with us and other daughter who lives quite close by. So, I do that with them sometimes."

(continued)

(continued)

Theme	Age	Participant answers
Expert friends (2)	Under 30 (1M)	M08: "I initiated by myself but then we [me and my friend] worked together."
	30–49	NA
	50 and over (1F)	F03: "if I hit something and I need something really precise and I have a good friend very good at wood working and she helps me to make a frame to mount 6 ipods. And we used old piece of bed, it was an old piece of oak. We spent a day, sanding."
People online (2)	Under 30 (2Ms)	M03: "Me: what kind of people did you consult? P: people at the Hackspace and people on the internet. Mostly. Occasionally my dad. He has done a lot of things himself." M07: "A couple of things I had, sort of active discussions with people who have more experience than me. Where there is any information that I was lacking, I would talk to people mostly online with whom knew what they are doing. But the majority of everything I've made, made my own."
	30–49	NA
	50 and over	NA
Depends on the project (2)	Under 30	NA
	30–49 (1M)	M02: "totally depends on the project."
	50 and over (1F)	F03: "It depends on what I am doing."

References

Akpomuvie, O. B. (2010). Self-help as a strategy for rural development in Nigeria: A bottom-up approach. *Journal of Alternative Perspectives in the Social Sciences, 2*(1), 88–111.

Ali, N. S., Khairuddin, N. F., & Zainal Abidin, S. (2013). Upcycling: Re-use and recreate functional interior space using waste materials. In *15th International Conference on Engineering and Product Design Education,* Dublin.

Alkire, S., Bebbington, A., Esmail, T., Ostrom, E., Polski, M., Ryan, A., et al. (2001). *Community-driven development* (PRSP Source Book). Washington, DC: World Bank.

Anderson, C. (2012). *Makers: The new industrial revolution*. London: Random House Business.

Barriball, K. L., & While, A. (1994). Collecting data using a semi-structured interview: A discussion paper. *Journal of Advanced Nursing, 19*(2), 328–335.

Bebbington, A., Dharmawan, L., Fahmi, E., & Guggenheim, S. (2004). Village politics, culture and community-driven development: Insights from indonesia. *Progress in Development Studies, 4*(3), 187–205.

Bekin, C., Carrigan, M., Szmigin, I., & Seyfang, G. (2007). Growing sustainable consumption communities: The case of local organic food networks. *International Journal of Sociology and Social policy, 27*(3/4), 120–134.

Binswanger, H., & Aiyar, S. (2003). *Scaling up community-driven development* (policy research working paper No. 3039). Washington, DC.: World Bank.

Binswanger, H., & Nguyen, T. (2005). A step by step guide to scale up community driven development. *African water laws: Plural legislative frameworks for rural water management in Africa workshop,* Johannesburg.

Bowen, G. A. (2005). Local-level stakeholder collaboration: A substantive theory of community-driven development. *Community Development, 36*(2), 73–88.

Braun, V., & Clarke, V. (2006). Using thematic analysis in psychology. *Qualitative Research in Psychology, 3*(2), 77–101.

Ceschin, F. (2012). *The introduction and scaling up of sustainable product-service systems. A new role for strategic design for sustainability* (PhD thesis). Milan: Politecnico di Milano.

Danish, K. W. (1995). International environmental law and the "bottom-up" approach: A review of the desertification convention. *Indiana Journal of Global Legal Studies, 3*(1), 133–176.

Davies, A. R., & Mullin, S. J. (2011). Greening the economy: Interrogating sustainability innovations beyond the mainstream. *Journal of Economic Geography, 11*(5), 793–816.

de Haan, J. H., & Rotmans, J. (2011). Patterns in transitions: Understanding complex chains of change. *Technological Forecasting and Social Change, 78*(1), 90–102.

De Keersmaecker, A., Parmar, V., Kandachar, P., & Baelus, C. (2012). Towards scaling up grassroots innovations in India: A preliminary framework. In *UNESCO Chair International Conference-Technologies for Sustainable Development: A Way to Reduce Poverty?* Lausanne.

Dongier, P., Van Domelen, J., Ostrom, E., Ryan, A., Wakeman, W., Bebbington, A., et al. (2003). *Community driven development* (Poverty Reduction Strategy Paper Sourcebook. Vol. 1.). Washington, DC: World Bank.

El Asmar, J., Ebohon, J. O., & Taki, A. (2012). Bottom-up approach to sustainable urban development in Lebanon: The case of zouk mosbeh. *Sustainable Cities and Society, 2*(1), 37–44.

Ellen MacArthur Foundation. (2015). Circular economy. Retrieved from http://www.ellenmacarthurfoundation.org/circular-economy

EUED. (2015). What is EUED? Retrieved from http://www.eued.ac.uk/whatiseued

Frank, C. (2013). *Living simple, free & happy: How to simplify, declutter your home, reduce stress, debt & waste*. Georgetown, Ontario: Betterway Books.

Fraser, E. D., Dougill, A. J., Mabee, W. E., Reed, M., & McAlpine, P. (2006). Bottom up and top down: Analysis of participatory processes for sustainability indicator identification as a pathway to community empowerment and sustainable environmental management. *Journal of Environmental Management, 78*(2), 114–127.

Fuad-Luke, A. (2013). *Design activism: Beautiful strangeness for a sustainable world.* London: Earthscan.

Geels, F. W. (2011). The multi-level perspective on sustainability transitions: Responses to seven criticisms. *Environmental Innovation and Societal Transitions, 1*(1), 24–40.

Gillespie, S. (2004). *Scaling up community-driven development: A synthesis of experience.* (Social Development Paper No. 69). Washington, DC: World Bank.

Goldsmith, B. (2009). Trash or treasure? Upcycling becomes growing green trend. Retrieved from http://www.reuters.com/article/2009/09/30/us-trends-upcycling-life-idUSTRE58T3HX20090930#gitVsJx9UzRpQuvS.97

Grootaert, C. (2003). *On the relationship between empowerment, social capital and community driven development* (working paper No. 33074). Washington, DC: World Bank.

Gupta, A. K. (1995). People's knowledge for survival: Grassroots innovations for sustainable natural resource management. In *IFAD's International Conference on Hunger & Poverty* (pp 16–23). Brussels.

Gupta, A. K. (2000). Grassroots innovations for survival. *ILEIA Newsletter for Low External Input and Sustainable Agriculture (LEISA), 16*(2), 5–6.

Gupta, A. K., Sinha, R., Koradia, D., Patel, R., Parmar, M., Rohit, P., et al. (2003a). Mobilizing grassroots' technological innovations and traditional knowledge, values and institutions: Articulating social and ethical capital. *Futures, 35*(9), 975–987.

Gupta, D. M., Grandvoinnet, H., & Romani, M. (2003b). *Fostering community-driven development: What role for the state?* (policy research working paper No. 2969). Washington, DC: World Bank.

Gupta, D. M., Grandvoinnet, H., & Romani, M. (2004). State-Community synergies in community-driven development. *Journal of Development Studies, 40*(3), 27–58.

Heeks, R. (2012). IT innovation for the bottom of the pyramid. *Communications of the ACM*, 55 (12), 24–27.

Hiremath, R. B., Kumar, B., Balachandra, P., & Ravindranath, N. (2010). Bottom-up approach for decentralised energy planning: Case study of Tumkur district in India. *Energy Policy, 38*(2), 862–874.

Horwitch, M., & Mulloth, B. (2010). The interlinking of entrepreneurs, grassroots movements, public policy and hubs of innovation: The rise of Cleantech in New York city. *The Journal of High Technology Management Research, 21*(1), 23–30.

Kay, T. (1994). *Reiner pilz. Salvo, 23*, 11–14.

Krishna, A. (2003). *Social capital, community driven development, and empowerment: A short note on concepts and operations.* (working paper No. 33077). Washington, DC: World Bank.

Lang, D. (2013). *Zero to maker: Learn (just enough) to make (just about) anything*. Sebastopol, CA: Maker Media Inc.

Lavery, S. H., Smith, M. L., Esparza, A. A., Hrushow, A., Moore, M., & Reed, D. F. (2005). The community action model: A community-driven model designed to address disparities in health. *American Journal of Public Health, 95*(4), 611–616.

Longhurst, N. (2012). The totnes pound: A grassroots technological niche. In A. Davies (Ed.), *Enterprising communities: Grassroots sustainability innovations* (pp. 163–188). Emerald: Bingley.

Longhurst, N., & Seyfang, G. (2011). Harnessing grassroots innovations: Complementary currencies and sustainability. In *International Conference on Community and Complementary Currencies 2011,* Lyon.

MacDonough, W., & Braungart, M. (2002). *Cradle to cradle: Remaking the way we make things.* New York: North Point Press.

Mansuri, G., & Rao, V. (2004). Community-based and-driven development: A critical review. *The World Bank Research Observer, 19*(1), 1–39.

Manzini, E. (2015). *Design, when everybody designs: An introduction to design for social innovation*. London: The MIT Press.

Mathie, A., & Cunningham, G. (2003). From clients to citizens: Asset-based community development as a strategy for community-driven development. *Development in Practice, 13*(5), 474–486.

Mathie, A., & Cunningham, G. (2005). Who is driving development? Reflections on the transformative potential of asset-based community development. *Canadian Journal of Development Studies, 26*(1), 175–186.

Middlemiss, L., & Parrish, B. D. (2010). Building capacity for low-carbon communities: The role of grassroots initiatives. *Energy Policy, 38*(12), 7559–7566.

Monaghan, A. (2009). Conceptual niche management of grassroots innovation for sustainability: The case of body disposal practices in the UK. *Technological Forecasting and Social Change, 76*(8), 1026–1043.

Nesta. (2015). Top findings from the open dataset of UK makerspaces. Retrieved from http://www.nesta.org.uk/blog/top-findings-open-dataset-uk-makerspaces?utm_source=Nesta+Weekly+Newsletter&utm_campaign=dea994abb8-Nesta_newsletter_29_04_154_28_2015&utm_medium=email&utm_term=0_d17364114d-dea994abb8-180827681

Pastakia, A. (1998). Grassroots ecopreneurs: Change agents for a sustainable society. *Journal of Organizational Change Management, 11*(2), 157–173.

Platteau, J., & Gaspart, F. (2003). The risk of resource misappropriation in community-driven development. *World Development, 31*(10), 1687–1703.

Rayner, S. (2010). How to eat an elephant: A bottom-up approach to climate policy. *Climate Policy, 10*(6), 615–621.

Scott-Cato, M., & Hillier, J. (2010). How could we study climate-related social innovation? applying Deleuzean philosophy to transition towns. *Environmental Politics, 19*(6), 869–887.

Seyfang, G. (2008). *Grassroots innovations in low-carbon housing.* (No. ECM 08-05). Norwich: The Centre for Social and Economic Research on the Global Environment.

Seyfang, G. (2010). Community action for sustainable housing: Building a low-carbon future. *Energy Policy, 38*(12), 7624–7633.

Seyfang, G., Hielscher, S., Hargreaves, T., Martiskainen, M., & Smith, A. (2014). A grassroots sustainable energy niche? Reflections on community energy case studies in the UK. *Environmental Innovation and Societal Transitions, 13*, 21–44.

Seyfang, G., & Smith, A. (2006). *Community action: A neglected site of innovation for sustainable development?* (No. EDM 06-10). Norwich: The Centre for Social and Economic Research on the Global Environment.

Seyfang, G., & Smith, A. (2007). Grassroots innovations for sustainable development: Towards a new research and policy agenda. *Environmental Politics, 16*(4), 584–603.

Sharma, M. P. (2004). *Community-driven development and scaling-up of microfinance services: Case studies from Nepal and India* (No. 178). Washington, DC: International Food Policy Research Institute.

Smith, J. L. (2008). A critical appreciation of the "bottom-up" approach to sustainable water management: Embracing complexity rather than desirability. *Local Environment, 13*(4), 353–366.

Stern, P. (2000). Toward a coherent theory of environmentally significant behavior. *Journal of Social Issues, 56*(3), 407–424.

Sung, K. (2015). A review on upcycling: Current body of literature, knowledge gaps and a way forward. In *17th International Conference on Environmental, Cultural, Economic and Social Sustainability* (Vol. 17, No. 4, pp. 28–40). Venice.

Sung, K., & Cooper, T. (2015). Sarah Turner–Eco-artist and designer through craft-based upcycling. *Craft Research, 6*(1), 113–122.

Sung, K., Cooper, T., & Kettley, S. (2014). Individual upcycling practice: Exploring the possible determinants of upcycling based on a literature review. In *Sustainable Innovation 2014 Conference* (pp. 237–244) Copenhagen.

Sung, K., Cooper, T., & Kettley, S. (2015). An exploratory study on the links between individual upcycling, product attachment and product longevity. In *Product Lifetimes and the Environment 2015 Conference* (pp. 349–356), Nottingham.

Szaky, T. (2014). *Outsmart waste: The modern idea of garbage and how to think our way out of it.* San Francisco, CA: Berrett-Koehler Publisher Inc.

UK Government. (2015). 2010 to 2015 government policy: Greenhouse gas emissions. Retrieved from https://www.gov.uk/government/publications/2010-to-2015-government-policy-greenhouse-gas-emissions/2010-to-2015-government-policy-greenhouse-gas-emissions#issue

UK Hackspace Foundation. (2015). UK hackspace foundation. Retrieved from http://www.hackspace.org.uk/view/Main_Page

van den Bosch, S. J. M. (2010). *Transition experiments: Exploring societal changes towards sustainability* (PhD thesis). Rotterdam: Erasmus University Rotterdam.

Viswanathan, M., Sridharan, S., Ritchie, R., Venugopal, S., & Jung, K. (2012). Marketing interactions in subsistence marketplaces: A bottom-up approach to designing public policy. *Journal of Public Policy & Marketing, 31*(2), 159–177.

Walker, G. (2011). The role for 'community' in carbon governance. *Wiley Interdisciplinary Reviews: Climate Change, 2*(5), 777–782.

Walker, G., Hunter, S., Devine-Wright, P., Evans, B., & Fay, H. (2007). Harnessing community energies: Explaining and evaluating community-based localism in renewable energy policy in the UK. *Global Environmental Politics, 7*(2), 64–82.

Author Biographies

Kyungeun Sung is a Ph.D. candidate in the School of Architecture, Design and the Built Environment, Nottingham Trent University, UK, with support from the RCUK Energy Programme's funding for the Centre for Industrial Energy, Materials and Products. Her research broadly deals with sustainable design while paying close attention to sustainable production and consumption by upcycling. She received her Master's degree in Strategic Product Design at Delft University of Technology in the Netherlands, and Bachelor's degree in Industrial Design at Korea Advanced Institute of Science and Technology in South Korea. She has industrial work experiences in Samsung Electronics, LG Electronics and Enviu; and academic work experience in the Industrial Design & Technology Department, University of Botswana.

Tim Cooper is a Professor of Sustainable Design and Consumption at Nottingham Trent University. His research interests are multidisciplinary, embracing design, consumer behaviour, public policy and environmental ethics. He has a specialist expertise in product longevity and is contributing editor of *Longer Lasting Products* (2010). Tim has managed and contributed to a wide range of research projects, funded by the European Commission, Council of Europe, EPSRC, ESRC, Defra, WRAP and industry and third-sector organisations. He has acted as research evaluator for the Research Council of Norway, Irish Environmental Protection Agency and Belgium Federal Science Policy. He is currently Head of the Sustainable Consumption research group, Co-Director of the Centre for Industrial Energy, Materials and Products and a member of the Product Design academic team.

Sarah Kettley is a Reader in Relational Design within the Product Design subject area at Nottingham Trent University. Her research is concerned with how networks of things can be designed for networks of people. She is a member of the Design for Health and Wellbeing, and the Digital Craft and Embodied Knowledge Research Groups at NTU, and represents the University as a member of the steering group for the international Arcintex research network. She is an elected council member of the Design Research Society, and convenes the tentSIG special interest group in tangible, embedded and networked technologies. She received her doctorate from Edinburgh Napier University in 2008, and is a Fellow of the Higher Education Academy.

Strategic Decisions and Sustainability: Nobody Wanted Heavy Trains, But Is that What We Have Got?

A Case Study of UK Rolling Stock Procurements Over Time: Strategic Decisions in Nationalised and Privatised Contexts

Michael J. King

Abstract

This research investigates the relationship between strategic decisions and the environmental outcomes of those decisions. The motivation for this work began with secondary data showing UK trains getting heavier from 1960 to the present day. Heavy trains mean more fuel consumption (diesel or electric power), more emissions and higher maintenance costs. The specification and procurement of new rolling stock (trains) is a highly complex strategic decision, with multiple stakeholders, long time-frames and typically costing several hundred million Euros and the resulting 30–40 year asset. During the procurements for these new trains it is highly unlikely that anyone said "Make sure they are heavy." So how could this have happened? This article first seeks to verify this apparent increase in weight. Then it seeks to isolate potential contributing factors, specifically whether a privatised or nationalised industry context can help to explain changes over time. To achieve this, the research will look at the characteristics over time for the same class of train—electric multiple units designed for duty as a commuter train. For this class of commuter train, four distinct tranches of rolling stock operating in the UK will be analysed, from the 1970s through to current day. The first two tranches are located within a nationalised rail industry, whereas the second two are post-privatisation. Strategic decisions to specify, procure and build new trains will be assessed in terms of the weight and other characteristics of the vehicles. Some early work will be done to gain insight into the strategic decision process and context that produced these outcomes. Theoretical frameworks drawn upon include the decision making literature, Social Issues in Management literature and insights from Foucault regarding power and influence over the decision or the silences within that decision. The

M.J. King (✉)
School of Environmental Sciences, University of Liverpool,
Jane Herdman Building, Liverpool L69 3GP, UK
e-mail: michael.king@liverpool.ac.uk; michael.j.king@innovys.co.uk

W. Leal Filho (ed.), *Sustainable Development Research at Universities in the United Kingdom*, World Sustainability Series,
DOI 10.1007/978-3-319-47883-8_13

findings provide support for the original chart and apparent increase in the relative weight of trains over time. There also appears support that weight increased in a privatised industry context compared to a nationalised setting. The reasons behind this are explored—with social and political characteristics appearing more important than simple changes in the vehicle formats over time. There are some signs that the most recent batch of trains (yet to enter service) may have addressed weight, as measured by the weight of the carriages. Perhaps lessons have been learned?

Keywords
Sustainability · Strategic decisions · Strategic decision making · Procurement · Climate change · Carbon · Energy · Transport · Trains · Rolling stock · Weight

1 Introduction

The motivation for this work began with a chart produced by Rail Research UK[1] showing UK trains getting heavier from 1960 to the present day. This chart showed high-speed inter-city trains from the 1970s weighing approximately 850–900 kg per seat, whereas Pendolino and Voyager high speed trains, introduced in the mid-2000s are 1050–1100 kg per seat.

Heavy trains should mean more fuel consumption, more emissions and higher maintenance costs. Given the threat of climate change and the need to reduce greenhouse gas emissions this appears to be a set of strategic decisions that have produced questionable outcomes for the environment.

The specification and procurement of new trains is a complex strategic decision: it has large economic value, multiple stakeholders, a decision process that can take several years and assets (trains) with a life span of 30–40 years. When these trains were being specified, procured and built, it would seem highly unlikely that anyone said: "Make sure they are heavy." The chart also shows Japanese trains getting lighter over the same period, providing support for the view that it was not technically inevitable for trains to get heavier, because of the introduction of electric doors, air conditioning and so on. So how has this happened?

The objectives of this research article are:

1. Seek to validate the apparent increase in weight through a quantitative assessment of GB rolling stock over time—looking at weight, number of seats and other characteristics;

[1] http://rruka.org.uk/.

2. Determine if there are differences between the rolling stock procured and built within the nationalised GB railway of the 1970s to early 1990s and the subsequent privatised railway; and
3. Attempt to isolate factors that may have contributed towards any weight increase over time—for example train construction materials;
4. Discuss how these different contexts and other factors could contribute towards positive or negative environmental outcomes, and specifically the potential impact upon the issue of climate change.

2 Review of Literature

The focus here is upon a series of strategic decisions within GB rail that have delivered new rolling stock over time. To help understand this I will look for guidance from the literature regarding:

1. How to understand decisions and strategic decisions in particular; and
2. How organisations produce outcomes in the environmental and social domain beyond a narrow economic definition of their purpose; and
3. How to understand the relationship between the external environment and context of strategic decisions and the resulting outcomes.

The key lessons from the literature are summarised below.

2.1 The Nature of Strategic Decisions

Defining what a "decision is, when it is made, and who makes it have all, at times, turned out to be problematic" (March et al. 1993). Decision making is described as "an unstructured process" (Mintzberg et al. 1976) and even "the idea of a 'decision' can also be elusive" (March 1993).

Strategic decision making is simply that which is "important, in terms of the actions taken, the resources committed, or the precedents set" (Mintzberg et al. 1976). Attempts to describe and model strategic decision making range from a "'rational-comprehensive' approach" (Bourgeois III and Eisenhardt 1988) through to something that is more of a "science of 'muddling through'" (Mintzberg 1973). The former sees "a highly rational, proactive process" (Fredrickson and Mitchell 1984) with "systematic analysis, particularly in the assessment of the costs and benefits of competing proposals" (Mintzberg 1973) within "a goal-driven appraisal and decision-making process" (Flyvbjerg et al. 2003). The latter views organisations as a "coalition" (Cyert and March 1992) with "political incrementalism" (Bourgeois III and Eisenhardt 1988) moving forward towards competing objectives.

A review of the dominant research paradigms (Eisenhardt and Zbaracki 1992) concluded with a compromise between these two extremes and stated that "strategic decision making is best described by an interweaving of both boundedly rational (March et al. 1993) and political processes." This will be the view adopted for this article.

2.2 Organisations and the Environmental Outcomes of Their Decisions

Trains procured and built over time are the outcome of strategic decisions over time. Weight is one characteristic of these trains that has both economic and an environmental implications. Weight influences fuel consumption costs for the train operator, track maintenance costs for the infrastructure operator and possibly ticket prices for the traveller or freight customer. From an environmental perspective weight could impact society at large through greenhouse gas and other emissions.

In complex strategic decisions, such as the specification and procurement of new rolling stock, there are always like to be trade-offs across different potential objectives and outcomes. This article will not examine in detail the specific decisions processes in each procurement to search for these trade-offs in the discourse. However the literature can provide guidance regarding the competing expectations of organisational responsibilities.

It is generally non-controversial that public sector agencies should seek deliver social and environmental outcomes—even if some question their effectiveness at achieving these aims. However, there is a debate regarding the objectives of private sector organisations. The literature around Corporate Social Responsibility can help here. Effectively this body of work is about the increasing expectation of organisations to respond to "social ills" (Margolis and Walsh 2003) faced by society at large.

There is a long history (Bowen 1953) to the arguments regarding the ability and necessity for private sector organisations to take on a broader role in society. One side of this debate recognises that, although business has multiple constituencies, it must maintain a singular focus upon economic returns for investors. This is often characterised by Milton Friedman's statement (1970) that the "social responsibility of business is to increase its profits." However, even authors who argue for this narrow role recognise that 'externalities', such as pollution, create limitations where "the arguments for profit maximisation break down" as the firm imposes "costs on others which are not easily compensated through an appropriate set of prices" (Arrow 1973).

The position beyond this narrowly defined economic domain argues that business has responsibilities to a wider set of stakeholders (Freeman 1984), who are not merely passive observers. Organisations already face "considerable latent conflict of goals" (Cyert and March 1992), even within the economic domain. Converting the "rhetoric of corporate responsibility into meaningful action" (Ackerman 1973) is a learning process and it may be possible to align goals by converting "social needs and problems into profitable business opportunities" (Drucker 1984). Many studies (for example: Waddock and Graves 1997; Griffin and Mahon 1997; Roman et al. 1999) have investigated the relationship between the economic, social and environmental performance. This has been termed the 'business case for sustainability'—measuring Corporate Social Performance (Wood 1991) across a "triple bottom line" (Elkington 1997) of economic, environmental and social expectations and outcomes. The outcomes of these studies are mixed. Some show a positive relationship: companies with 'better' performance on social and/or environmental issues have better financial performance. Others show negative relationships or no relationship at all.

This brief tour of the debate regarding the role of private sector organisations in the response to social and environmental issues illustrates the challenges and provides support that it is at least feasible, although some would question whether it is desirable. The search for the business case does not necessarily help with understanding how trade-offs are managed—leading to demands (Margolis and Walsh 2003) for research that releases the "grip of economic assumptions."

2.3 The Context of Strategic Decisions

The final area that this research will explore in the literature relates to the context in which these decisions take place. The focus of this article is upon the **outcome** of many strategic decisions to specify and procure new trains over time. There will be multiple influences over these strategic decisions that are internal and external to the organisations involved. The stimuli to begin the process to buy a train can come from various sources, but for the purpose of this research it is taken as a given that the decision has already been made to buy new rolling stock. Once the decision is underway, this article is interested in understanding how that specification process and procurement process is designed and operated, in particular, how weight is incorporated—if at all.

If a decision can be considered "less a theory of choice than a theory of attention" (March et al. 1993), then there is a need "to understand which social ills garner attention by which firms" (Margolis and Walsh 2003) and how this attention is influenced. According to Flyvbjerg et al. (2003) strategic decisions, such as the procurement of rolling stock, are "defenceless in the face of power" owing to their scale, value and political visibility. The literature on power is worthy of further exploration.

Foucault (1982) takes a pervasive view of power and says that it is "rooted in the system of social network" and is "a way in which certain actions may structure the field of other possible actions." With this view, individuals are "the vehicles of power, not its points of application" (Gordon 1980) and power can be understood by looking at practices as "embodied in a certain number of local, regional, material institutions." Institutions and practices are a guide to understanding power, rather than its source and location. For example, in a society in which rational ideology is dominant, "organisations gather information and conduct analyses because that is what proper organisations and proper decision-makers do" (March et al. 1993).

In the same way that Foucault saw the physical and social layout of hospitals as a reflection of the underlying power regime that had changed over time—"the new form of the hospital was at once the effect and the support of a new type of gaze" (Gordon 1980)—then the institutional structures supporting the strategic decision to buy and build new rolling stock should also reflect an underlying ideology.

A final key insight to highlight here from the work of Foucault (1979) is that silence is "an integral part of the strategies that underlie and permeate discourses." So when asking how heavy trains could have come about, consideration should also be given to the silences that may have contributed to this outcome. Given that climate change is a commons problem, then who speaks for the commons and was their voice silenced or drowned out?

Weight and other characteristics of the trains are the focal point of this analysis, rather than the actual decision process itself. From Foucault one can expect that these outcomes are the result of **practices**, which reflect different regimes of power, which are also identifiable in different institutions and people. Within the scope of this article it is not possible to study the actual underlying practices in detail, but it is possible to look at the outcomes and seek to draw inferences regarding how these outcomes could have been achieved and what this could say about the underlying practices. In a similar way to Cyert and March's (1992) desire to "study the decisions by studying the process", the aim here is to **study the outcomes, but then begin to ask what process could have produced these outcomes?**

3 Research Method

This research seeks to verify the apparent increase in the weight of UK trains and to isolate certain factors, specifically whether a privatised or nationalised industry context can help to explain any changes over time. To achieve this, a specific class of train operating in GB rail has been selected—electric multiple units (EMUs) designed for duty as a commuter train. This gives a consistent type of train for comparison over time. Freight is not assessed here, as the original chart was focused upon passenger rail. Although, GB rail is a mixed use network, passenger rail is by far the dominant traffic.

Four distinct tranches of UK rolling stock will be analysed, from the 1970s through to current day. The first two tranches are within a nationalised rail industry, whereas the second two are post-privatisation. This structure is depicted in Fig. 1.

The first two tranches shown in Fig. 1 span from the 1970s to 1990s and occur when British Rail was the nationalised operator of the railway—owning and managing the infrastructure, stations and vehicles, as well as operating all services. The second two tranches occur within a privatised context: with a regulated infrastructure owner, some 20+ Operators and various other organisations. Rolling

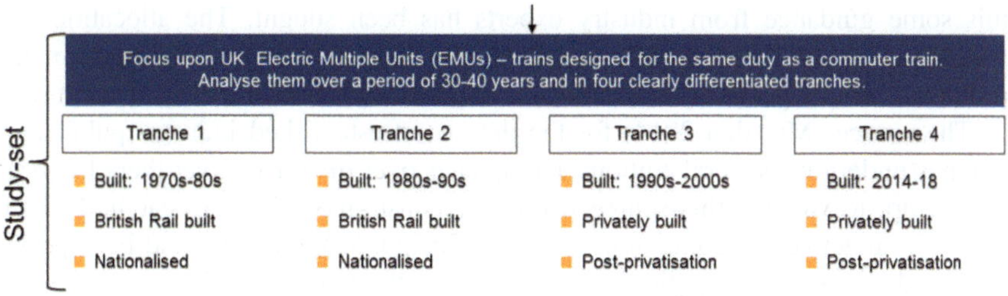

Fig. 1 The study set: four tranches of trains

stock is identified in the UK by its class and sub-class. The mapping of specific vehicle classes to each of the four tranches is shown for reference in Table 1.

Data on railway vehicles has been taken from an established industry source (Marsden 2014). The primary data to be captured is the year of introduction, total vehicle weight and seating capacity—to give a measure of mass per seat over time. Other characteristics of the vehicles to be captured include the construction materials used for the body, the power of the engines and also the maximum speed. Any differences between the tranches will be explored to identify possible explanations in the build and construction of the trains.

This investigation of the data and physical characteristics of the vehicles supports objectives 1–3 of this research, as stated in the introduction. The fourth objective builds upon these findings, and begins to explore the different contexts of these strategic decisions and how these outcomes might have come about.

3.1 Limitations of This Work

The main limitations of this study relate to the data.

The train classes have been allocated to the four tranches shown in Table 1. Tranches 1 and 2 contain trains built during the nationalised railway, whereas tranches 3 and 4 are for trains built after privatisation. This is a simple rule to implement. However the split between trains in tranche 1 versus tranche 2 and trains in tranche 3 versus tranche 4 is not so clear. This is more about judgement and for

Table 1 Rolling stock classes mapped against study tranches

	Classes included within the tranche listed by class identifier
Tranche 1	313/0, 313/1, 314, 315, 317/3, 317/5, 317/6, 317/7, 317/8, 318, 319/0, 319/2, 319/4, 321/3, 321/4
Tranche 2	365, 465/0, 465/1, 465/2, 465/9, 466
Tranche 3	332-4 car, 332-5 car, 333, 334, 357/0, 357/2, 360/0, 360/2, 375/3, 375/6, 375/7, 375/8, 375/9, 376, 377/1, 377/2, 377/3, 377/4
Tranche 4	345, 700/0 (8-car) RLU, 700/1 (12-car) FLU

this some guidance from industry experts has been sought. The allocations are visible in Table 1. In terms of the main results this is not expected to cause major problems, as it is primarily the nationalised-privatised context that is of interest.

The source (Marsden 2014) for the data is an established industry publication that has undergone several editions and updates over time. However there has been no attempt to validate this source within the constraints of this research. There is however an active rail community that uses this publication and could be expected to provide a source of verification and audit for the validity of data.

The measure of kg/seat that is used is this study is not presented as an ideal measure. Far from it, as one industry source has pointed out, the measure does not count standing area, which busy commuter trains often use. However, kg/seat will be kept for this research as the aim is to first validate the original study and explore further once that has been completed.

3.2 Constraints of This Work

This work is restricted to Great Britain Railways. Although some references to Japanese trains are used in this work it must be emphasised that Japanese trains operate in a different industry structure, culture and historical context that impacts the ability to generalise across countries.

This work has focused upon weight, but there is a complex relationship between weight, energy usage and resulting emissions. This relationship varies according to the different characteristics of urban, suburban and intercity services. With higher speeds "energy used overcoming aerodynamic drag becomes predominant" (Eickhoff and Nowell 2010) and so caution must be exercised to go beyond the focus upon weight and draw conclusions regarding energy and emissions. This would need further work.

4 Results and Discussion

Table 2 summarises the analytical results across the four tranches.

Table 2 Summary of the primary data for each tranche

	Total number of train-sets	Seating			Weight			kg/seat		
		Max	Min	Average	Max	Min	Average	Max	Min	Average
Tranche 1	388	319	194	265.4	144.5	102.0	130.3	744.8	400.9	499.6
Tranche 2	230	348	168	301.0	150.9	72.0	127.0	563.1	383.9	427.9
Tranche 3	500	359	174	254.7	214.8	122.4	165.2	1028.7	507.5	667.3
Tranche 4	180	666	400	497.7	410.0	278.0	329.3	750.0	615.6	672.2

The total number of 'train-sets' is shown in Table 2. A trainset consists of different numbers of rail carriages. Most EMUs in this study consist of four carriages, although tranche 4 and the newly ordered trains are 8 and 9-car sets. The carriage types vary depending upon if they contain the engine motor, have first class seating, disabled toilets or numerous other variations. The sets are shown here simply to give a feeling for the sample size and number of vehicles involved in each tranche.

In tranche 1—EMUs built during the 1970s–1980s under British Rail—there are 388 trainsets. The average tranche 1 trainset has 265.4 seats and weighs 130.3 tonnes. The average mass per seat (kg/seat) for a tranche 1 trainset is 499.6 kgs per seat.

Tranche 2 has 230 trainsets with an average of 301 seats per set. So tranche 2 EMUs are providing more seats on average, possibly through longer trains, more 5 car sets for example, or through changed seating layouts. The average weight of Tranche 2 trains has reduced and this combines with more seating to deliver an improved kg/seat figure of 427.9 kg/seat compared to Tranche 1.

Tranche 3 with 500 trainsets has the lowest average seating available of the four tranches. The average tranche 3 trainset has 254.7 seats and on average it weighs 165.2 tonnes. The average kg/seat reflects this and is 667.3 kg/seat for a tranche 3 trainset—more than 50 % heavier than the tranche 2 figure.

Finally, tranche 4 includes 180 trainsets which are currently on order and due to enter service in 2015–17. The data should be treated with some caution until they are formally released. However the increased length of these 8, 9 and 12-car EMUs is visible with a higher average seating of 497.7 seats per trainset, but also reflected in the increased weight of 329.3 tonnes per trainset. The relative measure shows the highest kg/seat of the four tranches at 672.2 kg per seat, but as said above this should be treated with some caution until the trains are formally released into service.

Therefore, with respect to objective 1 of this study: there is support that trains have got heavier over time relative to the number of seats.

This level of analysis is looking at the *average trainset* for each tranche. The next section breaks out the individual trainsets within each tranche. The data showing the mass per seat for each of the specific classes of trains within the tranches are shown in Fig. 2.

The chart shows data for individual sets within each tranche. The leftmost marks in the chart are for tranche 1 trainsets introduced between 1976 and 1989. Tranche 2 trainsets were introduced between 1991 and 1994, Tranche 3 between 1997 and 2004, and finally, Tranche 4 trainsets are to be introduced from 2015 to 2017.

Although the diagram shows variation with each tranche, there does appear to be visual support for an increase in relative weight in recent times—the various trainsets within tranches 3 and 4 appear relatively heavier than those in tranches 1 and 2. **This provides some evidence for objective 2 of this research—there do appear to be differences in the relative weight of trains in the nationalised tranches (1 and 2) compared to the privatised tranches (3 and 4).** We can quantify this further by referring back to Table 2, where we can see that tranche

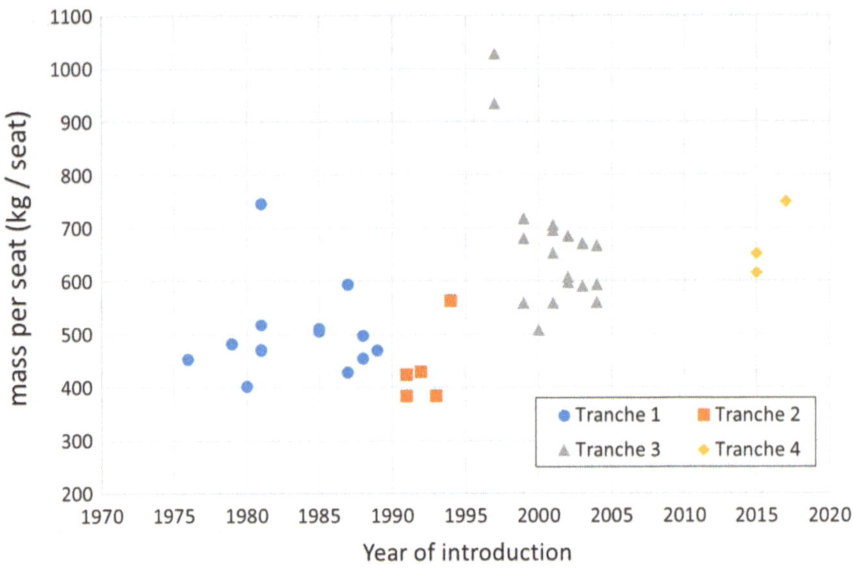

Fig. 2 Mass per seat for the study set of trains grouped by four tranches

Table 3 Weight of the carriage by tranche

	Weight per carriage (tonnes)		
	Max	Min	Average
Tranche 1	36.1	31.9	34.4
Tranche 2	37.7	33.4	34.7
Tranche 3	45.6	38.6	41.3
Tranche 4	34.8	33.3	34.1

1 and 2 trainsets weigh an average of 499.6 kg per seat and 427.9 kg per seat. This compares with 667.3 kg per seat for tranche 3 and 672.2 kg per seat for tranche 4.

Before exploring some vehicle characteristics that might have contributed towards these variations over time, I want to look at the absolute weight of the trains further. As stated above, trainsets consist of different formations: from 2 carriages per trainset through to some of the new tranche 4 trains that have 12 carriages. These carriage all have different layouts, components and materials as well. Table 3 shows the weight per carriage for each tranche i.e. ignoring the number of seats and just looking at the vehicle.

This view of the data shows tranche 4 trains actually with the lowest weight of 34.1 tonnes per carriage on average, with tranche 3 the heaviest trains. This is even more evident in Fig. 3, which shows the weight of individual classes within each tranche.

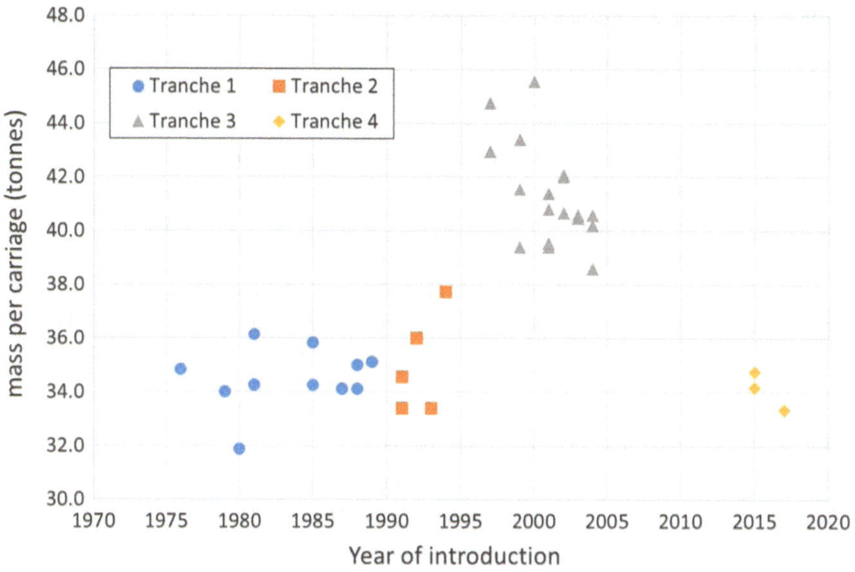

Fig. 3 Mass per carriage by tranche (tonnes)

Tranche 3 stands out as a much heavier group of trains, whereas tranche 4 appears similar to tranches 1 and 2. It is possible that changes in vehicle construction may have contributed to these changes and so this will be explored next.

4.1 Vehicle Characteristics as Possible Factors Contributing Towards Weight

A report by the Rail Safety and Standards Board (RSSB 2008) found that "the breakdown in mass of a typical multiple unit vehicle was presented as being roughly evenly split between five key areas: bogies, body shell, interiors, propulsion equipment and other." It is difficult in the space here to examine all of these areas, but it is possible to look quickly at the body-shell.

With the same source (Marsden 2014) for data we can see from Table 4 that there is an increasing use of aluminium in vehicle construction over time, which should reduce weight compared to steel.

This factor should advantage tranches 3 and 4 relative to the first two and so points towards other factors behind the apparent increase in relative weight.

Propulsion equipment is cited as one of the other key areas for weight by RSSB and this can be assessed to a certain extent. It isn't possible here to isolate the actual weight of the engine or other relevant equipment, but data is available regarding engine power and train speed as a feature of the vehicles. Perhaps increasing weight is needed to support extra speed and power to reduce journey times? The data for this is shown in Table 5.

Table 4 Body-shell construction across the tranches

	Body-shell construction
Tranche 1	• 11 out of 14 sub-classes within the tranche have steel bodies
	• 3 out of 14 sub-classes have steel frames with aluminium bodies
Tranche 2	• 3 out of 6 sub-classes within the tranche have steel bodies
	• 3 out of 6 sub-classes have steel frames with aluminium bodies
Tranche 3	• 4 out of 18 sub-classes within the tranche have steel bodies
	• 4 out of 18 sub-classes have aluminium bodies
	• 10 out of 18 sub-classes have aluminium bodies with steel cabs
Tranche 4	• All 3 sub-classes within this tranche use aluminium for the body shell

Table 5 Engine power and speed across the tranches

	Engine power and speed of tranches
Tranche 1	• Maximum speeds from 121 to 161 km/h
	• Engine power ranges from 657 to 1072 kW
Tranche 2	• Maximum speeds from 121 to 161 km/h
	• Engine power ranges from 657 to 990 kW
Tranche 3	• Maximum speeds from 121 to 161 km/h
	• Engine power ranges from 1000 to 2000 kW
Tranche 4	• Maximum speeds from 145 to 161 km/h
	• Engine power ranges from unknown to 5000 kW

Tranche 3 trains do have more power than tranche 1 and 2, but they don't show higher maximum speeds. It may be that track limitations constrain their top speed, but it is also possible that increased power supports faster acceleration and hence reduced journey times. Although there is limited data for the tranche 4 trains, they do shown a marked increase in power and maximum speed. This could possibly contribute towards any increase in weight for tranche 3 and 4, but would need further investigation to determine this for sure.

Other characteristics of the vehicles that could be different over time include the presence of 'special features'. Air conditioning is an example of this and is something that begins to appear in some of the tranche 3 vehicles, but is absent from tranche 1 and 2. This would add weight, but it is unclear how much. Other technology, such as electric doors could also be expected in more recent trains and not in the earlier ones, however it has not been possible to explore this further within this study.

From this analysis I summarise that newer trains should have benefited from the increased use of aluminium over steel. This may be reflected in the carriage weight of tranche 4 vehicles (Table 3), but does not show in the tranche 3 vehicles, nor does it show in the *relative weight* of tranche 3 and 4 compared to their earlier counterparts. It is possible that increased weight from engines for faster acceleration may be offsetting some of the weight savings from body shell construction, but this cannot be explored further here.

4.2 Context as Possible Factors Contributing Towards Weight

As this work is looking at rolling stock introduced in the 1970s through to modern day, then it is importance to remember that the "social context of meaning within which organisations operate" (March et al. 1993) changes over time—"as ideologies and world views change, organisations change, and vice versa." Tranche 1 and 2 are located in a nationalised industry context, with British Rail as a single publicly owned organisation. Tranche 3 reflects a very significant change with the privatisation of British Rail. Instead of one single organisation running the railway, the industry was split into:

- A single organisation owning and maintaining the physical infrastructure (track, signals, stations, etc.)—initially this was Railtrack, listed on the London Stock Exchange, but later replaced by Network Rail, effectively a public sector organisation;
- Some 20+ Train Operating Companies (TOCs) delivering passenger services against franchise specifications secured through a competitive bidding process;
- Three Rolling Stock Leasing Companies (ROSCOs)—effectively finance houses —owning and maintaining the rolling stock and leasing them to TOCs;
- Other agents included regulators and a range of suppliers, such as train manufacturers.

One of the arguments for privatisation was for improved efficiency and effectiveness through the benefits of specialisation. Organisations, such as Richard Branson's Virgin Group, could bring their expertise in customer service to passenger operations, financial institutions had specialist expertise in long-term asset management and leasing, while Railtrack focused upon infrastructure engineering and maintenance. However, in addition to any benefits of specialisation it replaced a single organisation with a complex industry structure and a "nexus of contracts" (Jensen and Meckling 1976) within and between organisations.

It seems hard to understand why any railway organisation would *want* heavy trains—with increased fuel bills and maintenance costs. This outcome could be interpreted as the result of potentially misaligned incentives. If an organisation can just pass the fuel cost on or they don't pay it all for some reason, then this could be another form of commons problem (Hardin 1968) and could open the possibility to *fixing* this by aligning incentives (Narayanan and Raman 2004) across the supply chain. The challenge of aligning incentives is known to the rail industry. A recent review (RSSB 2008) into technology for reducing mass found that: "In order for there to be take up of any technology that leads to mass reduction there needs to be clear incentive for the developer of the technology."

However, the use of incentives themselves reflects an underlying ideology. The direction of these incentives gives some insight into the underlying system, but, to borrow from Foucault, a lack of consideration could also be described as a silence, which is itself worthy of highlighting. A report (Schofield 2007) investigating the effect of train brake standards upon vehicle mass found “no consideration is given to mass at all.” Train standards are not set in the heat of a high value procurement, rather, they are set over a long period typically through consultation with numerous industry stakeholders. This apparent silence could be an accident or it could reflect a more deliberate process from an underlying system that allocates value and places little or no value upon environmental factors. This is an area for further study.

5 Conclusions

This analysis has provided further support that GB rolling stock has increased in weight relative to the number of seats. This increase in weight does not appear to have been inevitable owing to changes in vehicle composition. There is also some evidence (Network Rail 2009) that modern Japanese rolling stock has lower relative weight—for example the Shinkansen 700 Series weighs 480 kg per seat compared to 1050 kg per seat for the GB Pendolino. This comparison cannot be taken too far, but, if it is assumed that air conditioning, electric doors and so on have been introduced over time in Japan, then this also supports the view that increased weight was not *technically inevitable* because of modern ‘special features’ for train design.

The standout data here appears to be tranche 3 trains. They are markedly heavier in terms of their carriages (Fig. 3) and also relatively heavier per seat (Fig. 2). These are the first trains procured after privatisation with the new industry structure. Tranche 4 data provides some sign of improvement with reduced absolute and relative weight, but the jury must remain out until the trains are in active service.

From the literature review it is argued that it is reasonable to begin with an outcome (heavy trains) and infer underlying practices from this outcome. I do not believe that anyone wanted heavy trains, but that is what has happened. I conclude that **heavy trains reflect an underlying set of practices and social context which appeared to place limited value on weight**.

Therefore, this study concludes that, if weight (and potentially other environmental factors) are desirable outcomes from strategic decisions, such as the procurement of rolling stock, then they need to be adequately represented in the strategic decision making practices and context in which those decisions take place.

6 Further Study

There are several areas of further study highlighted here to provide some insight into how to manage strategic decisions more effectively for environmental outcomes.

One particular area is first to understand if increased weight is actually a problem? Modern trains have regenerative braking, which can capture some of the energy lost in earlier models. This can be explored through quantitative assessment of actual energy efficiency and the involvement of expert engineering domain knowledge.

Strategic decisions frequently involve trade-offs and this is a valuable area for further study. For example, increased engine power to provide greater acceleration and reduce journey times may be a desirable social benefit to encourage more people to travel by train. Similarly disabled access and on board toilets provide a social benefit, but the impact upon seating and weight is not known here. Understanding how strategic decisions, and the society in which these decisions take place, manage trade-offs such as this would be helpful.

Another area for exploration is to look further into the detail of the practices that produced these trains. The procurement of new rolling stock is a highly structured process, which reflects many elements of a "'rational-comprehensive' approach" (Bourgeois III and Eisenhardt 1988). These decisions are governed by European Union tendering processes, which ostensibly seek to open up competition and provide transparency of decision making. This is all part of the institutional infrastructure and "regime of practices" (Foucault et al. 1991). Environmental considerations are typically a standard part of these complex procurements and are often give 5–10 % of the scoring criteria—but this still has produced heavy trains. Studying the documentation and actors involved in these processes is a useful area for further study. This should also investigate the role of silence and how some actors or voices are excluded from the process.

Although it still remains unlikely that anyone said 'make sure they are heavy', it is no solace if this outcome was unintended. Strategic Decisions have important repercussions for environmental and social outcomes. We must do better.

References

Ackerman, R. W. (1973). How companies respond to social demands. *Harvard Business Review, 51*, 88–98.

Arrow, K. J. (1973). Social responsibility and economic efficiency. *Public Policy, 21*, 303–318.

Bourgeois, L. J., III, & Eisenhardt, K. M. (1988). Strategic decision process in high velocity environments: Four cases in the microcomputer industry. *Management Science, 34*, 816–835.

Bowen, H. R. (1953). *Social responsibilities of the businessman*. New York, NY: Harper.

Cyert, R. M., & March, J. G. (1992). *A behavioral theory of the firm* (2nd ed.). Oxford: Blackwell Business.

Drucker, P. F. (1984). The new meaning of corporate social responsibility. *California Management Review, 26*, 53.

Eickhoff, B., & Nowell, R. (2010). *Determining the benefit of train mass reduction—T712*. London: RSSB.

Eisenhardt, K. M., & Zbaracki, M. J. (1992). Strategic decision making. *Strategic Management Journal, 13*, 17.

Elkington, J. (1997). Cannibals with forks. *The Triple Bottom Line of 21st Century Business*. Oxford: Capstone Publishing Limited.

Flyvbjerg, B., Rothengatter, W., & Bruzelius, N. (2003). *Megaprojects and risk: an anatomy of ambition*. Cambridge: Cambridge University Press

Foucault, M. (1982). The subject and power. *Critical Inquiry, 4*, 777.

Foucault, M. (1979). *The history of sexuality [by] Michel Foucault; translated from the French by Robert Hurley*. London: Lane.

Foucault, M., Burchell, G., Gordon, C., & Miller, P. (1991). *The Foucault effect: Studies in governmentality; with two lectures by and an interview with Michel Foucault*. Chicago: University of Chicago Press.

Fredrickson, J. W., & Mitchell, T. R. (1984). Strategic decision processes: Comprehensiveness and performance in an industry with an unstable environment. *Academy of Management Journal, 27*, 399–423.

Freeman, R. E. (1984). *Strategic management: A stakeholder approach*. London: Pitman.

Friedman, M. (1970). The social responsibility of business is to increase profits. *New York Times*.

Gordon, C. (Ed.). (1980). *Power/knowledge: Selected interviews and other writings 1972–77*. Brighton: Harvester.

Griffin, J. J., & Mahon, J. F. (1997). The corporate social performance and corporate financial performance debate: Twenty-five years of incomparable research. *Business and Society, 36*, 5.

Hardin, G. (1968). The tragedy of the commons. *Science, 162*, 1243–1248.

Jensen, M. C., & Meckling, W. H. (1976). Theory of the firm: Managerial behavior agency costs and ownership structure. *Journal of Financial Economics, 3*, 305–360.

March, J. G., Simon, H. A., & Guetzkow, H. (1993) *Organizations* (2nd ed.). Cambridge Massachusetts USA: Blackwell Publishers.

Margolis, J. D., & Walsh, J. P. (2003). Misery loves companies: Rethinking social initiatives by business. *Administrative Science Quarterly, 48*, 268–305.

Marsden, C. J. (2014). *ABC traction recognition*. Birmingham: Ian Allan Publishing.

Mintzberg, H. (1973). Strategy-making in three modes. *California Management Review, 16*, 44–53.

Mintzberg, H., Raisinghani, D., Theoret, A. (1976). The Structure of "unstructured" decision processes. *Administrative Science Quarterly, 21*, 246–275.

Narayanan, V. G., & Raman, A. (2004). Aligning incentives in supply chains. *Harvard Business Review, 82*, 94–102.

Network Rail. (2009). *Comparing environmental impact of conventional and high speed rail*. Network Rail

Roman, R. M., Hayibor, S., Agle, B. R. (1999) The relationship between social and financial performance. *Business and Society, 38*, 109.

RSSB. (2008). *Technology transfer for reducing mass*. London: RSSB.

Schofield, K. (2007). *Investigating how the requirements for train brake systems affect vehicle mass: Review of Standards*. London: Rail Safety & Standards Board.

Waddock, S. A., & Graves, S. B. (1997). The corporate social performance-financial performance link. *Strategic Management Journal, 18*, 303.

Wood, D. J. (1991). Social issues in management: Theory and research in corporate social performance. *Journal of Management, 17*, 383.

Sustainable Communities: University-Community Partnership Research on Social Dimensions of Sustainable Development

Carolyn Kagan, Rebecca Lawthom, Ornette Clennon, Jenny Fisher, John Diamond and Katy Goldstraw

Abstract

The aim of this paper is to examine the role that university research can play in sustainable development at the level of community. The methodological approach is action research undertaken in collaboration with voluntary and community organisations, addressing their needs and including high levels of participation. An outline of the nature of sustainable communities and engaged university research, is followed by case examples of five collaborative research projects, each generating different types of qualitative data, which inform sustainable development of communities in diverse ways. The studies indicate that university-community partnerships can provide understanding of the challenges facing people in communities; encourage innovative local action for sustainability; and contribute to policy development at different levels.

C. Kagan (✉) · R. Lawthom · O. Clennon · J. Fisher
Social Change and Community Wellbeing Research Group,
Manchester Metropolitan University, Birley Campus, Manchester M15 6GX, UK
e-mail: c.kagan@mmu.ac.uk

R. Lawthom
e-mail: r.lawthom@mmu.ac.uk

O. Clennon
e-mail: o.clennon@gmail.com

J. Fisher
e-mail: j.fisher@mmu.ac.uk

J. Diamond · K. Goldstraw
Institute for Public Policy and Professional Practice, Edge Hill University,
Ormskirk, Lancashire L39 4QP, UK
e-mail: diamond@edgehill.ac.uk

K. Goldstraw
e-mail: goldstrk@edgehill.ac.uk

W. Leal Filho (ed.), *Sustainable Development Research at Universities in the United Kingdom*, World Sustainability Series,
DOI 10.1007/978-3-319-47883-8_14

They do this via the creation of ecological 'edges'. The challenges facing universities doing this kind of research are highlighted.

Keywords
Sustainable communities · Inclusion · Governance · Fairness · Participation · Action research · University-community partnerships

1 Introduction

Economic, social and environmental factors, taken in their cultural context constitute the three pillars of sustainability. In this paper we consider the role of universities in contributing to the socio economic elements of sustainability via their community partnership research, framed in particular, by the concept of sustainable *communities*.

We will draw on five examples of our own research praxis to understand, promote and transform social aspects of sustainability at community level—sustainable communities. A focus on sustainable communities is a crucial part of wider sustainable development, as it is only by problematising neoliberalism and its negative structural effects on *communities* (Coburn 2004) that the vision of a sustainable future can be fully realised. The implications of our research for university-community research partnerships for sustainable development at community level will be discussed in terms of creating and working at the ecological 'edge'.

HEFCE, the higher education funding body, has a strategy for sustainable development, applied to all aspects of university activity, including research (HEFCE 2014). It is worth noting that although the HEFCE strategy is for sustainable development, in their vision they refer to sustainability. This is an important distinction, as one of the things that university thinking and practice can do is to problematise the very notion of sustainable **development**. In terms of sustainable communities the process of *development* (as in community development; community organising and so on) remains important. However, when we set sustainable communities alongside those economic and environmental elements of sustainability, a different picture emerges. *Development*, and all that it implies in terms of advanced capitalism, which by its very nature depends on the extraction and allocation of limited natural resources, thereby jeopardising the natural environment, has to be challenged. Indeed the role of university based, intellectual endeavour in foregrounding the socio economic interdependence between neoliberalism (the role of the markets), consumerism (market agency) and sustainability (the relationship between the agent, the market and resource allocation) is gaining ground (see for example the collection by D'Alisa et al. 2014).

2 Sustainable Communities

In the UK, the concept of sustainable communities preceded the 2008 economic crash and emerged from a Government sponsored Sustainable Development Commission. It coincided with the growth of concern for environmental degradation, climate change and the need to reduce carbon emissions and was precipitated by the recognition of a housing crisis which threatened the viability of neighbourhoods (Power 2004). Sustainable communities are those which

> meet the diverse needs of existing and future residents, their children and other users, contribute to a high quality of life and provide opportunity and choice. They achieve this in ways that make effective use of natural resources, enhance the environment, promote social cohesion and inclusion and strengthen economic prosperity. (Egan 2004: 18)

Whilst all communities differ in terms of their specific circumstance in time and place, sustainable communities are places that embody the principles of sustainable development insofar as they

- *Balance and integrate the social, economic and environmental components of their community*
- *Meet the needs of existing and future generations*
- *Respect the needs of other communities in the wider region or internationally also to make their communities sustainable* (Geographical Association 2015).

Sustainable communities are places where people want to live and work, now and in the future, meeting the diverse needs of existing and future residents within the wider context of economic and environmental security. They are safe, inclusive and cohesive, strong in social capital and offering opportunities for participation in decisions and governance; they enable human flourishing and wellbeing, are well served, well connected and fair for everyone; they have strong community and voluntary associations and are knowledgeable about and sensitive to protection of the environment (see Coote 2015).

Egan (2004: 19) summed up the key dimensions of sustainable communities in a diagram, adapted in Fig. 1.

The social and cultural dimensions envision sustainable communities that support: a sense of community identity and belonging; tolerance, respect and engagement with people from different cultures, background and beliefs; friendly, co-operative and helpful behaviour in neighbourhoods; and social inclusion and good life chances for all.

Good governance leads to sustainable communities that are well run and enjoy: representative, accountable governance systems which enable inclusive, active and effective participation; effective engagement with the community at neighbourhood level, including capacity building to develop the community's skills, knowledge and confidence; strong, informed and effective partnerships; a strong, inclusive, community and voluntary sector; and sense of civic values, responsibility and pride.

Fig. 1 Dimensions of sustainable communities (adapted from Egan 2004)

Well served sustainable communities enable people to reach their potential through: a good range of accessible, affordable, integrated and high quality public, community, voluntary and private services; service providers who think and act long-term and beyond their own immediate geographical and interest boundaries, and who involve users and local residents in shaping and co-producing their policies and practice.

A flourishing and diverse local economy leads to thriving sustainable communities featuring: a wide range of jobs and training opportunities; local work opportunities that offer opportunities for life-long learning; dynamic social enterprise and business creation, with benefits for the local community—a focus on people not profit; a strong business community with links into the wider economy; and economically viable and attractive town centres.

The interdepartmental, coordinated policy arena of sustainable communities has gone off the political boil in the UK. Support for communities is the responsibility of the Department of Communities and Local Government, whilst the sustainable development agenda lies with the Department for Environment, Food and Rural Affairs, with a focus almost exclusively on the environment. Despite this fragmentation, the dimensions of sustainable communities remain an important cluster of priorities for the wider sustainability agenda and Bichard (2014), for example, illustrates the ways in which some communities have made progress towards sustainability, of alternative ways of living and co-operating through the building of community capacity and nurturance of the environment. It is at the community level of sustainability that many of the actions needed for sustainable futures will be implemented.

The research we report here talks to social justice and fairness, as well as to the social and cultural, governance, services, and economic dimensions of sustainable communities. Each of the examples is also an example of university-community partnership research.

3 The Research Approach

The research we are reporting is collaborative, born of strong university-community partnerships.

3.1 University Community Partnership Research

Formal and informal partnerships between universities and the community and voluntary sector generally falls under the umbrella of public engagement, a broad set of activities characteristic of an engaged university (NCCPE 2015).

These partnerships range from local, specific partnerships to inter-agency strategic partnerships, to networked partnerships linking projects or agencies (Kagan and Duggan 2009). The key features of our partnerships are that they are characterised by:

- Being values led
- Starting with the concerns of the community or voluntary organisation
- Highlighting the identification of assets and capacity building
- Achieving reciprocity and attention to power issues
- Ensuring participation, inclusion and engagement
- Adopting a systems approach that reflects a multi-layered understanding to change.

These partnerships can help communities move towards more sustainable futures, become more resilient and enhance the wellbeing of those who work and live in them.[1]

[1]We recognise many different kinds of communities. In this context we are talking of communities of place—geographical areas with which people identify and have a sense of belonging (see Kagan et al. 2011a for further discussion of the concept of community).

Table 1 Research needs, data and analysis of case examples

	Case example	Research need of community partner	Type of data collected	Method of analysis
1	Forced labour and Chinese migrant workers	To understand the drivers and consequences of forced labour in order to provide appropriate services to undocumented workers	Interview accounts (conducted in Chinese and translated)	Thematic and narrative analyses
2	Resilience and disabled people	To understand and work with disabled people to build resilience and support inclusion	Personal accounts; observation; interviews; focus group	Life story analysis; thematic analysis; toolkit testing
3	Capacity building for sustainable communities	District wide concern about how to enhance participation in governance and build capacity for participation and inclusion	Policy analysis; ethnographic data	Case study
4	Sustainable African Diaspora enterprises	To map the challenges facing African Diaspora community organisations and develop capacity for enterprise development	Ethnographic participant observation; document records; meetings; interviews	Organisational case studies
5	Evaluation of volunteers supporting vulnerable families	To understand the impact of volunteering expertise and time to enable stronger community cohesion and more appropriate services	Participant observation; focus groups telephone and face to face interviews, questionnaire	Thematic analysis based on objectives of the organisation

3.2 Action Research and the Identification of Research Needs

The research approach in all the case examples was action research, with high levels of participation where possible (Kagan et al. 2008). Action research is a process and methodological approach rather than a methodology per se. Each of our examples addresses a need identified by our community partners for research which will enhance their sustainability journeys, and different kinds of qualitative data were collected and analysed through a variety of methods. Table 1 summarises the research undertaken.

The qualitative approach was not exploratory, but rather a way of tapping into deep meaning for participants, and dealing with complexity in their lives.

4 Case Examples

The research we are reporting is collaborative, underpinned by strong ethics of partnership working, reciprocity, stewardship and a commitment to sustainable development. Our community partners are actively trying to build communities that support human flourishing. The case studies offered here illustrate just some of the possibilities for such research.

4.1 Forced Labour and Migrant Chinese Workers

Globalisation has led to an increase in migratory flows as people in areas of poverty and worklessness seek work away from home, in order to support their families. This has led to a complex web of workers, travel facilitators and people traffickers, gangmasters and employers, in which migrant workers can be caught up in situations of forced labour and vulnerable work in communities in the host country. This link from the global to the local undermines the sustainability of communities by putting strain on local employment opportunities, weakening community cohesion and threatening people's sense of belonging and identity.

We worked with a local voluntary organisation that offered services to and supported Chinese people living in the North West of England: they had noticed an increase in undocumented workers, who spoke little English and were often living and working in very vulnerable situations. They had no access to public services and were unable to exercise employment rights to decent working conditions. We were commissioned by a social policy funding body to use a co-researcher approach to design, implement and analyse research data in the UK (Kagan et al. 2011b; Lawthom et al. 2015). We collected accounts of Chinese migrant workers' experiences of travelling to the UK, often by circuitous routes and usually entering the UK without relevant papers (although some had come on some kind of visa which they then overstayed); of finding work and of working conditions. We explored the role that family, either in China or in the UK had played in the decisions people made. It was clear that people made active decisions to travel and to stay in particular jobs or not, linked closely to their responsibilities to their families. They were working in precarious situations, paid well below the minimum wage, with long working hours, no holidays or sick pay and frequently bullying in the workplace. They were unable to participate in their local communities due to little leisure time, lack of speaking English and a lack of confidence due to their unauthorised status. Most of the people we spoke to had made applications to remain the UK, either through the asylum system (which they did not understand) or through other immigration channels. Our study was one in a programme of studies that informed the development of new legislation, the Modern Slavery Act 2015.

The knowledge gained from the study addressed sustainable communities in a number of ways. It:

- strengthened the capacity of the voluntary organisation we collaborated with to develop services (such as English classes) to support migrants and help them participate in their local communities and gain a sense of belonging in their new countries;
- exposed the workings of an 'alternative' economy in which employers exploited migrant workers, reducing the availability of decent jobs and thereby economic viability;

- revealed some of the ways in which global labour chains and precarious status weakens and undermines sustainable communities in both communities of origin and host communities and thus the unfairness of the migratory labour system.

4.2 Disabled People and Resilience Across the Life Course

In times of increasing austerity in the United Kingdom, the underpinning rhetoric is often given a sustainability angle, in that metaphors of balancing and resilience are drawn upon to justify cuts. Within this austere climate, marginalised groups are often more vulnerable and 'at risk' from the impact of cuts and this in turn stands to threaten the stability of cohesion within communities. A leading disability charity was keen to investigate how disabled people demonstrated resilience across the life course, and how best to build resilience amongst different groups of disabled people. This led to a partnership between researchers from a university in the north-west of the UK and the charity (https://disabilityresilience.wordpress.com) to carry out the relevant research (Runswick-Cole and Goodley 2013). In our definition (in line with Ungar 2011) we positioned the resilience of disabled people as being linked to sustainability, both of relationships and communities, and not the individual traits or coping skills of individuals. Rather than seeing resilience as being a property of individuals, it is, instead, derived from the *networks* of material resource, relationships with people and participation in communities.

The research consisted of phases including a life story approach, a 'community of practice' analysis and the development of a toolkit. The partnership between the university and the charity was further enhanced by a reference group of disabled people who participated in and advised the project.

The project yielded rich information which made justice and resource distribution key to understanding disability and resilience. The complex relationships between resource allocation, power and identity of disabled people were illustrated by the life stories. They showed that:

- networks afforded disabled people are inextricably linked to welfare benefits, accessibility of transport and social systems;
- health and social care systems were positioned as sites of struggle for resources that were needed to create resilience (or not);
- advocacy and social justice were reference points for disabled people (across the life course) to become and remain members of the community; and
- resilience is a relational, social, community and networked phenomenon which requires resources, and coordinated services to develop and support networks that respond to community members' needs, at different points in life.

Through the research we developed a community of practice consisting of disabled people, academics, practitioners, young people and parents/carers to both generate new ways of thinking and in itself help build resilience. This co-researcher approach creatively engaged and shared knowledge, and underlined the value of peer support and the importance of place in building networks of resilience. As part of the research we developed a participative and accessible toolkit which was taken up by the organization to use with their membership and stakeholder groups.

The knowledge gained from the study addressed sustainable communities in a number of ways. It:

- strengthened the capacity of the partner voluntary organisation to work with disabled people to co-produce knowledge and to develop good health and social care services supporting the development of networks which form the foundation of resilience and participation;
- provided an example of good practice in strengthening the social and cultural inclusion of disabled people in everyday life and exposed some of the obstacles to inclusion at different life stages;
- contributed to capacity building and new practices, enabling disabled people to make active and positive contributions, strengthening participation, respect and community cohesion.

4.3 Capacity Building for Sustainable Communities

Universities are well placed to examine the impact of policy on communities and to communicate lessons from this. They are also there for the long haul: able to pick up threads from research findings and apply them to new situations of different communities. Community development, renewal, and regeneration have featured in public policy in different forms for decades. Diamond (2004) showed how whilst the language might change, the underlying conceptual thinking and social goals remain remarkably similar and the gaps between the policy goals (for localisim, partnership working and sustainable development) and the reality on the ground are shared. He compared approaches to neighbourhood regeneration in two large cities in the UK and found that the ways in which regeneration partnership schemes operated meant that local people were defined as 'dependent' and that local agencies tended to marginalise alternative views. Furthermore, local partnerships, dominated by the local authorities, sought to co-opt local activists and individualise, rather than collectivise the experience of local communities. Diamond was able to identify ways in which local people could participate in governance of their communities in more meaningful ways.

An alliance was formed, some years later, with community residents of another local authority. There, a programme of participatory action research was agreed, with the social justice goal, to enable action; and the social justice functions of attending to power relationships, being non-extractive fully collaborative, and involving research participants in the research process (Goldstraw et al. 2015: 9). In this research, university researchers working with community residents have introduced a programme of capacity building and community leadership, to enable residents to act as advocates, mentors or buddies to other residents experiencing difficulties (Diamond 2012). The training has at its core reflective thinking, and those residents who develop the interest and confidence to do so, go on to undertake small community based pieces of research, supported by university researchers. The kinds of research projects they have undertaken include: the recovery of social and cultural memory via a social history project about poverty in the area; experiences of residents in receipt of welfare benefits, culminating in a radio play, giving voice to those who are usually silenced; research into the advocacy work of the community group and the effect of volunteering on both volunteers and 'clients'; and the development of a piece of community drama about the researched experiences of, and involving older people living in the area in receipt of benefits. Taken as a whole, this programme of research includes policy critique and action research in the area of capacity building for community involvement in governance and the more effective delivery of services.

The knowledge gained from the study addressed sustainable communities in a number of ways. It:

- enhanced understanding of, and ways of developing community involvement in governance;
- encouraged effective engagement with community residents, building their capacity for community leadership incorporating skills of enquiry and critical reflection, advocacy and research and enhancing their skills, knowledge and confidence;
- facilitated a strong sense of community and belonging, tolerance, respect and co-operative behaviour in neighbourhoods.

4.4 Sustainable African Diaspora Community Enterprises

We have been engaged in an action research project involving local grassroots black and ethnic minority communities, who have been shown to be disproportionately affected by welfare reforms and cuts to local services (Khan 2015). An extensive programme of consultation and support, building relationships with local African Diaspora community groups has evolved. One part of this process was the support we were able to give to a local campaign to save an African Caribbean community centre, which was ultimately unsuccessful. Nevertheless, during the campaign we were able to galvanise local communities to think about the services and sustainable enterprises they wanted in the building and to develop a network of local agencies. What this stage of the research revealed, was the structural challenges that African Diaspora communities (in particular) face in protecting and sustaining their local community assets.

As a result of our partnership building during the campaign, we worked with local and voluntary sector support and capacity building organisations, one of which commissioned wider, national research into community asset mapping and the black and ethnic minority communities. We held a reception in the university for the dissemination of the research, which affirmed the widespread structural challenges facing African Diaspora groups and the "insecurities facing many BAME-led organisations trying to safeguard community assets" (Field et al. 2015, p. 6). These challenges include trying to secure grants/council funding, trying to win council or public sector contracts via local authority micro commissioning and also trying to win contracts for council asset transfers.

A new partnership was formed with the London based group that conducted the research, in order to co-facilitate a newly formed BAME enterprise forum to examine these issues of sustainability more locally in Manchester, especially in the context of major regional policy developments linked to regional devolution (known popularly as Devo Manc and the Northern Powerhouse initiative). Members of the forum are engaged in various forms of action, which we are following, reporting and informing subsequent actions. Here are two brief case examples of the forum support that we are giving to community enterprises that are specifically engaging with the sustainability agenda, as outlined earlier.

Waste Not is a small Ghanaian-owned community recycling project, which collects donations of children's 'pre-loved' and unwanted new or used items, such as pens, pencils, clothes, toys, calculators, books. The company sends the donated items to support young mothers and families who need them. The company also sends its donated items to Ghana to support nursery and primary school libraries, and contributes to humanitarian agencies that support refugee camps and orphanages. Our partnership work with *Waste Not* has helped identify their capacity building needs and helped them think about the most effective governance structure for their future progress.

Project Hermes is a community wifi initiative run by the *The Mbari Group,* an activist collective. They initiate and support projects that explore and address social and cultural equity. They use the arts, community building, history, politics, the environment, economics and technology in their practice. They are working on building a free wifi access to a local housing estate. We are currently exploring the possibility of using wifi extenders to extend our university's public wifi access to the housing estate. We are also working with a Telecoms provider who has had previous experience of the benefits of providing free wifi for other local deprived communities (Dawood 2013).

Through this multi-dimensional action research process we are informing sustainable communities by:

- enhanced understanding of the structural barriers to effective participation and inclusion by marginalised African Diaspora groups;
- building on and growing partnerships bringing expertise to contribute to the capacity building of local community social enterprises;
- facilitated a strong sense of community and belonging, tolerance, respect and co-operative behaviour.

4.5 Evaluation of a Project Supporting Marginalised and Vulnerable Families

A multi-professional team of researchers from the university worked in partnership with a local voluntary organisation, Home Start, to evaluate their services across a number of urban areas (O'Neill et al. 2014). The aim was to identify ways in which they could improve their support for families and demonstrate their impact for funders.

The Home Start model is centred on targeted, local volunteer support for families experiencing difficulties in the UK and in other countries too. Isolated, struggling parents, trying to do their best, often living poverty, severed from extended family support, find it difficult to participate fully in their communities, and all family members are under stress and fall short of realising their potential. Families are referred to the service by professionals including health visitors, teachers, social workers and can also self-refer. Volunteers, who themselves have been parents, are sought from the local community and are matched with families. They undertake training including child development, signposting to services and child protection (or safeguarding). They then provide whatever help and support is needed in the domestic space for one of two hours per week.

We worked in close partnership with the organisation to develop the most appropriate research design, reaching all stakeholders. We undertook focus groups with volunteers, participant observation at meetings and training events, interviews with families and trustees, as well as an online questionnaire for referrers. At the heart of the research was the parents' and volunteers' experiences of being involved. It was clear that parents found the volunteers' support beneficial and it enabled them to cope with difficult times: it helped them increase their confidence and find a renewed sense of purpose. The volunteers told us about how their views of struggling families had changed and they were now able to advocate for families experiencing difficulties. They provided a range of social and practical support for the parents, and, crucially, as they were neither a professional nor a friend were seen as people they could trust and talk with openly. The research was funded by Manchester Metropolitan University and has been extended, with a PhD studentship, to examine the family support model, nationally, in the context of austerity.

In undertaking the evaluation, we addressed sustainable communities in these ways, to:

- Strengthen the capacity of the voluntary organisation's ability to provide appropriate and inclusive support to local families;
- Show how sharing of time and expertise and activities could reduce stigmatisation and build local social capital;
- Demonstrate the role the university could play, as a community anchor organisation, in strengthening the voluntary sector.

5 Discussion

These projects have worked in one way or another to build networks and alliances and increase empowerment. They have done this through complex working to strengthen: insight and identities through building organisational capacity; human, cultural and social capital; and wellbeing. They have only been able to do this through the operation of coherent and well managed collaborative research partnerships. In so doing, they have contributed to sustainable communities that are fair, harmonious and inclusive, well run, with good quality services, a flourishing economy and sensitive to the environment. The co-produced knowledge, insight, action and understanding is central to the transformation process and is what distinguishes research for sustainable development from other change processes.

All of the case examples were of participative research. Community partners identified the need, and participated in the design, implementation and interpretation of findings. They are not short term 'cut and run' projects, but build on relationships formed over a number of years, often with excluded groups, and which continue beyond the specific project. Such partnerships are difficult to form and sustain from a university base, as partnership working is rarely factored into workloads. Constant vigilance and pressure is required to ensure that university systems, local and central, enable rather than obstruct this kind of research. One of the pressures on the university researchers is to ensure that in addition to benefits for the community partners, engaged research also meets the needs of the university (and the assessments of research excellence that takes place nationally). Whilst this is not always easy, there is a requirement to produce impact case studies (HEFCE 2015), and one kind of impact might be the contribution made to the development of sustainable communities.

All of the projects have interfaced with local, national or global policies. They have both informed policy developments in favour of more sustainable communities or have exposed the ways in which social policies (particularly, in recent years those of austerity) obstruct human flourishing. Furthermore, in different ways they have contributed to the advancement of sustainability literacy (Davies 2009), both amongst the teams of researchers as they endeavour to understand the obstacles and progress towards sustainable communities, and amongst our community partners as they struggle to find better ways of supporting people to live respectful, co-operative and fulfilling lives.

Figure 2 summarises the complex processes by which these lead to enhancing sustainable development and contribute, ultimately to a sustainable, viable and what Rutherford and Shah (2006) refer to as a 'good society'.

When we have thought about the ways in which university-community partnerships contribute to sustainable communities, we have drawn on concepts from ecology and sustainable agriculture: in particular, the ideas of complex systems, fields and edges (Burton and Kagan 2015; Kagan and Duggan 2009). As a field we are considering a terrain that has a boundary and within which interactions happen. Interactions within any field of activity have a structure and complexity that cannot

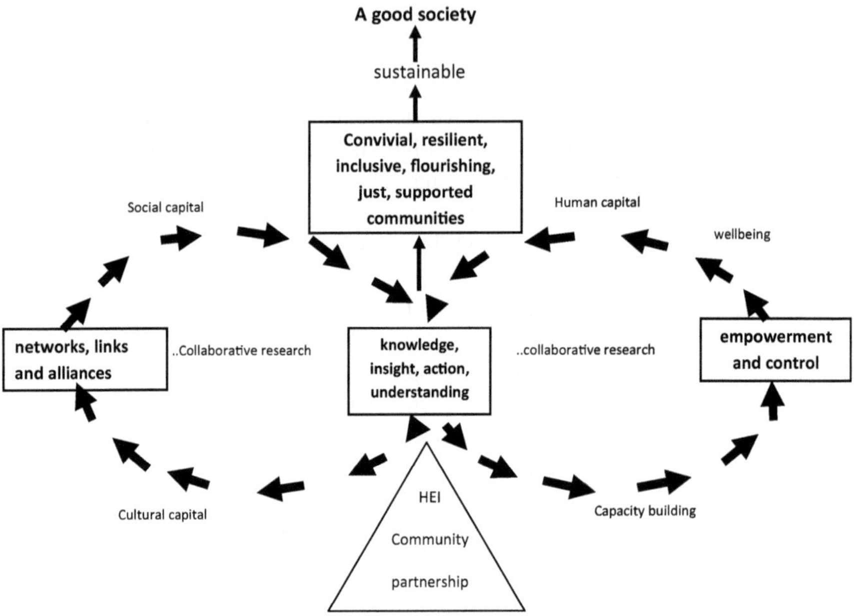

Fig. 2 HEI-community engagement supporting the development of sustainable communities

simply be reduced to the sum of those interactions. Furthermore, fields do not have fixed boundaries, they interact with and influence adjacent fields or ecosystems. The area where two ecosystems meet is called the 'ecotone' or ecological 'edge', and contains elements of both contributing fields. The edge can be applied to social systems to maximise resources.

As the 'edge' has characteristics of both ecosystems, it results in a richness of natural resources—both species and energy transactions. We have found it useful to use the concept of 'edge' to think about how to maximise available resources for sustainable development.

All of the projects worked across boundaries, and pooled the resources of different disciplines and professions as well as those of both the universities and community partners. They could be said to have created an ecological 'edge'. Working to create an ecological edge in research is an efficient way to generate and use resources and is a more sustainable way of working than within boundaries.

6 Conclusion

The concept of *sustainable communities* is a useful imaginary and organizing framework for university research into the understanding and enhancement of social aspects of sustainability.

It is possible to articulate the different dimensions of sustainable communities, but research which is capable of addressing complexity can usefully highlight their intersections.

Action research, with high degrees of participation, is an approach that is able to handle complexity, and enables meaningful community based research needs to be met. This requires an explicit value position, time, commitment and an interdisciplinary stance from researchers. One limitation of this approach is that it is time consuming for all concerned and because research questions evolve and cannot always be identified at the outset, funding can be difficult to attract. This makes it even more important that resources of the university and community are combined and maximized.

There is huge potential for university work to inform and contribute to the development of sustainable communities. To do this in a meaningful way they must commit to community engagement as an important subset of public engagement and work to sustain what is good and change what is not at a community level (Benneworth et al. 2010; Benneworth 2013). We have been able to show how engaged action research can contribute to sustainable development. However, we are aware that this is only touching on the possibilities for university research contributing to sustainable communities and that there are many different kinds of research approach which can be valuable.

Sustainability will only be achieved through the actions of people in families, neighbourhoods, communities and workplaces, and university research at the community level can help the transformation journey to more sustainable futures.

References

Benneworth, P. S., Charles, D. R., & Madnipour, A. (2010). Universities as agents of urban change in the global knowledge economy. *European Planning Studies, 18*(10), 1611–1630.

Benneworth, P. S. (Ed.). (2013). *University engagement with socially excluded communities*. London: Springer.

Bichard, E. (2014). *The coming of age of the green community: My neighbourhood, my planet*. London: Routledge.

Burton, M., & Kagan, C. (2015). Theory and practice for a critical community psychology in the UK. *Psicología, Conocimiento y Sociedad, 5*(2), 182–205.

Coote, A. (2015). *People, planet, power. Towards a new social settlement*. London: New Economics Foundation. 64p.

Coburn, D. (2004). Beyond the income inequality hypothesis: Class, neo liberalism, and health inequalities. *Social Science and Medicine, 58*(1), 41–56.

D'Alisa, G., Demaria, F., & Kallis, G. (Eds.). (2014). *Degrowth: A vocabulary for a new era*. London: Routledge. 220p.

Davies, K. (2009). A Learning Society. In A. Stibbe (Ed.), *The handbook of sustainability literacy: Skills for a changing world* (p. 220). Dartington: Green Books.

Dawood, S. (2013, July 18). *Free Wi-Fi scheme helps ease burden on poor families in Bury*. Retrieved from The Guardian: http://www.theguardian.com/uk-news/the-northerner/2013/jul/18/free-wi-fi-lampposts-bury-broad-oak

Diamond, J. (2004). Local regeneration initiatives and capacity building: Whose 'capacity' and 'building' for what? *Community Development Journal, 39*(2), 177–189.

Diamond, J. (2012). Learning about leadership in the third sector: From activist grassroots to managerial conformity. In M. Sotarauta, et al. (Eds.), *Leadership and change in sustainable regional development*. London: Routledge.

Egan, J. (2004). *The Egan Review: Skills for sustainable communities*. London: ODPM. Available from http://resources.cohesioninstitute.org.uk/Publications/Documents/Document/DownloadDocumentsFile.aspx?recordId=157&file=PDFversion. Last Accessed March 3, 2016

Field, Y., Murray, K., & Chilangwa-Farmer, D. (2015). *A place to call home: Community asset ownership in the African Diaspora community*. London: Locality.

Geographical Association. (2015). *Building sustainable communities*. http://www.geography.org.uk/projects/buildingsustainablecommunities/onlinecpdunit/stimulus/activity1/. Last Accessed March 3, 2016

Goldstraw, K., Chilcot, H., & Diamond, J. (2015). *Rochdale community champions. Building community Knowledge, developing community research* (56p.). Ormskirk, I4P, Edge Hill University. https://www.edgehill.ac.uk/i4p/files/2014/09/Cs-1756-Rochdale-56pp-A4-2015.pdf. Last accessed March 17, 2016.

HEFCE. (2015). *The nature, scale and beneficiaries of research impact: An initial analysis of Research Excellence Framework (REF) 2014 impact case studies*. London: Kings College. http://www.hefce.ac.uk/pubs/rereports/Year/2015/analysisREFimpact/. Last Accessed March 26, 2016.

HEFCE. (2014). Sustainable development in higher education. HEFCE's role to date and a framework for its future actions. http://www.hefce.ac.uk/media/hefce/content/pubs/2014/201430/HEFCE2014_30.pdf (Retrieved March 3, 2015).

Kagan, C., & Duggan, K. (2009). *Breaking down barriers: Universities and communities working together. Community cohesion thematic evaluation report*. Manchester: RIHSC. 126p.

Kagan, C., Burton., M., Duckett, P., Lawthom, R., & Siddiquee, A. (2011a). *Critical community psychology*. London: Wiley-Blackwell.

Kagan, C., Lo, S., Mok, L., Lawthom, R., Sham, S., Greenwood, M., et al. (2011b). *Experiences of forced labour amongst Chinese migrant workers*. York: Joseph Rowntree Foundation.

Kagan, C., Burton, M., & Siddiquee, A. (2008). Action research. In C. Willig & W. Stainton-Rogers (Eds), *The Sage handbook of qualitative methods in psychology* (pp. 32–53). Sage, London

Khan, O. (2015). *The 2015 budget: Effects on black and minority ethnic people*. London: Runnymede Trust.

Lawthom, R., Kagan, C., Baines, S., Lo, S., Sham, S., Mok, L., et al. (2015). Experiences of forced labour amongst UK based Chinese migrant workers: Exploring vulnerability and protection in times of Empire. In L. Waite, G. Craig, H. Lewis, & K. Skrivankova (Eds.), *Vulnerability, exploitation and migrants. Insecure work in a globalised economy*. London: Palgrave McMillan.

NCCPE. (2015). *What is public engagement*? National Coordinating Centre for Public Engagement Bristol. http://www.publicengagement.ac.uk/explore-it/what-public-engagement. Last Accessed March 6, 2016.

O'Neill, T., Mitchell-Smith, Z., Fisher, J., Lawthom, R., McLaughlin, H. (2014). '*The Spice of Life'. An evaluation of Home-Start Manchester South*. Manchester: RIHSC. http://www.rihsc.mmu.ac.uk/projects/docs/HSMS%20report.pdf. Last Accessed April 15, 2016

Power, A (2004) *Sustainable communities and sustainable development: A review of the sustainable communities plan. CASE report 23*. London: Centre for Analysis of Social Exclusion, London School of Economics and Political Science. http://sticerd.lse.ac.uk/dps/case/CR/CASEreport23.pdf. Last Accessed October 3, 2015

Runswick-Cole, K., & Goodley, D. (2013). Resilience: A disability studies and community psychology approach. *Social and Personality Psychology Compass, 7*(2), 67–78.

Rutherford, J., & Shah, H. (Eds.). (2006). *The good society. Compass programme for renewal.* London: Lawrence and Wishart. 98p.
Ungar, M. (2011). The social ecology of resilience. Addressing contextual and cultural ambiguity of a nascent construct. *American Journal of Orthopsychiatry, 81*, 1–17.

Authors Biography

Professor Carolyn Kagan holds an Emerita Chair in Community Social Psychology at Manchester Metropolitan University where she was the Director of the Research Institute for Health and Social Change. Throughout her career she has worked on action research projects in community settings, in pursuit of greater social justice with those marginalised by the social system. Much of Carolyn's work is action oriented, with projects extending over several years, addressing sustainable communities, and complexity in community and human service systems. Her work is collaborative and interdisciplinary and she has worked with colleagues in a number of different countries. She is founding editor of the journal *Community, Work and Family* and a co-author of the groundbreaking text, *Critical Community Psychology*.

Professor Rebecca Lawthom is a Professor in Community Psychology at Manchester Metropolitan University where she leads the Social Change and Community Wellbeing Research Group. Rebecca uses feminist, qualitative, creative and community participatory methods in her research which is focused on areas such as migration, disability and community engagement, all of which contribute to the sustainable communities agenda. She has a range of international collaborations for research. She is the Editor of the journal *Community Work and Family* and co-author of the acclaimed texts: *Critical Community Psychology; Qualitative Methods in Psychology;* and *Researching Life Stories*. She teaches on a number of undergraduate and postgraduate courses using dialogic methods.

Dr. Ornette D. Clennon is a Visiting Enterprise Fellow at Manchester Metropolitan University, UK. Ornette writes for Media Diversified and Open Democracy and is a Public Engagement Ambassador for the National Co-ordinating Centre for Public Engagement (NCCPE). He is also a community activist at local and national levels. Ornette's enterprise and activism work has been recognised with the 2011 NCCPE Beacons New Partnerships Award. Ornette is widely published and his previous books include *Alternative Education and Community Engagement: Making Education a Priority* (2014) and *Urban Dialectics, The Market and Youth Engagement: The Black Face of Eurocentrism?* (2015).

Dr. Jenny Fisher is a Senior Lecturer in Health and Social Care at Manchester Metropolitan University (MMU). Prior to working at MMU, she gained a wealth of experience in community development and area regeneration. Her PhD was concerned with place based sense of belonging, and she has expertise in community capacity building and community engagement. Jenny led a pioneering Community Organisers' project which tested different ways of engaging people from deprived areas in sustainable community practices. Other recent projects include personalisation and social care reform for older people, evaluation of a community based dementia programme and explorations of social eating with older people.

Professor John Diamond is the Director of the *Institute for Public Policy and Professional Practice* at Edge Hill University (UK). In 2015 he was invited to give the Annual Keib Thomas Memorial Lecture in London. He is currently the national chair of the not for profit *Association for Research with Voluntary and Community Organisations* (ARVAC). He has over 25 years experience of working as an external evaluator for a range of funded agencies and charitable organisations. He acts as a critical friend to CEOs working in the not for profit sector. In 2014 he was a co-researcher on a national study funded by the Webb Memorial Trust which examined the role of Fairness Commissions and is working on a follow up study. He is, also co-editor of the journal *Teaching Public Administration* and co-editor of an annual series—*Critical Perspectives on International Public Sector Management* (published by Emerald and launched in 2012). He is responsible for the *Collaborations Across Boundaries* module on the MSc Leadership and Management Development programme at the University.

Dr. Katy Goldstraw is a Research Assistant on a project funded by the Webb Memorial Trust, researching concepts of a Good Society for the Institute for Public Policy and Professional practice at Edge Hill University. She holds a Masters Degree in Poverty Reduction and Development Management and is currently undertaking a PhD, studying the effects of austerity policies on the Voluntary Sector in Manchester, UK. Katy is also the director of a small social enterprise, The *Volunteer Training Company*, which is a voluntary sector research and training provider. Katy's primary research interests are in participatory research, feminist research and Livelihoods Analysis.

Sustainability Research as Presented in UK University Sustainability Policies

Katerina Kosta

Abstract

Sustainability policies are an important tool for mainstreaming Sustainability in Higher Education (SHE) as they entail management level commitments. In 2013, following government request, all UK universities have an (environmental) sustainability policy publicly available on their website, which normally includes provision for education, research, estates and engagement. Given their recent existence, few studies to date have explored the nature and content of these documents while even less have focused on how sustainability research is conceptualised in the policies. The present paper aims to address this gap through an exploration of UK universities' sustainability policies and their conceptualization of sustainability research. References to research are compared with references to teaching, estates and engagement in order to explore the tendency identified by previous literature of the estates dominating the SHE discourse. The sample used consists of the sustainability policies of the thirty 'greenest' universities in the UK, according to the People and Planet University League 2015. The method used is content analysis assisted by QSR NVivo 10 software (QSR International Pty Ltd). The findings indicate that estates and operations are the most popular themes in the policies, with research occupying a secondary position. References to research mostly focus on the creation of new sustainability research centres and the generation of impact and funding. Regional government legislation is seen as affecting the content of sustainability policies as regions with ESD implementation legislation enjoy higher sustainability uptake in their universities. UK higher education institutions may find a comparative analysis of sustainability policies useful as these documents

K. Kosta (✉)
Faculty of Humanities and Social Sciences, Oxford Brookes University, Headington Campus, Oxford OX3 0BP, UK
e-mail: katerina.kosta-2016@brookes.ac.uk

W. Leal Filho (ed.), *Sustainable Development Research at Universities in the United Kingdom*, World Sustainability Series,
DOI 10.1007/978-3-319-47883-8_15

constitute part of their CSR profile and they are now widely scrutinized by sustainability assessment organisations and other interested parties.

Keywords
Sustainability research · Sustainability policies · Sustainability in higher education · Sustainability reporting

1 Introduction

For decades, higher education institutions (HEIs) have been monitoring and scientifically recording environmental degradation, remaining mostly passive observers of the situation. However, the end of the decade 2005-2014, which was defined by the United Nations as the Decade of Education for Sustainable Development (DESD) finds HEIs attempting to include sustainability in their curricula, policies and practice. 'Greening' the campus and the curriculum is dictated not only by governmental policies and funding councils but also by an increasingly assertive student body that wishes to see SHE implemented in their higher education experience (Drayson and Taylor 2015). Moreover, sustainability literate graduates are reported as being more attractive to prospective employers who wish to safeguard their corporate social responsibility profiles (Sterling 2011). Skills in sustainable development are expected to be overwhelmingly important for employment in the future: between 80 and 90 % of third year students rank the majority of skills as important or very important in terms of employment (Drayson and Taylor 2015). As a result, Sustainability in Higher Education (SHE) is emerging as a fast-growing movement.

The beginning of this movement can be located in 1990 when University Leaders for a Sustainable Future (ULSF) signed the Talloires Declaration (Tilbury 2011). The declaration invited universities to lead societal transition to sustainability. Twenty-five years later, the SHE movement is gaining momentum and is establishing itself as a separate research field. It is considered as a distinct specialization within sustainability scholarship and a subset of educational research (Lidstone 2014). UNESCO gave this movement a boost when it invited universities to co-deliver the Decade of Education for Sustainable Development, encouraging sustainability knowledge creation (Barth and Rieckmannn 2016).

Universities have been credited with the moral obligation to promote sustainability as they constitute the training grounds for future leaders and decision makers (Lidstone 2014; Lukman and Glavic 2007). For White (2014) universities are societies in microcosm and cultural changes that take place there can later be scaled to other settings. Moreover, universities are uniquely positioned for this role as they have the academic freedom to explore and test new ideas (White 2014).

However, the implementation of sustainability in university settings has been nothing but simple as HEIs are complex organisations that depend on various stakeholders with conflicting interests. In addition, academic disciplines are largely self-regulating and self-sustaining, which means that top-down implementation approaches are sometimes met with resistance (Lidstone 2014). Yet, as Paul Rowland contends, there needs to be 'a shift from universities serving the disciplines to universities serving society' (2013: x). Despite universities being notoriously resistant to change, SHE *does* take place and HEIs *do* strive to incorporate it in their curricula, policies and practice. To demonstrate their commitment to sustainability universities publish their own sustainability policies, strategies and reports. The appearance of such documents is a 'recent trend' for higher education. In the UK the government in 1993 and again in 1997 recommends that all universities adopt and publish a sustainability policy. Yet, only in 2013 do all UK universities have a sustainability policy or report publicly available on their website. Given their short existence, these documents have only recently become the object of research.

In 2013, Lee, Barker and Mouasher conduct an analysis of all Australian universities' sustainability visions and missions to find that approximately 8 % of the universities did not possess such documents (Table 1). A year later, White (2014) explores 27 university sustainability plans in the US finding that operations is the most prominent element while research is much less common with only 41 % of the policies addressing it. Lidstone et al. (2015) conduct a similar analysis of sustainability policies in 21 Canadian HEIs discovering that only 50 % of the policies have goals related to sustainability research. Again in Canada, Vaughter et al. (2016) analyse the policies of 50 HE-FE institutions to find references to research in only 16 % of the documents. In all four studies research is under-represented in the documents (Table 1).

Sustainability policies of UK universities have not been extensively researched and more so with a focus on SHE research. The present study aims to address this gap by exploring the sustainability policies of UK HEIs and how they address SHE research. It geographically complements the previous four studies by answering the following research questions.

(a) How much emphasis is given to SHE research in the policies in relation to education, operations and engagement?
(b) How do sustainability policy statements conceptualize SHE research?

Table 1 SHE research in university sustainability policies (Author's own)

Lee et al. (2013)	White (2014)	Lidstone et al. (2015)	Vaughter et al. (2016)
Australia	US	Canada	Canada
	41%	50%	approx. 16%

2 Literature Review

The emerging phenomenon of university sustainability policies will be presented first. Then, various attempts to circumscribe SHE research will be presented followed by an exploration of the limited number of studies that have addressed university sustainability policies and SHE research.

2.1 Sustainability Policy Statements of Higher Education Institutions

Sustainability policies are a significant integrative tool for the institutionalization of sustainability as they have normative and coercive powers; 'If a policy says you must do X, then you must (should) do X. Not everyone is empowered to do these sort of statements' (Lidstone et al. 2015: 19). Moreover, they help institutions develop appropriate planning processes, set measurable targets and assign responsibility and accountability. Velazquez et al. (2006) contend that the creation of a sustainability policy could be one of the most important tasks for building a sustainable university and that absence of it, is likely to result in uncoordinated efforts and unfocused or short-lived outcomes. This agrees with McNamara (2010) who claims that high quality plans are correlated with progress on sustainability initiatives. Finally, policies facilitate the communication of sustainability initiatives to multiple stakeholders like students, staff and funding bodies. However, according to HEFCE (2009) there has been no evidence of a systematic relationship between the existence of such plans and policies and the strength of sustainability activity within universities; some without such plans are relatively active, while others are not. Yet, where an institution promotes a sustainable development plan across its activities, that activity is more likely to have a greater coherence (HEFCE 2009). Overall, it seems that the existence of a sustainability policy is aligned with more coordinated sustainability efforts.

However, while policies provide an insight into university sustainability activity, they do not necessarily reflect what is happening 'on the ground'. Velazquez et al. (2006) report that raising levels of awareness is an important function of those policies but note that improved awareness does not automatically translate into successful implementation. They thus agree with Lidstone et al. (2015) who state that in many cases policies do not have much impact in guiding daily university activities. White (2014) also reports low levels of implementation between the specifics of plans and ensuing development, while Lidstone et al. (2015) notice that implementation is enhanced when policies contain SMART targets, timelines and accountability for the tasks. Finally, implementation success seems higher in cases with greater stakeholder involvement in the planning process (White 2014). The successful uptake of the sustainability policies is an issue that merits further exploration.

The literature also points to how the creation of university sustainability policies has been accelerated by the appearance of sustainability assessment tools (SATs). SATs like the People and Planet University League in the UK or STARS[1] in the US record and evaluate sustainability activity taking place in higher education institutions. White (2014) reports that STARS may have encouraged American universities publish a sustainability policies by offering extra credits if universities have such a document publicly available on their website. Along the same lines, Lidstone et al. (2015) report that in Canada having a sustainability policy is strongly related to having completed an assessment like STARS. Similarly in the UK, AUDE (Association for Universities' Directors of Estates) state that People and Planet University League has led to the development of many 'useful and effective sustainability policies' (2016: 2). There thus appears to be reciprocity between the existence of sustainability assessment systems and the development of a sustainability policy.

2.2 SHE Research

Research is one of the defining characteristics of universities and sustainability research is expected to be one of the defining characteristics of a sustainable university. Yet, Vaughter et al. (2016) note that there seems to be little agreement as to 'what counts as sustainability research' (34). This could be seen as a consequence of the lack of a widely accepted definition of sustainability in higher education. Most institutions adopt the 'triple bottom line' approach to sustainability, which provides for environmental, social and financial viability. This model of sustainability also called *People, Planet, Profit* does not come without its criticisms. Kopnina (2014) claims that business as usual will continue despite educating citizens about the need to treat the planet with respect as long as the P for 'Profit' dominates the ESD discourse. Vaughter et al. (2016) note that when applied to research, this three pillar approach might mislead researchers or institutions into thinking that 'they are doing sustainability research, if they just address any one of economic, social or environmental aspects' (35). The 'floating signifier' of sustainability has also been criticised by Brooks and Ryan (2008) who state that the instability and flexibility of the term make it 'liable to various interpretations, serving various interests' (6). And while the SHE definition is debated, SHE research is one of the unique contributions universities can make to a sustainable world.

Tracing the characteristics of SHE research, Tilbury (2011) sees it as a new wave of academic research that breaks away from disciplinary boundaries and seeks to transform rather than merely inform, aiming at societal rather than simply academic impact. Similar characteristics of sustainability research have been identified by Waas et al. (2010) who conduct an extensive analysis of the literature and international SHE declarations, accompanied by a workshop at the university of

[1]Sustainability, Tracking, Assessment and Rating System.

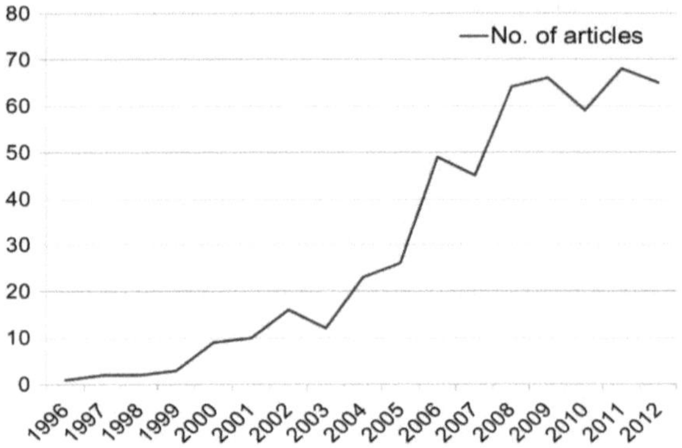

Fig. 1 Published articles from 1992 to 2012 (Barth and Rieckmannn 2016)

Antwerp called 'University Research for Sustainable Development'. Tracing the characteristics of SHE research, Waas et al. (2010) end up with a definition of SHE research as 'all research conducted within the institutional context of a university that contributes to sustainable development' (630). Interestingly, while adopting a definition that includes research in all disciplinary areas, the authors exclude research accomplished by non-higher education institutions as they contend that the mission of higher education is the service of the public interest while companies for example, serve the private interests of their shareholders (630). Thus, SHE research emerges as a new multidisciplinary genre that aims at social rather than simply academic impact, being primarily subject to 'societal peer review' (Waas et al. 2010; 633).

Criteria that promote sustainability research slowly make their way into high profile assessment systems like the Research Excellence Framework (REF), compliance with which directly affects funding allocation to universities. First implemented in 2014, the REF introduces real-world impact as one of the criteria of high quality academic research. Environmental impact is one of the eight impact types[2] defined by the REF, pushing environmental sustainability higher up the universities' research agenda. Assessment systems like the REF begin to influence the type of research undertaken by institutions and the introduction of the impact criterion might further incentivize sustainability research production.

The content of sustainability research remains to be explored. To identify trends in existing SHE research Barth and Rieckmann (2016) performed a systematic literature review of all research papers published in 110 different journals from 1996 to 2012 (Fig. 1). Their analysis reveals a noticeable increase in research output after 2005 which signifies the beginning of UNESCO's Decade of Education for Sustainable Development (DESD) in 2005. This illustrates the impact global initiatives like the DESD can have on the evolution of SHE research. Barth and Rieckmann (2016) also analyze the content of these articles to find that one third of them focus on curriculum development while only 1.8 % address research in Higher Education for Sustainable Development (HESD) (Table 2). Up to 2012 only

[2] The eight impact types according to the REF 2014: political, legal, health, cultural, technological, societal, economic, environmental.

Table 2 Thematic focus of research papers on SHE

Content focus	n	%
Curriculum development	170	33.4
T&L approaches	100	19.6
Organisational change/learning	64	12.6
Student view/ lecturer view	55	10.8
Research in HESD	9	1.8

Adapted from Barth and Rieckmann (2016)

9 articles explicitly focused on the newly established area of SHE or HESD research, illustrating that this is an emerging field that has not yet been extensively explored. There are calls in the literature for researchers to focus on this emerging field (Vaughter et al. 2016) and the present study can be seen as a response to these calls.

3 Methodological Design

The purposive sample used is the 30 'First-Class' sustainability universities, according to the *People and Planet University League* 2015. It was chosen under the assumption that since they fulfill many of the University League criteria, these institutions are extensively engaged in sustainability and thus more likely to have a comprehensive sustainability policy containing references to sustainability research. People and Planet University League was selected as this student-led NGO provides a transparent and detailed assessment methodology, which makes the existence of an (environmental) sustainability policy a prerequisite (People and Planet 2014). The analysis of each policy as found on the organisation's website was used as a reference point for this analysis.

For the purposes of the study, the conceptualization of sustainability policies offered by White (2014) has been adopted. Sustainability policies are defined as comprehensive management tools different from environmental management systems (EMS) in that they are not restricted to operations but encompass a wide array of issues (White 2014). This 'definition' of a sustainability policy was used to inform the data collection process, which took place in June 2015. Content analysis was the method chosen to thematically categorize the content of the documents and allow for themes to emerge relevant to SHE research. The same method had been used by the four previous studies in Australia, the US and Canada.

For the identification of the policies, the universities' websites were searched using the phrase 'sustainability policy'. This made obvious that while certain universities have a single comprehensive sustainability policy others have a

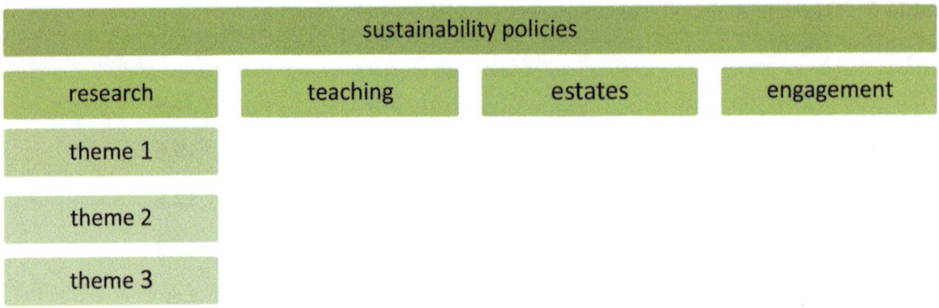

Fig. 2 Diagram of content analysis steps (Author's own)

separate environmental sustainability and SRS[3] policy or even multiple smaller policies like sustainable transport, procurement or energy. Given the recent existence of these documents, universities are not consistent in which ones they make available. Instead of a policy, a strategy, mission, vision, plan or a report might be available or any combination of the above with policies and reports mainly referring to current practices while strategies, missions, visions and plans being more future oriented. All terms described above were taken into account when searching to make sure no documents were omitted.

The coding strategy was informed by the research questions, using a combination of predetermined and emerging codes (Creswell 2009). Themes already extant in the literature were followed by themes emerging from the data analysis. Initially, every document was searched for four codes; *operations, research, teaching* and *engagement*. These first order codes represent the four key areas of sustainability in higher education as widely presented in the literature (Fischer et al. 2015). Yet, a policy might refer to these four terms differently through synonyms or alternative terms. It was thus decided to add the items *curriculum* and *courses* to *teaching*, *estates* and *campus* to *operations* and outreach to *engagement*. *Research* is not easily substituted by a synonym and when it does not appear in a policy, it might be inferred that sustainability research is probably not yet within the university's stated priorities. The underlying construct this keyword text search query aimed to illuminate is the emphasis SHE research attracted in the policies compared to teaching, estates and engagement.

Second level coding took place to identify emerging themes in the texts referring to sustainability research. The segments coded under 'research' were revisited while repeated thematic patterns were identified, making sure their definitions were as mutually exclusive as possible. During theme generation, staying faithful to the wording of the texts was a priority as the more one moves towards abstract categories and inferencing, the more reliability may be compromised as the researcher's agenda imposes itself on the process (Cohen et al. 2011). A visualization of the data analysis steps can be seen in Fig. 2.

[3]Social Responsibility and Sustainability.

4 Results

4.1 Comparison of Four SHE Areas

Research receives a lot of emphasis in the documents and a separately run word frequency query shows that it is the ninth most frequently used item. This is further supported by the text search query where the four areas of SHE are used as keywords (Fig. 3) to illustrate the frequency of appearance of the four basic SHE elements in the policies. Estates and operations appear as the most common theme, followed by research, teaching and engagement. The domination of the SHE discourse by estates and operations themes is a tendency that has been identified by numerous previous studies (Cotton and Winter 2010; White 2014; Vaughter et al. 2013; Yarime and Tanaka 2012; Fischer et al. 2015). This is a striking fact given that education and research are the main services universities are known for.

It should be mentioned however that the 'space' each of the four areas occupies in the matrix coding query depends on the keywords inserted in the search box. So the chart of Fig. 3 would be slightly different, if the term campus had been omitted from the campus/estates/operations keyword cluster. Text queries are sensitive to the nature and number of keywords and conclusions are provisional on the characteristics of the search.

4.2 Research Themes

The close reading of the texts containing references to research, revealed some repeated thematic patterns, the most common of which can be seen in Table 3. Universities were keen to report the establishment of new sustainability research

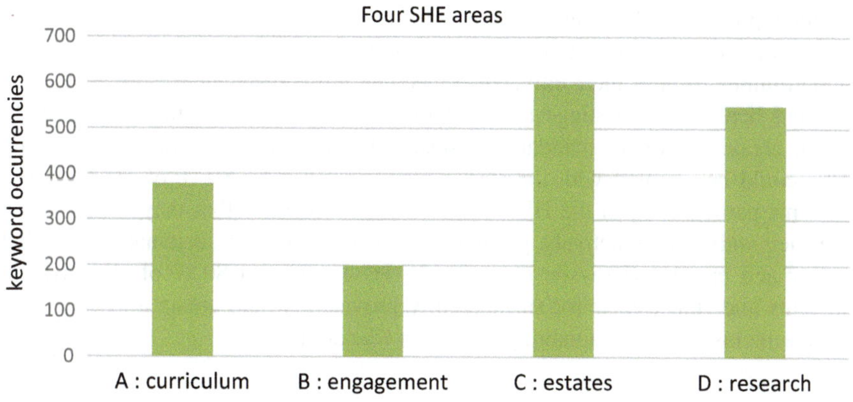

Fig. 3 Four main SHE areas in the policies (Author's own)

Table 3 Cross study comparison of research themes in the policies

Research themes of 30 policies (2016) UK	Vaughter et al. (2016) Canada	Lidstone et al. (2015) Canada	White (2014) US
Creation of new research centres	✓	✓	
Social impact			
Proving funding	✓	✓	
Knowledge transfer partnerships	✓	✓	
Using the campus as a 'living lab'	✓	✓	✓

Author's own

centres, a tendency identified by Vaughter et al. (2016) and Lidstone et al. (2015) in Canada. Producing sustainability research that has real-world impact was the second most common theme and it seems that it is specific to the UK HE system as it has not been identified by previous studies in Canada, Australia or the US. This may be a reflection of the way research is assessed in the UK with real-world impact becoming an evaluation criterion.

Attracting and providing funding was another common theme followed by the creation of knowledge transfer partnerships for sustainability research both of which topics have also been identified by the Canadian studies. Finally, using the campus as a living lab for modelling and researching sustainability appears as a cross-cutting theme in this and all previous studies. This could be an indication of the growing popularity of the 'living lab' concept as a means of bridging the gap between estates and educational departments.

Provision of staff training for sustainability research was another common theme and a specific initiative is highlighted here as important. The Staff Sustainability Skills survey carried out by the UWTSD[4] can be seen as part of the overall staff training provision. The survey revealed that 78 % of the UWTSD and Cardiff Metropolitan University staff were interested in sustainability, while 79 % saw opportunities to help their universities become more sustainable. This instance is isolated here with the suggestion that more surveys on academic staff attitudes might create a stronger mandate for sustainability from the academic staff body. This could be coupled with the strong mandate given by 80 % of the third year students participating in the HEA-NUS survey who stated that they would like to see their universities actively promoting and incorporating sustainability in their policy and practice (Drayson et al. 2013). Having almost 80 % of all university students and staff asking for sustainability provision would create a stronger case for the inclusion of sustainability in higher education.

[4]University of Wales Trinity Saint David.

4.3 Regional Government Legislation

Differences in regional government legislation seem to be reflected in the content of the university sustainability policies. Among the thirty 'greenest' universities in the UK, three are located in Wales, which is a large percentage given that Welsh universities constitute only 5 % of the UK HE population. This may not be a coincidence as in 2006 the Welsh government made Education for Sustainable Development and Global Citizenship (ESDGC) an integral part of the Welsh higher education curriculum. Moreover, the Higher Education Funding Council for Wales (HEFCW) has increased its requirements for HEIs to report on their delivery of sustainable development. As a result, sustainability policies issued by Welsh universities contain detailed descriptions of sustainability research delivery including specific operationalization, naming initiatives, outcomes and impact. Vaughter et al. (2016) have identified a similar trend in Canada where provinces with ESD implementation legislation enjoy higher sustainability uptake in their universities.

5 Discussion

Based on the above findings, several issues are highlighted here for discussion. The domination of estates and operations in the SHE discourse has been identified by previous researchers (Fischer et al. 2015; White 2014; Vaughter et al. 2016) who have given several explanations for it. White (2014) claims that campus operations are predictably the focus of university sustainability as efforts in this area 'produce cost savings which is a motivating factor for any institution' (235). Drawing on the Canadian HE sector, Vaughter et al. 2016 note that the operational focus seems to be spurred on by national policies which promote operations-related aspects. Similarly, in the UK the Higher Education Funding Council for England (HEFCE) incentivized the reduction of carbon emissions, through schemes like the Revolving Green Fund while initiatives for the promotion of SHE research were not similarly supported. Fischer et al. (2015) state that the focus of SHE activity is implicitly shaped by sustainability assessment tools (SATs) which set indicators that focus on these aspects of SHE. In the UK, the People and Planet University League has been up to now a major driver of sustainability activity at universities. Of its fourteen indicators in 2015, eleven are estates and operations oriented, with only one indicator addresses research for sustainable development.[5] Thus, universities that wish to excel in the League need to focus more on estates and operations rather than research. This shows how sustainability assessment systems can influence the content of sustainability in HE. Another explanation for the domination of estates and operations put forward by Vaughter et al. (2016) is that university sustainability policies are typically produced by officers in the estates or environmental management departments, which have little interaction with educational departments

[5]https://peopleandplanet.org/navid17492.

and this might be resulting in the policies' discourse being more estates and operations oriented. However, Tilbury contends that estates activity mostly driven by estates directors and their teams rarely makes an impact on students' formal learning opportunities (Tilbury 2011). A connection between estates activities with research, as exemplified by the promising 'using the campus as a living lab' approach might be a way to bring together estates and educational or research activity.

Impact and funding are among the top topics for sustainability research in UK universities' sustainability policies and this could be connected to the introduction of real-world impact as a new criterion in judging the quality of academic research in the UK. Finally, the appearance of regional government legislation as a motivating factor for the incorporation of sustainability in HEIs favourably compares with previous findings (Vaughter et al. 2016) pointing to important role the wider political context can play in the promotion of SHE research. An interesting issue to be explored by future research is whether the Scottish governments' 2015 Order[6] under the Climate Change Act—which makes sustainability reporting compulsory for public bodies—will result in more comprehensive sustainability reporting by Scottish universities.

6 Limitations

The findings of this research need to be critically appraised against some limitations resulting from the study's design. By exploring the sustainability policies as found on the universities' websites the study literally presents the 'virtual reality' of the phenomenon, providing limited information of what is happening 'on the ground'. This is a limitation, as a fairly consistent finding in studies of policy implementation is low levels of correlations between the specifics of plans and ensuing development (White 2014; Vaughter et al. 2013). The implementation of university sustainability policies might constitute an area for future research.

Another limitation of the study is that the thematic categories of SHE research are created by a single coder, the author. Krippendorf (2013) suggests that two or more coders are used in content analysis, as multiple coders provide multiple perspectives and reduce discrepancies. To compensate for the lack of a second coder, the thematic categories were contrasted with themes identified by researchers that have conducted similar studies in the past (Lidstone et al. 2015; Vaughter et al. 2016).

The purposive sample which was chosen due to its special characteristics cannot be said to be representative of the whole population. Thus, findings cannot be generalised to all higher education institutions in the UK. A systematic sample can be used by future studies in order to achieve greater representativeness.

[6]http://www.gov.scot/Topics/Government/sustainabilityperformance/reporting/sustainabilityreporting.

Finally, some important SHE research initiatives are absent from the policies but present on sustainability websites or separate departmental websites. Activity not included in the policies, is not recorded by the present study which focuses exclusively on the content of the documents.

7 Conclusion

Despite these limitations the study hopes to have identified tendencies in the content of university sustainability policies. By delineating how SHE research is depicted in the policies, in comparison to estates, teaching and engagement the study responds to calls in the literature for extra focus on the overlooked area of SHE research (Waas et al. 2010; Fischer et al. 2015). By focusing specifically on UK universities the paper addresses a geographical gap as this type of research has taken place in Canada, the US and Australia but not in the UK, where universities are actually assessed for the sustainability impact of their research. A description of 'what counts' as SHE research in the policies might be seen as critical at a time when delivery on sustainability is becoming the core of many European and UK funding initiatives.

Higher education institutions may find a comparative analysis of sustainability policies beneficial as these documents constitute part of their CSR profile, which is now widely scrutinized by sustainability assessment organisations and other interested parties. Sustainability reporting is growing into an important agenda, especially after the Paris COP21 agreement, where it was decided that overall sustainability reporting processes should be put in place to monitor progress. Universities could be leaders in this area by developing exemplary sustainability reporting.

References

AUDE. (2016). Association for University Directors of Estates: HE sector sustainability evaluating methodology summary report. Leeds: ASAC, AUDE, REPI.

Barth, M., & Rieckmannn, M. (2016). State of the art in research on higher education for sustainable development. In M. Barth, G. Michelsen, M. Rieckmannn, & I. Thomas (Eds.), *Routledge handbook of higher education for sustainable development*. New York: Routledge.

Brooks, C., & Ryan, A. (2008). *Education for sustainable development: Strategic consultations among English HEIs*. Higher Education Academy. https://www.heacademy.ac.uk/resource/education-sustainable-development-strategic-consultations-among-english-heis

Cohen, I., Manion, L., & Morrison, K. (2011). *Research methods in education*. Oxon: Routledge.

Cotton, D., & Winter, J. (2010). 'It's not just bits of paper and light bulbs': A review of sustainability pedagogies and their potential for use in higher education. In P. Jones, D. Selby, & S. R. Sterling (Eds.), *Sustainability education: Perspectives and practice across higher education* (pp. 39–54). London: Earthscan.

Creswell, J. W. (2009). *Research design: Qualitative, quantitative and mixed methods approaches*. Thousand Oaks: Sage.

Drayson, R., Bone, E., Agombar, J., & Kemp, S. (2013). *Student attitudes and skills for sustainable development*. Higher Education Academy/NUS. http://www.heacademy.ac.uk/assets/documents/sustainability/ESD_student_attitudes_2013_v4.pdf

Drayson, R., & Taylor, C. (2015). In W. Leal Filho, L. Brandli, L. Kuznetzova, & A.M.F. de Paço (Eds.), *Integrative approaches to sustainable development at University Level: World sustainability series*. Switzerland: Springer.

Fischer, D., Jenssen, S., & Tappeser, V. (2015). Getting an empirical hold of the sustainable university: A comparative analysis of evaluation frameworks across 12 contemporary sustainability assessment tools. *Assessment & Evaluation in Higher Education, 41*(1), 1–16. doi:10.1080/02602938.2015.1043234

HEFCE. (2009). *Sustainable development in higher education—2008 update to strategic statement and action plan*. http://www.hefce.ac.uk/pubs/hefce/2009/09_03/

Kopnina, H. (2014). Revisiting education for sustainable development (ESD): Examining anthropocentric bias through the transition of environmental education to ESD. *Sustainable Development, 22*, 73–83. doi:10.1002/sd.529

Krippendorff, K. (2013). *Content analysis: An introduction to its methodology* (3rd ed.). Thousand Oaks: Sage.

Lee, K., Barker, M., & Mouasher, A. (2013). Is it ever espoused? An exploratory study of commitment to sustainability as evidenced in vision, mission and graduate attribute statements in Australian universities. *Journal of Cleaner Production, 48*, 20–28. doi:10.1016/j.jclepro.2013.01.007

Lidstone, L. (2014). *A content analysis of sustainability policies and plans from stars-rated Canadian higher education institutions*. Master's thesis. Halifax, Nova Scotia: Dalhousie University. http://dalspace.library.dal.ca/xmlui/handle/10222/54034

Lidstone, L., Wright, T., & Sherren, K. (2015). An analysis of Canadian STARS-rated higher education sustainability policies. *Environment, Development and Sustainability, 17*(2), 259–278. doi:10.1007/s10668-014-9598-6

Lukman, R., & Glavič, P. (2007). What are the key elements of a sustainable university? *Clean Technologies and Environmental Policy, 9*(2), 103–114. doi:10.1007/s10098-006-0070-7

McNamara, K. H. (2010). Fostering sustainability in higher education: A mixed-methods study of transformative leadership and change strategies. *Environmental Practice, 12*(1), 48–58. doi:10.1017/S1466046609990445

People and Planet. (2014). *People and planet university league; the 2014 guide*. http://peopleandplanet.org/dl/greenleague/2013/guide.pdf

QSR International Pty Ltd. (2012). *QSR NVivo 10 qualitative data analysis software*. Doncaster, Australia: QSR International Pty Ltd.

REF. (2014). *Assessment criteria and level definitions*. http://www.ref.ac.uk/panels/assessmentcriteriaandleveldefinitions/

Rowland, P. (2013). Foreword. In L. F. Johnston (Ed.), *Higher education for sustainability, vii-x*. New York: Routledge.

Sterling, S. (2011). *The future fit framework; An introductory guide to teaching and learning for sustainability in HE*. Higher Education Academy. https://www.heacademy.ac.uk/system/files/future_fit_270412_1435.pdf

Tilbury, D. (2011). Higher education for sustainability: A global overview of commitment and progress. In GUNI (Ed.), *Higher education in the World 4. Higher education's commitment to sustainability: From understanding to action* (pp. 18–28). Barcelona: Palgrave.

Vaughter, P., McKenzie, M., Lidstone, L., & Wright, T. (2016). Campus sustainability governance in Canada. *International Journal of Sustainability in Higher Education, 17*(1), 16–39. doi:10.1108/IJSHE-05-2014-0075

Vaughter, P., Wright, T., McKenzie, M., & Lidstone, L. (2013). Greening the Ivory Tower: A review of educational research on sustainability in post-secondary education. *Sustainability, 5*, 2252–2271. doi:10.3390/su5052252

Velazquez, L., Munguia, N., Platt, A., & Tadde, J. (2006). Sustainable university: What can be the matter? *Journal of Cleaner Production, 14*, 810–819. doi:10.1016/j.jclepro.2005.12.008

Waas, T., Verbruggen, A., & Wright, T. (2010). University research for sustainable development: Definition and characteristics explored. *Journal of Cleaner Production, 18*(7), 629–636. doi:10.1016/j.jclepro.2009.09.017

White, S. S. (2014). Campus sustainability plans in the United States: Where, what, and how to evaluate? *International Journal of Sustainability in Higher Education, 15*(2), 228–241. doi:10.1108/IJSHE-08-2012-0075

Yarime, M., & Tanaka, Y. (2012). The issues and methodologies in sustainability assessment tools for higher education institutions: A review of recent trends and future challenges. *Journal of Education for Sustainable Development, 6*(1), 63–77. doi:10.1177/097340821100600113

Author Biography

Katerina Kosta is a new Ph.D. student at Oxford Brookes University funded by the *150 Anniversary Scholarships* award. She has recently completed a research internship at the Environmental Association for Universities and Colleges (EAUC) where she explored sustainability assessment and reporting systems in the UK tertiary education sector. She is also a graduate of the M.Sc. in Educational Research by the University of Exeter, where she used quantitative and qualitative research approaches to explore the fast-growing field of Sustainability in Higher Education (SHE). Her M.Sc. research focused on the influence of curriculum content on student attitudes to sustainability, student engagement in SHE and academic staff views on the inclusion of sustainability in higher education. She is currently exploring how higher education institutions report on their sustainability performance.

The UK Sustainable Development Research Network—Bridging the Sustainability Science/Policy Divide

Gary Kass, Ben Shaw and Fred Steward

Abstract

Between 2001 and 2014, the UK Government funded the Sustainable Development Research Network (SDRN) as an interface between the SD research community and the policy community. The authors were involved with the SDRN for many years: Shaw and Steward as members of the SDRN coordinating team and Kass as a member of the Advisory Committee. This paper will explore the work of the SDRN especially its effectiveness as a bridge between the research and policy communities in the area of SD research. The paper will examine the evolution of the SDRN throughout its lifespan, tracking the shifts in its remit and activities and focussing on a review of the SD research landscape undertaken by SDRN (Steward et al. in Mainstreaming sustainable development research in an age of austerity: SDRN review of the UK sustainable development research and policy landscape. Policy Studies Institute, 2013). It will explore current opportunities for spanning the boundaries between SD research, policy and practice, situating SDRN in the context of an increasing recognition of 'sustainability science' as a challenge-led, problem-oriented co-production approach involving researchers, policy-makers and practitioners, described by Future Earth as a 'novel way of doing research'. The paper also consider the impact of the 'impact agenda' and recent adoption of the Sustainable Development Goals and whether SDRN may have a useful role to play in bringing SD research more to the fore.

G. Kass (✉)
Natural England, Nobel House 17 Smith Square, London SW1P 3JR, UK
e-mail: gary.kass@naturalengland.org.uk

B. Shaw · F. Steward
Policy Studies Institute, University of Westminster, 35 Marylebone Road, London NW1 5LS, UK
e-mail: b.shaw@psi.org.uk

F. Steward
e-mail: f.steward@psi.org.uk

W. Leal Filho (ed.), *Sustainable Development Research at Universities in the United Kingdom*, World Sustainability Series,
DOI 10.1007/978-3-319-47883-8_16

Keywords
Sustainability · Science · Policy · Boundary · Organisation

1 Introduction

Following the United Nations Conference on Environment and Development in 1992, the UK Government published its first sustainable development strategy in 1994 (DETR 1999). This was updated in 1999 and included a principle of "using scientific knowledge" which was articulated as follows: "when taking decisions, it is important to anticipate early on where scientific advice or research is needed, and to identify sources of information of high calibre. Where possible, evidence should be reviewed from a wide-ranging set of viewpoints" (HM Government 1999).

Further, in 2001, the Department for Environment, Food and Rural Affairs (Defra), established the Sustainable Development Research Network (SDRN) in order to "contribute to sustainable development in the United Kingdom by facilitating the better use of evidence and research in policy-making" (Eames 2002). In 2002, the then Environment Minister, Michael Meacher, expressed the motivation for establishing the SDRN, recognising that achieving sustainable development was a "formidable task" and stating "We have some idea of what it will involve, including an increasing awareness of the environmental limits within which we need to operate. But we still have a very imperfect idea of how to get there." This, he said, underlined "the central role that research (and researchers) have in sustainable development." Consequently, Meacher stated that funding the SDRN was "one of the government's contributions to sustainable development research" (Eames 2002).

SDRN has been convened by Policy Studies Institute (PSI), part of the University of Westminster from its outset to the present day. In 2014 Defra removed its funding from the Network. Between 2001 and 2014, SDRN was funded through four contracts or 'phases', with the transition between phases used as an opportunity to review and revise as necessary the Network's aims, objectives and activities. Table 1 summarises the key details of each phase.

The SDRN's overarching objectives remained static from 2003 to 2014. In Phase IV, SDRN had a more extensive set of objectives (Defra 2011a) including:

1. Contribute to knowledge transfer of evidence relevant to the work of mainstreaming SD across Government.
2. Attain a broader remit across the Green Economy and Strategy programme and other policy areas such as Big Society, Sustainable Food and the Natural Environment.
3. Refresh the Advisory Committee membership to improve alignment and co-ordination with other external evidence initiatives.

Table 1 The four phases of SDRN

Phase	Dates	Aims
I	2001–2003	Specific aims include • Monitoring and mapping research relevant to the UK Sustainable Development Strategy • Fostering a network of organisations with an interest in sustainable development research; facilitating the flow of information about current and planned activities; and • Promoting sustainable development research activity by influencing funders and research organisations …to promote research collaboration across disciplinary boundaries… actively involves natural scientists, engineers and social sciences alike (Eames 2002)
II	2003–2006	… • Facilitate the provision of research and evidence to policymakers • Engage government policymakers, scientists and members of the research community • Promote sustainable development in the research and academic communities • Work with funding bodies to encourage relevant research • Advise the Defra Sustainable Development Unit on sustainable development research issues
III	2007–2011	As Phase II (see SDRN 2007) Phase III additional objectives 1. Increase the profile of SDRN across government departments and the devolved administrations 2. Enhance the engagement and involvement of policy-makers across government in the work of SDRN 3. Refresh the Network's communications approach and activities to enhance the accessibility and value of SDRN outputs to its target audiences in policy and research. (SDRN 2011)
IV	2011–2014	As Phase III (with minor changes, SDRN 2011)

4. Increase the profile among researchers of the need to incorporate consideration of inevitable changes in the earth's climate into SD thinking.
5. To facilitate a series of meetings in areas such as: (a) Climate risk perception; (b) Climate risk communication; (c) Risk and organisational psychology; (d) Community based ways to deal and lessen climate impacts; and (e) Social impacts of a changing climate, including impacts on equity.
6. To inform government of policy-relevant evidence and for government researchers to outline policy-relevant topics likely to have greatest impact.
7. Contribute to work by the Government's Chief Scientific Adviser to strengthen cross-disciplinary work on low carbon/green growth.

These more extensive objectives coincided with the formation of the Coalition Government, following the 2010 UK General Election. SDRN recognised the new agenda as a "challenge" the main driver of which was the new Government's commitment to "mainstreaming sustainable development" (Defra 2011b)—shifting from SD as a discrete policy area to "focus on how SD can be pursued and consolidated through cross-departmental policies on wider priority issues, such as the Green economy, the Big Society, the natural environment, food security, wellbeing and fairness" (SDRN 2011).

Despite these shifts, SDRN's basic 'business model' remained focused on:

1. undertaking and publishing research and evidence reviews,
2. organising and delivering an annual Sustainable Development Research Conference,
3. delivering seminars and workshops on specific topics,
4. producing the SDRN email newsletter and maintaining its website (www.sd-research.org.uk).

These activities have been used to develop the network over time. Membership of the network is free and open to all those with a professional interest in UK SD research and policy. SDRN has provided added-value 'services' to members, including news/research collation, summary and distribution; dissemination of evidence requests from policy-makers, events and networking fora; conducting evidence reviews and enabling members to communicate their work to their peers. It has also targeted engagement of policy-makers and researchers on specific issues. Membership has been developed by targeted direct recruitment, word of mouth and general network communication activities and currently stands at over 2500.

SDRN reaches out to numerous communities beyond Defra and academia, such as businesses, NGOs, the devolved administrations and to other government departments. It has engaged with more than 500 UK organizations, representing a diverse group spread across academic, business and public spheres (Steward and Piterou 2012). While SDRN has played an important and valued role at the interface of SD research and policy it faces obvious limitations. The resources available to the Network necessarily restrict what it can do: a small part-time coordinating team can't engage with all of government to address its SD evidence needs. Pressures on policy-makers and researchers also limit their ability to engage. Policy/evidence agendas and processes in government can evolve unpredictably and at a pace research finds hard to respond to effectively. Identifying agendas of mutual interest to specific researchers and policy-makers, while retaining relevance to the wider membership are perennial challenges. The Network's position at the research-policy interface presents difficulties in managing relationships between research and policy on a 'many-to-many' basis when only a few can actively be pursued.

In this context is possible to situate SDRN within the wider landscape around evidence and policy and the role of institutions in bringing evidence and policy closer together (e.g. Solesbury 2001; Mulgan and Puttick 2013). Guston (2001) describes 'boundary organizations' as meeting three criteria:

- they provide the opportunity and sometimes the incentives for the creation and use of boundary objects and standardized packages
- they involve the participation of actors from both sides of the boundary, as well as professionals who serve a mediating role
- they exist at the frontier of the two relatively different social worlds of politics and science, but they have distinct lines of accountability to each (Guston 1999, 2000).

SDRN satisfies Guston's criteria for a 'boundary organization'; producing 'boundary objects' in the form of rapid research reports and email newsletters; involving policy-makers, practitioners and researchers in meetings, workshops and conferences; and sitting at the interface between research, policy and practice—enabling the exchange and brokering of research-based evidence to inform decision-making. According to Guston, the operation of boundary organizations "gives both the producers and the consumers of research an opportunity to construct the boundary between their enterprises in a way favourable to their own perspective". In considering the relationship between politics and science, Guston is relaxed, stating that "boundary organizations suggest that the old idea that politics and science should be neatly cleaved should be abandoned in favor of the newer attempt to mix the interests of both".

SDRN can also be located within the territory of 'sustainability science' (Komiyama and Takeuchi 2006; Clark 2007; Kates 2011; Kass 2015). Here, the endeavour aims for a "new kind of new type of science that links disciplines, knowledge systems and societal partners to support a more agile global innovation system" (Future Earth 2014). Sustainability science aims to be "defined by the problems it addresses rather than by the disciplines it employs" (Clark 2007). As SDRN seeks to bring the necessary disciplines to bear and a wide range of partners into the process, it can be said to be a practitioner and broker of sustainability science.

This paper does not provide a detailed account of the history and achievements of the SDRN nor a detailed analysis of SDRN as a boundary organisation or knowledge broker in sustainability science. Rather, this paper illustrates how SDRN has provided value as a focal point at the interface between SD research, policy and practice. The paper concentrates on a key piece of SDRN's work, its 2013 review of the UK sustainable development research and policy landscape (Steward et al. 2013); reflecting on the review's conclusions and recommendations and looking forward to where SDRN might go in light of recent developments: the removal in 2014 of Defra's funding for SDRN; the 'impact' agenda in the 2014 Research Excellence Framework (King's College and Digital Science 2015); the formation, in November 2015 of the UK National Committee for Future Earth (Royal Society 2015); and the publication of the Nurse Review of the Research Councils (Nurse 2015).

2 The SDRN Review—*Mainstreaming Sustainable Development in the Age of Austerity*

The key aim of the review was to assess "the degree to which existing research activities meet the needs of policy-makers working on sustainable development" (Steward et al. 2013). SDRN had undertaken previous reviews of the SD research landscape (Eames 2006) and the 2013 review aimed to build on these, spanning the period from SDRN's formation in 2001 to the date of publication (mid-2013). The report highlighted seven key messages:

(a) **Growing diversity of actors**

SD is an increasingly multi-actor area of policy and research, with a greater diversity of users, funders, and researchers evident since the previous review in 2006. Growing concerns about climate change in particular have led to new users such as the Department for Energy and Climate Change (DECC) and the Committee on Climate Change, and a new agenda of a low-carbon transition.

The business sector has become a more prominent research funder, with 45 independent non-profit organizations funding research were identified. Sustainability researchers cover an increasingly broad field, with leading research centres scattered among seven different disciplinary panels in the 2008 Research Assessment Exercise.

(b) **Policy action at multiple levels**

This growing diversity is accompanied by an increasingly multi-level character of policy action on sustainability. The sub-national level is particularly important. The devolved administrations of Scotland, Wales and Northern Ireland have new responsibilities and many cities are active members of. Supra-national institutions such as the European Union and the UN are also sites of key policy initiatives. SD researchers frequently engage with these policy actors rather than with national government.

(c) Mainstreaming sustainable development in policy

These broad trends have been accompanied by specific policy developments since 2010. Centred on its Mainstreaming Sustainable Development policy of 2011, the UK coalition government (Defra 2012) redefined its commitments and priorities in relation to SD. The significance of these for research has also been influenced by its broader policy agenda: limiting public expenditure and making growth a priority. Policy statements have particularly highlighted the transition to a green economy and an increased requirement on all government departments to address sustainability.

(d) Sustainable development in an age of austerity

The mainstreaming of SD in an age of austerity has been accompanied by a sharper political contestation of economic and environmental purposes while attempting to ensure that sustainability is mainstreamed across government departments; highlighting the difficulties of strategic choices under public funding constraints. Controversies over the institutional coordination of SD across government and the meaning of SD in the national planning policy framework are examples.

(e) Coordinating and prioritising research funding

The review identified a need for more effective research funding coordination and priorities:

- While some interdisciplinary programmes have developed across research councils, their effectiveness in addressing SD needs appraisal. Research for SD should be more coordinated, and further opportunities pursued for new initiatives to involve users in research design.
- Multi-level coordination between national, regional/local and international research activities needs greater focus, revisiting priorities
- Greater attention should be given to policy-relevant research themes such as transition to a green economy; SD in land use planning; and wellbeing and SD.
- The viability of a SD policy/research community depends the interdisciplinary identity in organisations and peer review; influence on the research impact debate; new contributions to sustainability competence and learning for policy practice; and a growing role of social science for sustainability in research and policy.
- The diversity of policy/research interfaces requires different approaches such as more engagement with local and international spheres; a new emphasis on sustainability goal-setting; new collaborative models for knowledge synthesis and co-production; and a new focus on practice and experiment to inform policy.

(f) **A shared identity**

The growing diversity and interdisciplinarity of SD research needs a more favourable institutional context for professional learning and careers. The SD research community needs a stronger shared identity in the face of growing diversity and specialisation. Its blend of interdisciplinarity must be meaningful and relevant for a core network of researchers and practitioners.

(g) **Conclusions from the SDRN review**

In its conclusion, the review stated that "there remains considerable potential for enhancing the boundary capabilities for knowledge exchange with users in both policy and innovation, while the shift in innovation policy from a science-driven to a challenge-led approach and a new emphasis on transformative systems innovation needs to be consolidated and translated into new modes of policy relevant research".

The review concluded that new activities which span the boundary between research, policy and practice need to be built on and extended, taking advantage of the "growing permeability between policy and innovation" and the "opportunities for challenge-led practice-based research with a multiplicity of public and private stakeholders at difference governance levels."

3 Discussion

3.1 What Is Sustainable Development Research?

The contested and dynamic nature of the term 'SD' or 'sustainability' (Dobson 1999) is both a help and hindrance to attempts to set out and deliver an agenda for SD research. The ambiguity of the term allows the scope of SD and/or sustainability research to be drawn widely and dynamically, changing as priorities shift and the scope flexes. But this can also be a hindrance, as presenting a stable long-term agenda for SD research becomes difficult and has to be continually negotiated and defined. In essence, despite its ambiguity and lack of a precise or stable definition, the scope of the SD agenda has continually sought to encompass positive economic, social and environmental outcomes. Consequently, SD research could be expected at least to comprise research activities within and between the various disciplines focused around these outcomes from the social sciences, economics, humanities and life sciences, natural sciences and engineering.

However, a further enduring question in SD has been what are (and should be) the relationship between the three elements of economy, society and environment? Are they 'pillars' or 'legs of a stool'? are they 'strands in a triple-helix'? Are they 'nested circles' working outwards from economy, to society to environment? There is no clear, stable and unambiguous answer here and it is not the goal of this paper to attempt to present such an answer. It is unsurprising that agendas for SD research have been similarly amorphous and flexible but SDRN has maintained an enduring objective as an interface between SD researchers, policies and practitioners, even as its objectives and focus of its activities changed throughout its evolution. In particular, in its final phase of government funding, SDRN adapted to the more radical shift towards 'mainstreaming sustainable development'. Here, SDRN coped with both the relative reduction in the value of the currency of the terms SD and sustainability within government and their reframing within agendas such as green economy, resilience and wellbeing.

The nature of the concept of SD and way it frames issues may be part of the problem:

- SD tends address large-scale, complex and transformative challenges often with ill-defined or long-term endpoints for example in economic and energy systems or enhancing natural capital.
- SD cuts across traditional policy areas and priorities; being seen as conflicting rather than as supporting.
- Public policy debate is not conducive to addressing long-term transformative challenges. Many accept or support many sustainability challenges and the need for action. However, public debate takes place in a less reflective and supportive environment which trends not to encourage exploration, experimentation and learning.
- While formally the responsibility of all government, there is a danger of SD being no-one's responsibility. It can be seen as an additional burden under conditions of reduced resources, ands increased pressure to deliver outputs and short-term impacts on deficit-reduction and economic growth.
- SD requires an evolutionary or iterative response. Solutions need to be worked towards over time, through learning and the engagement of different societal actors.

These criticisms can easily be used to suggest a diminishing value in SD as an underpinning or cross-cutting framework for policy. However, equally well they can be reformulated into aspects which could be considered as relevant to good policy-making or research and necessary to address. For example, Waas et al. (2010) suggest a set of characteristics (Table 2) that could be used to conceptualise SD research that need not all be met but which can help with framing research better able to contribute to SD.

3.2 Issues Arising for Sustainable Development Research

This paper has provided an overview of the activities of the Sustainable Development Research Network in the UK (SDRN), an initiative funded by UK government bodies between 2001 and 2014. Throughout its life-span, SDRN acted as boundary organization (Guston 2001), bridging the communities of SD research,

Table 2 Preliminary set of characteristics of university research for sustainable development proposed by Waas et al. (2010)

Content	Process
1. Different levels of scale (local–global)	1. Action oriented
2. Different time perspectives (short, medium and long term)	2. Collaboration (international and sectoral)
3. Distribution aspects	3. Continuity
4. Multidimensionality (economy, environment, institutional, social)	4. Environmental, safety and security management
5. North–south	5. Independence
6. Precautionary principle	6. Knowledge transfer
	7. Multi-/interdisciplinarity
	8. Normativity
	9. Participation (including local knowledge)
	10. Proactive
	11. Problem oriented
	12. Public interest
	13. Societal peer review
	14. Impact monitoring
	15. Relevance check
	16. Transparency

policy and practice. In this role, SDRN helped to facilitate two-way interaction across these fluid boundaries, both bringing the findings of SD research into the policy and practice spheres in one direction while also seeking to encourage the research community to work on issues relevant to SD policy-making and implementation. Such activities are conducted in necessarily dynamic, fluctuating and, at times, contested contexts, with the boundaries being continually redrawn between research, policy and practice, between institutions, and between different framings of SD.

This paper has focussed on the key messages arising from the SDRN's 2013 review of the SD research landscape. The review noted significant changes including a greater diversity of users, funders, and researchers, policy action on sustainability occurring at a range of governance levels, especially at the sub-national level. In addition, the review highlighted the effect of the *Mainstreaming Sustainable Development* drive brought in by the Coalition Government following the 2010 General Election. Here, the key shift was to seek to embed SD within a broad range of policies, rather than view SD as a separate policy issue. However, the incoming government also heralded an 'age of austerity' characterised by sharper political contestation of economic and environmental purposes. Here, the terms SD and sustainability became further contested, with issues such as green economy, resilience and wellbeing being offered as alternative framings where SD became less favourable or less salient as a term.

Within the research community itself, the SDRN review called for a number of critical shifts to embed SD firmly. The review called for greater effectiveness in research funding coordination and priorities, including between national, regional/local and international research and greater attention paid to policy-relevant themes. The review also suggested the need to strengthen inter-disciplinarity, shared identity and competence for researchers, policy-makers and practitioners, including collaboration on knowledge synthesis and co-production, enhancing 'boundary capabilities'; and experimental approaches to inform policy and practice. Furthermore, the review called for SD researchers to influence the emerging debate around research impact, shifting from a science-driven to a challenge-led approach.

The review concluded that new activities which span the boundary between research, policy and practice need to be built on and extended, taking advantage of the "growing permeability between policy and innovation" and the "opportunities for challenge-led practice-based research with a multiplicity of public and private stakeholders at difference governance levels."

3.3 Recent Developments

In the two years following the SDRN review, there were a number of important developments that affect the landscape for SD research. In 2014, Defra's removal of funding for SDRN resulted in significant reduction of SDRN's activities but more extensive activities are currently being considered. Also in 2014, the international Future Earth initiative published its 2025 vision and negotiations began to set up UK committee. In 2014 the Research Excellence Framework (REF) assessment process took place and the Government launched a fundamental review of the Research Councils to be conducted by Sir Paul Nurse, the President of the Royal Society (the UK's National Academy of Science).

In 2015, the results of the REF were published, with much attention on its key innovation, the assessment of research impact (King's College and Digital Science, 2015). In particular, the 2014 REF involved 154 universities and higher education institutions submitting 6975 impact case studies across 36 'units of assessment' based on broad disciplines (http://www.ref.ac.uk/panels/unitsofassessment/). The database of impact case studies (http://impact.ref.ac.uk/CaseStudies/search1.aspx) provides a search facility enabling the impact case studies to be interrogated by keywords. Table 3 summarises the numbers of impact case studies and units of assessment using keyword searches on the range of familiar terms associated with 'sustainable development' referred to above. In the REF, the units of assessment were arranged into four Main Panels (A life sciences; B physical sciences and engineering; C social sciences; and D humanities). Table 3 shows how the impact case studies under each of these search terms breaks down by Main Panel.

Across this range of search-terms, "green economy" is significantly the least frequent, suggesting that this term was not well developed at the time when impact case studies were being prepared (2013–2014), not supporting the SDRN 2013 review's suggestion that SD had become reframed as green economy. By contrast, "sustainability" appears considerably more frequently than "sustainable

Table 3 Prevalence of search-terms related to 'sustainable development' within the REF 2014 impact case studies

Search term	Main panel A	Main panel B	Main panel C	Main panel D	Total
Sustainable development	11	23	106	24	164
Sustainability	72	133	271	91	567
Resilience	33	57	98	35	223
Wellbeing	143	30	129	72	374
Green economy	1	4	8	2	15

development" suggesting that this framing was considerably more attractive. The terms resilience and wellbeing were also more popular than "sustainable development".

Looking across the Main Panels, it is evident that the social sciences (Main Panel C) accounted for the majority of the case studies relevant to sustainable development. Main Panel C included a number of more 'obviously' SD-related units of assessment, (approximately 65 % of the case studies referencing "sustainable development" and 48 % of those "sustainability" case studies). The SDRN review suggested that social sciences were becoming more prominent in SD research and the REF 2014 impact case studies reflect this.

This simple analysis is not a precise assessment of the prevalence of SD within the REF impact case studies. As these terms are both ambiguous and contested, further assessment would be needed to examine the specific records to understand the context of their use. Also, these figures are likely to represent an element of double-counting as more than one term may have been used in any individual case study. Nevertheless, this initial analysis provides a useful starting point to gauge prevalence.

Within policy circles, towards the end of 2015, the United Nations adopted the Sustainable Development Goals (SDGs) and the Future Earth UK Committee was established and met for the first time. In December 2015, Sir Paul Nurse published the results of his review of the research councils (Nurse 2015) with the government agreeing to take the recommendations forward (HM Treasury 2015) and Defra scientists responding (Boyd 2015).

The Nurse Review called for a greater coordination between those involved in the 'research endeavour' in driving excellent research to help meet social, economic and environmental objectives. A key recommendation within the Nurse Review was to create a new body, Research UK, to develop UK research strategy and to enable multi-and interdisciplinary research and on 'challenge-led' research, bringing researchers and research-users together in co-production partnerships.

3.4 Relationship of Research, Knowledge, Policy and Practice

The relationship between research, knowledge, policy and practice is often overlooked or not made explicit in thinking about the mechanisms of research and policy interactions. However, it is important for effective exchange between research and policy. Best and Holmes (2010) propose three generations of thinking about how to bridge evidence and policy/practice—or knowledge to action: linear models, relationship models and systems models, suggesting three phases of development in an organisation's approach to knowledge transfer and promoting impact: firstly a simple linear-model of communication and dissemination: research is conducted, then communicated and then used. Secondly, developing knowledge transfer opportunities through building relationships, networks and stakeholders: introducing a dynamic between research and use. Thirdly, organisations moving towards a more whole-systems approach, addressing organisational issues including learning, information systems and leadership. Importantly, Best and Holmes argue that each subsequent phase adds to rather than replaces the former so that systems approaches are complemented by communication systems and networks with stakeholders.

Similarly, Young et al. (2014) emphasise the importance of models of research and policy interaction in relation to biodiversity: 'Many initiatives exist to improve communication, but these largely conform to a 'linear' or technocratic model of communication in which scientific 'facts' are transmitted directly to policy advisers to 'solve problems'. While this model can help start a dialogue, it is, on its own, insufficient, as decision taking is complex, iterative and often selective in the information used.'

3.5 Improving the Interaction of Sustainable Development Research and Policy

While there are numerous debates in research and policy in general, these also play out in SD research. This should seek to improve links between research and policy-makers working on SD and secondly ensure SD is seen as a relevant and useful construct for wider policy debates and government priorities. To address these issues Young et al. emphasise the need to:

(a) Frame research and policy jointly;
(b) Promote inter- and trans-disciplinary research and 'multi-domain' working groups, including scientists and policy makers from various fields and sectors;
(c) Establish· structures and incentives supporting interactive dialogue in the long-term.

An observation made by one of this paper's authors drawn from the 2014 SDRN conference[1] is that there are many related but parallel debates being had on government policy and research to support policy which could be better integrated. Processes to identify shared questions, analyses and objectives for future research could add value to current and future policy development.

Addressing these issues is a challenge for SDRN to consider but also for government and others (e.g. the research councils, Future Earth, Horizon 2020): to develop appropriate channels of research/policy interaction to focus on sustainability challenges, requiring multiple and interlinked organisations.

4 Conclusion—Where Next for SD Research and the SDRN?

Nurse's calls for a greater focus on challenge-led, interdisciplinary, co-production approach to research (through Research UK) chimes strongly with the Future Earth 2025 Vision. Therefore, with the recent formation of the Future Earth UK Committee, a timely opportunity arises to bring SD research and the SD research community more to the fore. Challenges to be taken forward by Research UK could be interpreted and framed in the context of the SDGs, progressing integration across disciplines, across the boundaries between researchers, policy-makers and practitioners; and across levels of governance, from local to international—all issues highlighted by SDRN in 2013.

SDRN occupies a unique point as a boundary organisation that can act as a focus for strengthening both coordination among SD researchers and brokering SD research into policy and practice. Such opportunities include:

- supporting Future Earth UK as a ready-made stakeholder community, focusing attention on the science/research aspects of the SDGs.
- supporting the Stakeholder Forum (http://www.stakeholderforum.org/index.php) and its UK Policy Advisory Panel and other actors in driving UK strategy for delivering on the goals
- coordinating a SD-research focused response to Nurse Review, especially in relation to the role of Research UK in multi- and inter-disciplinary, challenge led co-production approaches to research, innovation and knowledge exchange. The UK Government has said that, subject to legalisation, it will take forward the recommendations of the Nurse Review and the SD research community can play a strong role in this.

[1]See http://www.sd-research.org.uk/latest/sdrn-annual-conference for details of conference agenda and presentations.

These ideas are suggested as possible ways that SD research and policy capacity could be built. They are not the only ways this could be done. However, we are clear that there is a need to reinvest in the institutional capacity and architecture to address these SD issues across research, policy and wider society. We offer these perspectives and ideas in the hope of stimulating debate, and vitally, the action necessary if the substantial challenges we are currently facing on SD are to be resolved.

References

Best and Holmes. (2010, May). System thinking, knowledge and action: towards better models and methods. *Evidence & Policy: A Journal of Research, Debate and Practice*, *6*(2), 145–159 (15).

Boyd, I. (2015). The nurse review: A response from Defra's scientists. https://ianlboyd.wordpress.com/

Clark, W. (2007, February 6). Sustainability science: A room of its own. *PNAS*, *104*(6), 1737–1738. http://www.pnas.org/content/104/6/1737.full.pdf

Defra. (2011a). *Sustainable development research network phase IV—Science and research projects*. SD1223. http://sciencesearch.defra.gov.uk/Default.aspx?Menu=Menu&Module=More&Location=None&Completed=0&ProjectID=17928

Defra. (2011b). *Mainstreaming sustainable development—The Government's vision and what this means in practice*. Department for Environment, Food and Rural Affairs. February 2011. https://www.gov.uk/government/uploads/system/uploads/attachment_data/file/183409/mainstreaming-sustainable-development.pdf

Defra. (2012). *Defra and DECC social science expert panel*. http://sciencesearch.defra.gov.uk/Default.aspx?Menu=Menu&Module=More&Location=None&Completed=0&ProjectID=18289

DETR. (1999, May). *A better quality of life, Cm 4345*. London: The Stationery Office.

Dobson, A. (Ed.). (1999). *Fairness and futurity: Essays on environmental sustainability and social justice*. Oxford, New York: Oxford University Press.

Eames, M. (2002). *A new agenda for UK sustainable development research*. London: Policy Studies Institute.

Eames, M. (2006). Securing the future: The role of research. interdisciplinary, cross-cutting and strategic research needs to support the UK sustainable development strategy. An SDRN report to the Environment Research Funders Forum. SDRN. http://www.sd-research.org.uk/sites/default/files/publications/Securing%20the%20Future%20The%20Role%20of%20Research_0.pdf

Future Earth. (2014). *Future Earth 2025 vision*. Future Earth Interim Secretariat, Paris. http://www.futureearth.org/sites/default/files/files/Future-Earth_10-year-vision_web.pdf

Guston, D. (1999). Stabilizing the boundary between US politics and science: The role of the Office of Technology Transfer as a boundary organization. *Social Studies of Science, 29*(1), 87–112.

Guston, D. (2000). *Between politics and science: Assuring the integrity and productivity of research*. New York: Cambridge University Press.

Guston, D. (2001, Autumn). Boundary organizations in environmental policy and science: An introduction. *Science, Technology, & Human Values, 26*(4), 399–408. Sage Publications, Inc. Stable URL: http://www.jstor.org/stable/690161

HM Government. (1999). *Sustainable development: The UK strategy, Cm 2426*. January 1994. London: HMSO.

HM Treasury. (2015). Spending review and autumn statement 2015. https://www.gov.uk/government/publications/spending-review-and-autumn-statement-2015-documents/spending-review-and-autumn-statement-2015

Kass, G. (2015). *Implementing sustainability science in UK nature conservation policy and practice*. Paper presented at Global Sustainability Research Conference, Cambridge, Cambridge, January 26–27, 2015.

Kates, R. (2011, December 6). What kind of a science is sustainability science? *PNAS*, *108*(49), 19449–19450 (Robert W. Kates). http://www.pnas.org/content/108/49/19449.full.pdf

King's College and Digital Science. (2015). The nature, scale and beneficiaries of research impact: An initial analysis of Research Excellence Framework (REF) 2014 impact case studies. Research Report 2015/01. King's College London and Digital Science, March 2015. http://www.hefce.ac.uk/media/HEFCE,2014/Content/Pubs/Independentresearch/2015/Analysis,of,REF,impact/Analysis_of_REF_impact.pdf

Komiyama, H., & Takeuchi, K. (2006, October). Sustainability science: Building a new discipline. *Sustainability Science*, *1*(1), 1–6. First online: 24 August 2006. http://link.springer.com/article/10.1007/s11625-006-0007-4

Mulgan, G., & Puttick, R. (2013). Making evidence useful: The case for new institutions. *NESTA*. http://www.alliance4usefulevidence.org/assets/MakingEvidenceUseful.pdf

Nurse, P. (2015). *Ensuring a successful research endeavour: Review of the UK research councils by Paul Nurse*. Department of Business, Innovation and Skills, 19 November 2015. BIS/15/625. https://www.gov.uk/government/uploads/system/uploads/attachment_data/file/478125/BIS-15-625-ensuring-a-successful-UK-research-endeavour.pdf

Royal Society. (2015). What on Earth is future Earth? https://blogs.royalsociety.org/in-verba/2015/11/20/what-on-earth-is-future-earth/

SDRN. (2007). Sustainable development research network (phase III). http://www.psi.org.uk/site/project_detail/sustainable_development_research_network_phase_iii

SDRN. (2011). Sustainable development research network (Phase IV). http://www.psi.org.uk/site/project_detail/sustainable_development_research_network_phase_iv

Solesbury, W. (2001). *Evidence based policy: Whence it came and where it's going*. ESRC UK Centre for Evidence Based Policy and Practice. Working Paper 1. http://www.lgsp.uz/old/publications/option_paper_training/ebp_when_it_came_and_where_it_is_going_eng.pdf

Steward., F, Mocca, E., Shaw, B., Watson, B., & Bicket, M. (2013). *Mainstreaming sustainable development research in an age of austerity: SDRN Review of the UK Sustainable Development Research and Policy Landscape*. Policy Studies Institute.

Steward, F., & Piterou, A. (2012). The UK sustainable development research network—Analysing the interaction of knowledge and policy. In *Evidence for Sustainable Development Conference*. Berlin, October 5–6, 2012.

Waas, T., Verbruggen, A., & Wright, T. (2010). University research for sustainable development: Definition and characteristics explored. *Journal of Cleaner Production, 18*(2010), 629–636.

Young, J. C., Waylen, K. A., Sarkki, S., Albon, S., Bainbridge, I., Balian, E., et al. (2014). Improving the science-policy dialogue to meet the challenges of biodiversity conservation: Having conversations rather than talking at one-another. *Biodiversity and Conservation, 23*, 387–404. Open access: http://link.springer.com/article/10.1007/s10531-013-0607-0

Understanding Attitudes Towards Native Wildlife and Biodiversity in the UK: The Role of Zoos

Adriana Consorte-McCrea, Alan Bainbridge, Ana Fernandez, Dennis Nigbur, Siri McDonnell, Aïssa Morin and Oksana Grente

Abstract

The present paper draws from a study of the role of zoos in forming attitudes towards biodiversity and native wild carnivores that are considered for reintroduction. The project is being developed by an interdisciplinary team (wildlife conservation, psychology, education) working towards the development of a questionnaire to investigate this topic in the UK. Research suggests that experiences with live animals in zoos may encourage empathy, through personal connection, which in turn facilitates greater concern towards biodiversity. Concomitantly, the reintroduction of wild carnivores to their native habitats may contribute to biodiversity by helping regulate ecosystem dynamics. Carnivores also carry a rich cultural and historical heritage. IUCN guidelines state the need for public support to establish a reintroduced population in the wild, therefore, carnivore restoration efforts benefit from the understanding of the human dimensions. A pilot study was carried out in Kent (spring 2015) using focus groups and interviews to investigate attitudes towards biodiversity, with particular focus on two species of carnivores native to the British Isles and currently considered for reintroduction (the European lynx *Lynx lynx* and the pine marten *Martes martes*) and the role of zoos in promoting support towards

A. Consorte-McCrea (✉) · S. McDonnell · A. Morin · O. Grente
Wildlife and People Initiative-Ecology Research Group, School of Human and Life Sciences, Canterbury Christ Church University-CCCU, Canterbury, Kent CT1 1QU, UK
e-mail: adriana.consorte-mccrea@canterbury.ac.uk

A. Bainbridge
Auto/Biography and Narrative Research and Knowledge Exchange Group, Faculty of Education, Canterbury Christ Church University-CCCU, Canterbury, UK

A. Fernandez · D. Nigbur
Society & Environment Group, School of Psychology, Politics and Sociology, Canterbury Christ Church University-CCCU, Canterbury, UK

W. Leal Filho (ed.), *Sustainable Development Research at Universities in the United Kingdom*, World Sustainability Series,
DOI 10.1007/978-3-319-47883-8_17

biodiversity conservation. Results suggest an association between seeing native wild carnivore species in the zoo and emotional responses such as 'breaking down fears', but also concerns about a disconnect between people and nature, and misunderstanding about the role of zoos in 'protecting' species. Below we offer a discussion of the themes that emerged from the analysis of focus groups and interviews in relation to biodiversity.

Keywords
Biodiversity • Environmental sustainability • Zoos • Wild carnivores • Attitudes • Narrative

1 Introduction: Environmental Sustainability and Biodiversity Targets

Since the Convention on Biological Diversity- CBD was created in the wake of the Global Forum Rio 92, international mobilization to address the biodiversity crises is still to meet the targets outlined by consecutive international agreements. In 2010, the worldwide CBD Strategic Plan 2011–2020 produced the Aichi Biodiversity Targets to be met by 2020:

> Meeting the Aichi Biodiversity Targets would contribute significantly to broader global priorities addressed by the post-2015 development agenda; namely, reducing hunger and poverty, improving human health, and ensuring a sustainable supply of energy, food and clean water. Incorporating biodiversity into the sustainable development goals(...) provides an opportunity to bring biodiversity into the mainstream of decision-making" (Secretariat of the Convention on Biological Diversity 2014:10).

The importance of promoting public awareness about the values of biodiversity and actions to support and to sustainably use it has been particularly recognised and comprises the first of the Aichi Biodiversity targets.[1]

A growing commitment to biodiversity conservation has also been reflected by zoos and aquaria directives. BIAZA (Britain and Ireland Association of Zoos and Aquariums) actions for the conservation of biodiversity are guided by WAZA (The World Association of Zoos and Aquariums), which formally supports the UN Decade on Biodiversity and has committed to develop a framework for guiding member zoos in meeting the Aichi Targets (WAZA 2005, 2011; Moss et al. 2015).

[1]Aichi Strategic goal A. Address the underlying causes of biodiversity loss by mainstreaming biodiversity across government and society; Target 1: By 2020, at the latest, people are aware of the values of biodiversity and the steps they can take to conserve and use it sustainably. https://www.cbd.int/decision/cop/?id=12268.

Within the context of biodiversity conservation, international conventions encourage the use of reintroduction to restore populations of native species (see Bern Convention (1979), Article 11(2); and CBD (1992), Article 9(c)). The IUCN's guidelines for reintroductions state that to establish a viable, free-ranging population in the wild it is necessary to enlist public support (IUCN/SSC 2013). Investigating the attitudes and understanding of diverse interest groups is therefore instrumental to inform strategies for the conservation of biodiversity.

2 Findings from Pilot Focus Groups and Interviews in Kent

2.1 Methodology

Focus group sessions were 90 min long and individual interviews were 60 min. Volunteers gathered at Canterbury Christ Church University- CCCU (8 for focus groups; "Mary" and "David" for interviews), Wildwood Trust (zoo specialised in native wildlife; 5 for focus groups) and Howletts Wild Animal Park (zoo mostly focussed on exotic wildlife; 6 for focus group and "Sadie" for interview). The call for zoo public volunteers was circulated on the Facebook page of respective zoos 2 weeks in advance of event, and participants were offered free family tickets to visit the zoo; for the CCCU based event, invitations were distributed in public spaces, cafes, notice boards in Canterbury and at CCCU. Each of the 22 volunteers (over 18) was offered a £10 high street gift voucher as reward. Once volunteers made contact they were sent a formal invitation to participate and a letter of information about the study. Some scripted questions were used to prompt discussion, but participants were free to elaborate and deviate. We used an extended model of The Theory of Planned Behaviour (Ajzen 1991) as our theoretical framework to plan focus groups' questions, to elicit the thoughts of participants about the protection, conservation and reintroduction of the focus species, responsibility, action, opportunities and risks posed by their presence. These were intercalated with some information about the species, their ecology, status and threats. During interviews each participant was asked a single stimulus question "Can you tell me as little or as much as you wish about yourself and your interest in the re-introduction of wild carnivores, such as the lynx and pine martin and biodiversity." When the participants had finished talking the interviewer would ask further prompt questions about the content of the response. Events were audio recorded and transcribed, with permission of participants. The qualitative analysis that followed aimed to identify key issues that emerged during the discussions, to inform the planning of questionnaires for the main body of this study. Comparisons between the samples in relation to site were not intended considering the scope of this pilot. The following themes relating to biodiversity emerged from this analysis. Quotes are taken from transcription, and the names of participants have been omitted or changed.

2.2 Restoring Biodiversity: Spatial Concerns and Co-Existence

At the beginning of all three sessions focus group participants voiced concerns about humans sharing space with reintroduced carnivores: "*if there were wolves introduced to Scotland, how long would it take them to come down towards the more urban areas? The more heavily populated areas of the country*?". Even participants who were enthusiastic about the return of wildlife to other countries had reservations about encountering native predators in their own backyard. There was an underlining sense of fear towards predators that people in the UK did not grow up with:

> (Adam) People psychologically reject wilfully bringing something in that has the potential to bite you.

Participants believed that most people would fear the reintroduction of large carnivores based on preconceptions and economic interests (game and farming). Concerns were also voiced about already established species—if there is not enough space to be shared among all species, we should focus on those already in the wild.

> it's also understanding the impact already on the existing wildlife and ecosystems by reintroducing another species that have been dead for… They're going to need another food source if you start reintroducing these other species….

Fears were also voiced concerning the safety of reintroduced animals, based on a long history of retaliations:

> (Charlotte) "…it would be a case of trying to protect us so that they (reintroduced animals) then stay protected."

After centuries of intensive persecution, wild carnivores started making a comeback during the last few decades. In spite of localised conflicts, most populations of large carnivores are increasing and spreading through mainland Europe, some aided by successful reintroduction initiatives. There may be concerns that although the reintroduction of carnivores may benefit biodiversity under appropriate conditions, it becomes more problematic in human-modified landscapes such as the ones in most of the UK (Linnell et al. 2005; Ray 2005). However, Scotland in particular has experienced large-scale reforestation accompanied by an increase in populations of woodland deer, which resulted in large connected areas of suitable habitat for many native carnivores, such as pine marten and for a viable populations of lynx[2] (Wilson 2004; Hetherington 2008; Hetherington et al. 2008). At the moment, herbivory has a high impact on the economy and on the biodiversity of the area, which studies suggest can be much improved by the reintroduction of the lynx.

[2]According to population viability analysis over 20,000 km^2 of Scottish habitat exists and it is suitable to support around 450 lynx (Hetherington et al. 2008).

2.3 Biodiversity and Wild Carnivores

Half-way through the sessions, focus groups were introduced to two news features (2.5 min each "Will wild lynx return to Britain?" and "Reds Return" (BBC News 2015; BBC Radio 2015) and to basic information about the ecology of the European lynx and pine marten. Although some participants had previous knowledge, for most the role of native carnivores in controlling ecosystems made sense then ("*Lynx helps to regenerate forest, which is good for people*"; "*Pine martens benefit the native wildlife by controlling grey squirrels*"). Beliefs were also voiced that increased biodiversity means less need for human intervention in management of populations; and that increased biodiversity benefits entire systems—extending to flora and geographical features (e.g., rivers; erosion):

> You could think, well... there is too many deer, they are eating our plants, why don't we just shoot them? That doesn't have the same effect, and part of the reason why it doesn't have the same effect is because the deer doesn't really understand being shot and they don't know how to avoid it, but because they understand how wolves predate on them this changes their behaviour in a different sort of way, so they stay away from certain areas, which regenerates.

In fact, wild predators require biodiverse habitats and also help to maintain their integrity by provoking cascading effects that affect the structure of communities; they are sensitive to impacts to ecosystems' integrity and provide food resources to other trophic groups (scavengers) (Jedrzejewska and Jedrzejewski 2005; Linnell et al. 2005; McShea 2005; see Sergio et al. 2006 for a review). In their absence ungulates overgraze and have a detrimental effect on biodiversity as they impact on plant species and consequently on birds, insects and mammals (see study by McShea 2005; Jedrzejewska and Jedrzejewski 2005). Therefore conservation strategies to protect them also meet the needs of many other species.

Nevertheless, proposals to restore ecological function by reintroduction of wild carnivores are often met by public misconceptions of what pristine environments should be like, modelled on the idea of parks and woodlands that have been dominated by large populations of deer and small predators for many generations (Steneck 2005).

Resigning to the fact that human influence cannot be excluded from natural systems in Europe, Linnell et al. (2005:393) suggest that the role of carnivores goes beyond their potential to support ecological function:

> Many view the return of carnivores as highly symbolic, almost as the ultimate test of human ability to coexist with biodiversity. In other words, although we cannot achieve wilderness (…), we can at least restore some of the wildness to the landscape.

3 Barriers to Understanding Why *Biodiversity* Matters

In spite of an understanding about benefits of a rich array of species and habitats for healthy ecosystems, the term *biodiversity* in itself was considered abstract and uninspiring by focus groups' participants. There was a belief that scientific terminology alienates general public and causes disconnection:

> The language around it needs to change because I actually think biodiversity sounds… I feel detached from that term.

> It's like we are trying to get to something that we have not had before, but actually we had biodiversity for centuries and centuries and centuries until the last fifty years and it was called nature. I wonder whether why we have to kind of keep inventing this language that actually means quite a lot people don't buy into it because they don't think it has anything to do with them and I think that a really important way forward to try and change the language so people can relate to it and feel a part of it.

> Biodiversity I think that most of the time it would just go over people's heads. They'd have to google what the word meant.

The open-ended narratives also provided a wide range of understandings as to the nature of biodiversity: from seemingly simplistic conceptions to complex notions of the place and role of humans in the wider environment. Sadie's narrative suggested less developed understanding of biodiversity: she was unclear as to what this might be and spent some time discussing wider issues of diversity including cultural and ethnic groups. She appreciated that biodiversity included a mix of animals and plants, that the animals needed to be happy and yet was concerned that even natural predation may upset animals. Equally, she was concerned that a return to a more diverse animal population would threaten both humans and domesticated animals.

Mary acknowledged a degree of confusion about what biodiversity might be but did have a well-established understanding of the complexity of interaction between all species, including plants and animals. She admitted to finding the reality messy as there was still '*lots that we do not know about*' but despite this was able to articulate about the role of carnivores/predators to 'keep things in check'. David's descriptions of biodiversity offer the most complex understanding and were quite clear that '*whether we are aware of it or not—everything is connected*' and that this involves the whole planet. From the very start of David's narrative the theme of interconnectedness was central to how he conceived biodiversity and he returned to and embedded this with his responses throughout. Both Mary and David articulated the place of humans within the natural environment.

Consistent patterns of biodiversity loss indicate that the message of urgency regarding the conservation of biodiversity is either not reaching the general public, or not engaging them. As highlighted above some of the barriers that may prevent people from engaging with biodiversity issues relate to difficulty in engaging with the term *biodiversity* itself: Novacek (2008) noted that not only there is a lack of familiarity with the word but it does not convey the intricate interconnectedness and interdependence between life forms and their environment.

4 Wild Carnivores as a Biodiversity Conservation Tool: Connecting People and Nature

During focus groups there was evidence that emotion plays a part in how people relate to biodiversity and to its loss. Some participants voiced feelings of guilt over harm caused by humans:

> I think it's always very sad when you hear that something else has become extinct, we are basically becoming poorer and poorer with various… whether it be other animals or plants or whatever, you just realise that your own species is having such a negative impact on the rest of the world and animal populations.

Other conveyed feelings of excitement:

> I think it makes the environment more exciting as well and I think... you know, you have got all these different species and stuff living free within the UK. I think it is quite exciting.

Michael J. Novacek, of the American Museum of Natural History (2008:1157) argues that it is essential to foster a connection between people and nature to engage them with the biodiversity crisis. He says:

> That linkage should be built from a clear and compelling message about the importance of biodiversity and what we risk in depleting it.

While studies suggest that a cognitive element bears an important role in attitudes towards the conservation of wild carnivores (Roskaft et al. 2007; Bath et al. 2008; see Consorte-McCrea 2011 for a review), knowledge is not enough to predict attitudes towards biodiversity.

In the Wildwood focus group, although some believed parents with young children feel more negative about danger of reintroduced predators in the countryside or towns, there were also considerations for the long term benefits for future generations:

> (Charlotte) "I think for the people with children they've got to look at the future of their children as living as part of the planet. That if we introduce these, it will benefit their children because there will be more forests and natural environments for our animals and show that as a positive thing for the adults of children, that's what they're going to grow up with. It's not going to be a major threat to your child, it's going to be a positive step for them in the future."

Such concerns fit in with findings in cognitive development which suggest that the development of active environmental concern may be influenced by early life experiences (Keliher 1997; Bjerk et al. 1998a, b). According to research, the development of appreciation and value towards wildlife in children and adolescents is mediated by frequent access to nature areas (including gardens, parks or wild places) in urban and rural settings; positive messages from relevant adults about wildlife; and opportunities to take part in varied wildlife related activities in a safe and supportive environment. Declining opportunities to engage with nature from childhood, on the other hand, promote a lack of interest in nature and commitment to biodiversity conservation, while misconceptions and negative messages about wildlife may foster negative perceptions and limit their interest in relation to wildlife (Pyle 2002; Kellert 2002; Velsor and Nilon 2006).

Wilson (1984) proposes we have a connection with nature, both developmentally (Kahn and Kellert 2002; Clayton and Myers 2009) and in the way nature affects us emotionally, and the best opportunities to promote our engagement with biodiversity are offered by direct experience with living organisms (Dingwall and Aldridge 2006; Weprin 2007; Novacek 2008). Nevertheless "learning, culture and experience" seem necessary to strengthen our innate bond with living organisms (Kellert 2002; Hinds and Sparks 2008:110).

Amongst focus group participants, biodiversity was also seen as enriching in itself:

> I think it enhances my life experience to know that if you go down to Worth you might be able to see a beaver. I probably never will, but it is the fact they exist.

There was also a sense that wild carnivores are charismatic species that can catalyse attention towards biodiversity ("*not just having some nice carnivores around*"), as there were beliefs that "*the lynx benefit biodiversity*", "*the lynx is a symbol of UK's biodiversity*" and "*pine martens are 'a force for good*'", which go beyond their direct benefits to ecosystem function.

The narratives provided by Mary and in particular David contained very complex thinking in relation to the role and position of humans within the natural world. Mary was unequivocal that all humans and animals are connected—even at the consciousness level—and that humans are part of biodiversity and consequently the food chain. She provides a compelling account of her decision to walk into the wilds of Canada and despite seeing the signs of bears she was prepared to take the risk of a possible bear attack. It is not as though Mary was not afraid that enabled her to take this risk but rather her deep connection with the natural world and her realization that the potential of becoming part of the food chain was an acceptable stance to take.

David provides a rich narrative that charts his shift from not appreciating the '*interconnectedness of everything*' to becoming aware of '*every vibration*' and the '*moveable energies and signs*' within the natural world. He talks about how this shift to a '*deep ecology*' originated from his past use of psychedelic drugs and encountering traveller communities. What David describes is how, before his drug use, the natural world gave him little or no pleasure but that since this period in his life he is now aware of the '*wonder*' of the interconnections. These views are not held with little awareness of the pragmatics of every day living, as David still grapples with the reality of re-introducing carnivores and their potential impact on both humans and animals. Just like Mary and Sadie, he is worried about the risks to humans and other animal species when re-introducing top predators, alongside knowledge that such animals have a role in maintaining a balance of nature. And also in agreement with Mary and Sadie he recognizes the impact of humans on depleting biodiversity and argues that humans have a responsibility to put right their wrongs.

Although David's understanding of biodiversity is one of deep ecological interconnectedness with humans as a natural part of this, he is also very aware that it is the action of humans that has put them at odds with the wider ecosystem. He goes as far to suggest that the action of humans, in a bid to '*rule*' the world, such as schools and buildings, have separated humans from their natural world. This act of separation is for David the root cause of the inability for many others to not see the wonder in nature that he does. David describes his relationship with the interconnectedness of the natural world as the source of meaning for his life. Indeed, he actively seeks to reconnect by spending time in the woods, by slowing down and listening to the birds and trees.

When support for the conservation of wild carnivores is considered, however, the experiences involved in a rural upbringing seem to play an important role. Negative attitudes towards carnivores in rural areas may be associated with the expectation that encounters with carnivores put themselves or their families in danger, and may result in financial loss, while positive may be associated to the expectations and with the excitement of seeing animals in the wild (Roskaft et al. 2007; Consorte-McCrea 2014). As well as facilitating connections, emotions also seem to reinforce intentions to engage with the natural environment (Hinds and Sparks 2008). A sense of connection with animal species may be a precursor to empathy, especially for species that are perceived as similar to us, and to an interest in taking action to protect them (Clayton et al. 2011, 2014).

5 The Role of Zoos in Connecting People, Carnivores and Biodiversity

Zoos were seen by participants of the focus groups as a place where safe contact with wildlife takes place. Within this context, seeing animals from a safe distance provides the sense of "wildness" suggested by Linnell et al. (2005), an illustration of reality of power, size, danger:

> (Elizabeth) "I think zoos surely are good things because they must bring up on most people that relationship of potential danger and you get more in touch with your natural – well the past probably when there were wolves around and things that could eat you."

Over 7000 million people visit WAZA member zoos and aquariums yearly, all over the world, and around 25 million visit BIAZA member zoos and aquariums in the UK alone (WAZA 2016; BIAZA 2016). Since the world's population has become mostly urban, for many people zoos provide the closest encounters they will ever have with wild animals, which may be powerful opportunities to connect with nature (Myers and Saunders 2002; Bowkett 2009; Clayton and Myers 2009; Packer and Ballantine 2010; Vanstreels and Pessutti 2010; Clayton et al. 2014). In turn, wild carnivores may help us engage with biodiversity by helping us glimpse into the dynamics and interconnections that are at its essence, as they "put some of the wild back into our lives" (Linnell et al. 2005:399).

Focus groups' participants were asked if seeing these animals face to face in a zoo made difference to their point of view about reintroduction. At Wildwood, were both focus species are kept, volunteers suggested that seeing the animals familiarizes people with them, as they believed native species are not usually exposed to the same visibility as exotic ones:

> (Ed) "...if they didn't see them in zoos they wouldn't really be aware of countryside animals that are under threat."
>
> (Charlotte) "Being able to see something and know what you're dealing with (...). The fear can be greater of the unknown than it can of seeing it and learning about it."
>
> Others related a break down on misconceptions: "The lynx is surprisingly small in real life"; (Diana) "...you just think it sounds scary but then you look at it and it's beautiful."

Gwynne (2008:51) suggests no media or museums have the "potential for moving people to care about an animal" in the same way that zoos have. As indicated by social development research, experiences with live animals in a zoo may encourage empathy, through a sense of personal connection, which in turn facilitates greater concern towards their conservation and ultimately for their native ecosystem, having an effect in the formation of lasting values (Myers and Saunders 2002; Falk et al. 2007; Clayton and Myers 2009; Clayton et al. 2011; Clayton et al. 2014).

Furthermore, research suggests that zoos still "support and reinforce" the positive values and attitudes of visitors who already have environmental identity and values (Falk et al. 2007:3; Sterling et al. 2007). While direct experience may promote a more affective evaluation of an object, repeated exposure to that object may strengthen the affective connections with it (Hinds and Sparks 2008). A sense of connection to animals and nature; an understanding of the ecological role of the reintroduction of carnivores; and support towards it, seems to increase with frequency of visits and membership to a zoo, suggesting a cumulative effect that builds on visitors' capacity for future learning and for action (Rounds 2004; Falk et al. 2007; Packer and Ballantyne 2010; Reading and Miller 2008; Clayton et al. 2014). The higher their sense of connection, the more visitors may use the zoo visit to reflect on their relationship with nature and concerns for the animals in the wild, suggesting that close associations with the zoo, through membership or frequent visits relates to positive emotional and cognitive responses to wildlife.

6 Linking Knowledge, Responsibility, and Action

Participants also recognised the education role of zoos, and there was an attribution of "good zoo" value to zoos according to their work in education and wildlife conservation: '*Good zoos*' educate visitors about conservation work and status in wild. '*Bad zoos*' just show off their animals.

> I have certainly been to other zoos, where the ethos is simply if you want to see the animals that you won't find in the wild, then we can show you some if you pay us enough money. So you see animals in cages, with no education going on

> I get the impression that some other zoos they've only got interest within themselves, it could be money making, it could be profit making.

Amongst focus group participants, seeing animals in the zoo was considered as an educational experience for children:

> There is probably a role for places like zoos, parks and things, to give children a face to face encounter with these other creatures. Partly to challenge some of their stereotypes they have already picked up. Challenging some of their ideas you know.

Sadie's narrative interview response also argued that humans need to prevent further species becoming extinct and saw this as one of the roles of 'good' zoos. During our interview that took place within Howletts Wild Animal Park, she made repeated links between the animals in captivity and successful breeding and re-introduction programmes. Howletts was a '*good*' zoo in Sadie's estimation due to their successes in re-introducing gorillas and rhinos and that the animals appeared to be happy enough to breed. Another strength identified by Sadie was the role of Howletts to provide quality information on each animal and their natural habitat.

Overall, zoos offer a wide range of learning experiences which can involve "reflecting, thinking, and acting" (West and Dickie 2007; Packer and Ballantyne 2010:31). This is particularly relevant in the current environmental climate when we consider that adults must engage with biodiversity issues right now and may have already left formal education streams. Visitors seem to particularly remember sights, sounds, smells, emotional affinity and connection, feelings of protectiveness, associated with being in close proximity of the animals, and the information about human caused threats to their survival (Packer and Ballantyne 2010). For some these resulted in reflection about their own responsibility and connectedness with nature and wider global issues and an increased desire to learn more, which impacted on their understanding, attitudes and behaviours towards the environment.

Some participants voiced a sense that humans need to work for biodiversity because we are responsible for its decline:

> I personally think it is about the bigger picture in a kind of way. I think, you know, we do share this world I suppose with many different species and many different animals. I think we have a right to conserve as much as possible, we probably do more damage than anyone... I think it's more about doing what I think needs to be done, not me personally thinking it should be done so we should do it… We are going to want to look out for our best interests but when are looking at animals that are going extinct, largely to what we have done, I think it is our right to help out as much as possible and we are in a place to do that.

A 'duty of care' reflected beliefs that humans must look after animals

> Us human beings do have a duty to look after these animals, you know, however many there are in the world or however many there are not.

This duty of care was reflected in the narrative interviews that shared a common theme between all three participants related to the impact of humans on biodiversity and as a consequence the responsibility for humans to correct the damage caused by their actions. Mary commented on how biodiversity and food chains are threatened by human egos and the whole planet was now in need of our help—she was quite strident that 'if we had ruined it then we should fix it'.

People's duty of care about environmental degradation and loss of wildlife may relate to a sense of responsibility and stewardship towards nature and concerns for future generations, which can be motivated by "aesthetic, ethical, patriotic, familial, and religious values". (USA Biodiversity Project, in Novacek 2008:11572). A *moral purpose* may be necessary to motivate *society* to meet challenges such as the ones presented by biodiversity loss, shifting the focus from individual moral choices to "our collective ability to recognize, reflect upon, and reasonably address the value questions we face." (Clayton and Myers 2009:53). Biodiversity loss affects people directly and play a great role in health, economy, migration and political stability—areas that may be more readily prioritised by society—rather than being in competition with them (Novacek 2008).

Misconception about the role of zoos in biodiversity conservation may lead to the belief that species are being '*saved by the zoo*' simply because individual animals are looked after and are '*protected*' from lack of food, attacks or threats they would face in their natural environment, whether caused by human pressure or by a natural fight for survival. There were worries that having species preserved in zoos may lead to lack of concern/ involvement with conservation as the public may see it as enough in itself:

> I guess the dilemma with the zoos is the implication in a way, that we have these animals packaged for you and that's because in zoos you don't have to go see them in the wild, or it doesn't really matter if they die off in the wild because we can still home them here.

Some concerns related to beliefs that caged animals can convey a sense of security—behaviours are not natural (e.g., hunting for food) and do not represent their ecological roles in living systems:

> I don't think zoos necessarily convey how difficult it is for species to survive and I think people are generally disconnected from any sort of understanding about their own survival.

It is uncertain the worth of assessing the value of individual species or groups, such as carnivores, considering the importance of interconnectedness and the interdependence of each species within whole functional ecosystems (Gascon et al. 2015). In view of the intricacies, the most valuable aspect of the conservation of wild carnivores may be its potential to promote the protection of all biodiversity.

7 Conclusions

Prompted by our schedule of questions, participants reflected extensively on the impact of people in the natural environment and their responsibility for the survival of species. Although the term biodiversity in itself was not familiar, connections were made between native wild carnivore species and their benefits to the natural environment and ultimately to people's lives. Fears emerged, in relation to the long absence of these carnivores and associated lack of knowledge and familiarity with their needs and impacts, as well as feelings of longing for a connection with the living world. Participants' responses to their own zoo experiences indicate that zoos can help to dispel 'fear of the unknown', but raise questions about captivity creating 'false ideas' about wilderness and wildness.

Our findings also raise questions regarding an apparent dichotomy between feelings of 'stewardship' towards the living world and feelings of connection and of 'belonging to nature', and the implications these may have on attitudes towards biodiversity, which require further investigation. Other areas for further investigation include associations between attitudes and proximity of residence to areas of species recovery; the role of frequent visits/membership to zoos as opposed to sporadic/no visits in local people's attitudes, as well as the role of interactions with the focus/native species rather than other/exotic species.

Our preliminary results support beliefs that humans have an innate emotional bond with living organisms, which can be nurtured by learning, culture and first hand experiences of wild animals in the zoo setting. Such experiences may foster empathy and an interest in finding out more about the animals. By incorporating rich affective experiences and social reinforcement with poignant information 'good zoos' can help people reflect on their own role in the natural environment, and wider biodiversity issues during and after visits. They can thus empower people to take action to protect biodiversity. Results will inform the design of questionnaires and interviews to be carried out in the UK to help identify key areas that must be addressed by plans to reintroduce native wild carnivores to benefit biodiversity, and to suggest ways in which zoos may support such plans.

Acknowlegments The authors would like to thank the staff of Wildwood Trust and of the Aspinall Foundation, in particular Hazel Ryan and Claire Capdevila-Wright, Tony King, Neil Spooner, Amanda McCabe and Bryony Hatcher for their support and invaluable time. Many thanks to all the volunteers who took part in the focus groups and interviews.

References

Ajzen, I. (1991). The theory of planned behaviour. *Organizational Behavior and Human Decision Processes, 50*(2), 179–211.

Bath, A., Olszanska, A., & Okarma, H. (2008). From a Human Dimensions perspective, the unknown large carnivore: public attitudes towards Eurasian lynx in Poland. *Human Dimensions of Wildlife, 13*(1), 31–46.

BBC News. (2015). Will wild lynx return to Britain?, Retrieved April 27, 2015 from http://www.bbc.co.uk/news/uk-32450051

BBC Radio. (2015). Reds Return: Could the return of the Pine Marten mean the end of the Grey Squirrel takeover? BBC Radio 4, Retrieved April 8, 2015 from http://www.bbc.co.uk/programmes/b05pn674

BIAZA. (2016). http://www.biaza.org.uk/about-biaza/. Accessed on January 11, 2016.

Bjerke, T., Ødegardstuen, T. S., & Kaltenborn, B. P. (1998a). Attitudes toward animals among Norwegian children and adolescents: Species preferences. *Anthrozoös, 11*(4), 227–235.

Bjerke, T., Ødegardstuen, T. S., & Kaltenborn, B. P. (1998b). Attitudes toward animals among Norwegian adolescents. *Anthrozoös, 11*(2), 79–86.

Bowkett, A. E. (2009). Recent captive-breeding proposals and the return of the ark concept to global species conservation. *Conservation Biology, 23*(3), 773–776.

Clayton, S., & Myers, G. (2009). *Conservation Psychology: Understanding and promoting human care for nature*. UK: Wiley-Blackwell. ISBN 978-1-4443-5641-0.

Clayton, S., Fraser, J., & Burgess, C. (2011). The role of zoos in fostering environmental identity. *Ecopsychology, 3*(2), 87–96. doi:10.1089/eco.2010.0079

Clayton, S., Luebke, J., Saunders, J., Matiasek, J., & Grajal, A. (2014). Connecting to nature at the zoo: Implications for responding to climate change. *Environmental Education Research, 20*(4), 460–475. doi:10.1080/13504622.2013.816267

Consorte-McCrea, A. (2011). Conservation of the maned wolf (Chrysocyon brachyurus): Carnivore and people relationships in the southeast of Brazil. Ph.D. thesis, UKC. http://create.canterbury.ac.uk/10662/

Consorte-McCrea, A. (2014). Relationships between the maned wolf and people. In A. G. Consorte-McCrea & E. F. Santos (Eds.), *Ecology and conservation of the maned wolf: Multidisciplinary perspectives* (pp. P35–52). London: CRC Press.

Dingwall, R., & Aldridge, M. (2006). Television wildlife programming as a source of popular scientific information: A case study of evolution. *Public Understand Sci, 15*, 131–152.

Falk, J. H., Reinhard, E., Vernon, C., Bronnenkant, K., Deans, N., & Heimlich, J. (2007). *Why zoos and aquariums matter: Assessing the impact of a visit*. Silver Spring, MD: Association of Zoos and Aquariums.

Gascon, C., Brooks, T. M., Contreras-MacBeath, T., Heard, N., Konstant, W., Lamoreux, J., et al. (2015). The importance and benefits of species. *Current Biology, 25*(10), R431–R438.

Gwynne, J. A. (2008). Inspiration for conservation: Moving audiences to care. In A. Zimmermann, M. Hatchwell, L. Dickie, & Chris West (Eds.), *Zoos in the 21st century, catalysts for conservation? conservation biology 15* (pp. P51–62). Cambridge, UK: Cambridge University Press.

Hetherington, D. (2008). The history of the Eurasian Lynx in Britain and the potential for its reintroduction. *British Wildlife, 20*(2), 77–86.

Hetherington, D. A., Miller, D. R., Macleod, C. D., & Gorman, M. L. (2008). A potential habitat network for the Eurasian lynx *Lynx lynx* in Scotland. *Mammal Review, 38*, 285–303. doi:10.1111/j.1365-2907.2008.00117.x

Hinds, J., & Sparks, P. (2008). Engaging with the natural environment: The role of affective connection and identity. *Journal of Environmental Psychology, 28*, 109–120. doi:10.1016/j.jenvp.2007.11.001

IUCN/SSC. (2013). Guidelines for Reintroductions and Other Conservation Translocations. Version 1.0. Gland, Switzerland: IUCN Species Survival Commission, viiii + 57 pp.

Jedrzejewska, B., & Jedrzejewski, W. (2005). Large carnivores and ungulates in european temperate forest ecosystems: Bottom-up and Top-down control. In J. Ray, K. Redford, R. Steneck, & J. Berger (Eds.), *Large Carnivores and the conservation of biodiversity* (pp. 230–247). D.C: Island Press. Washington.

Kahn, P. H., & Kellert, S. R. (2002). *Children and nature: Psychological, sociocultural, and evolutionary investigations*. London: MIT Press.

Keliher, V. (1997). Children's perceptions of nature. *International Research in Geographical and Environmental Education, 6*(3), 240–243.

KellertKellert, S. R. (2002). Experiencing nature: Affective, cognitive, and evaluative development in children. *Children and nature: Psychological, sociocultural, and evolutionary investigations*, 117-151. Cambridge, Massachusetts London: The MIT Press.

Linnell, J. D. C., Promberger, C., Boitani, L., Swenson, J. E., Breitenmoser, U., & Andersen, R. (2005). The linkage between conservation strategies for large carnivores and biodiversity: The view from the half-full forests of Europe. In J. Ray, K. Redford, R. Steneck, & J. Berger (Eds.), *Large Carnivores and the conservation of biodiversity* (pp. 381–399). D.C: Island Press. Washington.

McShea, W. J. (2005). Forest ecosystems without Carnivores: When ungulates rule the world. In J. Ray, K. Redford, R. Steneck, & J. Berger (Eds.), *Large Carnivores and the conservation of biodiversity* (pp. 138–154). Washington DC: Island Press.

Moss, A., Jensen, E., & Gusset, M. (2015). Evaluating the contribution of zoos and aquariums to Aichi biodiversity target 1. *Conservation Biology, 29*, 537–544. doi:10.1111/cobi.12383

Myers Jr, O. E., & Saunders, C. D. (2002). Animals as links toward developing caring relationships with the natural world. *Children and nature: Psychological, sociocultural, and evolutionary investigations*, (pp. 153–178). Cambridge, Massachusetts: The MIT Press.

Novacek, M. J. (2008). Engaging the public in biodiversity issues. *Proceedings of the National Academy of Sciences, 105*(Supplement 1), 11571–11578.

Packer, J., & Ballantyne, R. (2010). The role of zoos and aquariums in education for a sustainable future. *New Directions for Adult and Continuing Education, 2010*, 25–34. doi:10.1002/ace.378

Pyle, R. (2002). "Eden in a Vacant Lot: Special Places, Species and Kids in Community of Life". In P. H. Kahn Jr and S.R. Kellert (Eds.). Children and nature: Psychological, sociocultural, and evolutionary investigations (pp. 305–328). Cambridge, Massachusetts: The MIT Press.

Ray, J.C. (2005). "Large Carnivorous Animals as Tools for Conserving Biodiversity: Assumptions and Uncertainties" In J. Ray K. Redford R. Steneck & J. Berger (Eds.). Large Carnivores and the conservation of biodiversity. (pp. 34–56). Washington DC: Island Press.

Reading, R. P. & Miller, B. (2008). Attitudes and attitude change among zoo visitors. In A. Zimmermann M. Hatchwell L. Dickie & W. Chris (Eds.). Zoos in the 21st century, catalysts for conservation? conservation biology (15th ed., pp. 63–91). Cambridge, UK: Cambridge University Press,

Roskaft, E., Handel, B., Bjerke, T., & Kaltenborn, B. P. (2007). Human attitudes towards large carnivores in Norway. *Wildlife Biology, 13*(2), 172–185.

Rounds, J. (2004). Strategies for the curiosity-driven museum visitor. *Curator: The Museum Journal, 47*(4), 389–412. doi:10.1111/j.2151-6952.2004.tb00135.x

Secretariat of the Convention on Biological Diversity. (2014). Global Biodiversity Outlook 4. Montréal, 155 pages. https://www.cbd.int/gbo/gbo4/publication/gbo4-en.pdf Accessed on December 12, 2015.

Sergio, F., Newton, I., Marchesi, L., & Pedrini, P. (2006). Ecologically justified charisma: preservation of top predators delivers biodiversity conservation. *Journal of Applied Ecology, 43*, 1049–1055.

Steneck, R. S. (2005). An ecological context for the role of large carnivorous animals in conserving biodiversity. In J. Ray, K. Redford, R. Steneck, & J. Berger (Eds.), *Large Carnivores and the conservation of biodiversity* (pp. 9–33). Washington DC: Island Press.

Sterling, E., Lee, Jimin, & Wood, T. (2007). Conservation education in zoos; an emphasis on behavioural change. In A. Zimmermann, M. Hatchwell, L. Dickie, & Chris West (Eds.), *Zoos in the 21st century, catalysts for conservation? conservation biology 15* (pp. 37–50). Cambridge, UK: Cambridge University Press.

Vanstreels, R. E. T., & Pessutti, C. (2010). Analysis and discussion of Maned wolf Chrysocyon brachyurus population trends in Brazilian institutions: Lessons from the Brazilian studbook, 1969–2006. *International Zoo Yearbook, 44*(1), 121–135.

Velsor, S. W., & Nilon, C. H. (2006). A qualitative investigation of the urban african-american and Latino adolescent experience with wildlife. *Human Dimensions of Wildlife, 11*(5), 359–370.

WAZA. (2005). Building a future for wildlife—the world zoo and aquarium conservation strategy. ISBN 3-033-00427-X

WAZA. (2011). Communiqué, World Association of Zoos and Aquariums mobilized for the United Nations Decade on Biodiversity. United Nations Decade on Biodiversity. http://www.waza.org/files/webcontent/1.public_site/5.conservation/un_decade_biodiversity/waza_involvement/UNDOB%20press%20release.pdf. Accessed on February 02, 2016.

WAZA. (2016). "Zoos and Aquariums of the World" http://www.waza.org/en/site/zoosaquariums. Accessed on January 1, 2016

Weprin, A. (2007). "Planet Earth Delivers for Discovery". Available at http://www.broadcastingcable.com/news/programming/planet-earth-deliversdiscovery/30333. Accessed on January 11, 2016

West, C., & Dickie, L. A. (2007). "Introduction: is there a conservation role for zoos in a natural world under fire?" P 3-11. In A. Zimmermann, M. Hatchwell, L. Dickie, & Chris West (Eds.), *Zoos in the 21st Century, Catalysts for Conservation? Conservation Biology 15*. Cambridge, UK: Cambridge University Press.

Wilson, E. O. (1984). *Biophilia*. Cambridge, MA: Harvard Univ Press.

Wilson, C. J. (2004). Could we live with reintroduced large carnivores in the UK? *Mammal Review, 34*, 211–232. doi:10.1111/j.1365-2907.2004.00038.x

Authors Biography

Dr. Adriana Consorte-McCrea has an MSc in Conservation Biology by the Durrell Institute of Conservation and Ecology (DICE, UKC) and a PhD in Ecology by the University of Kent. She has worked with zoos and the conservation of wild canids for 30 years in Brazil and in the UK and has co-edited a pioneering book on the ecology and conservation of the maned wolf. She is particularly interested in the human dimensions of wildlife conservation. In 2006 she created the Wildlife and People Research Group, which she chairs (as part of the Ecology Research Group) at Canterbury Christ Church University, where she has taught undergraduate students since 2004 and also works as an Education for Sustainable Development Lead (Futures Initiative). She has been a member of the IUCN-SSC Reintroduction Specialist Group since 1997.

Dr. Alan Bainbridge is a Chartered Psychologist, Doctor of Clinical Science and Senior Fellow of the Higher Education Academy and began working in Higher Education in 2001 having previously taught in secondary schools for 18 years. He is interested in the contested space between psychoanalytic thought and practices to education in its widest sense. He has written on how educational professionals develop their professional practice, the nature of academic understanding and is currently exploring the fetish in education and how learning and the 'natural world' are interconnected. Alan has recently used narrative and biographical techniques to research the motivators and barriers towards a community engagement project and the attitude of individuals towards the re-introduction of native wild carnivores. He uses his experience as a UKCP registered psychoanalytic psychotherapist to inform his research and as such works qualitatively to seek to provide opportunities and spaces where participants can provide rich contextual data of their life experiences. He is the co-coordinator for the Faculty of Education Auto/biography and Narrative Research and Knowledge Exchange Theme Group.

Dr. Ana Fernández is a senior lecturer in Psychology at Canterbury Christ Church University. She received her PhD in Psychology from the University of Kent in 2007, which focused on the effects of emotion on visual attention. Since 2009 she has been involved in research into environmental sustainability focusing on attitudes, public engagement and climate change discourses. She has recently become involved in conservation psychology as well as community psychology, working with independent dog rescues in Kent. As part of her undergraduate teaching she has developed a Psychology of Sustainability module and she also teaches on the postgraduate module Society & Environment. She is currently supervising a PhD student who is investigating sustainability in the NHS, and an MSc project in conservation psychology.

Dr. Dennis Nigbur is a senior lecturer in Psychology at Canterbury Christ Church University, with particular interests in the social psychology of culture and sustainability. He graduated from the University of Kent with a BSc in Social Psychology in 1999 and conducted his doctoral studies on national identity, receiving a PhD from Royal Holloway in 2004. He conducted his post-doctoral research at the University of Surrey about predictors of kerbside waste recycling, where he became a Research Fellow. He up a lectureship at Canterbury Christ Church University in 2007, where he is currently the director of the B Sc programme in Psychology and the Chair of the Faculty Ethics and Governance Committee.

Siri McDonnell has completed a BSc in Ecology and Conservation at Canterbury Christ Church University. She joined the Wildlife and People Research Group in January 2013 to research attitudes towards the conservation of native wildlife and biodiversity. After she graduated in 2013 she became research assistant in this project, and as so she has collaborated in its development, in preparing research conference presentations, as well as organising and assisting the running of focus groups. In 2015 Siri begin working as Events Co-Ordinator for the RSPB.

Aïssa Morin has a Bachelor's degree in Environmental and Animal Sciences from the University of Rennes 1 (France), obtained in 2015, during which she has studied for one year at the University of Canterbury. She is currently doing a Master's degree in Ecological Management, Biodiversity and Evolution at the National Natural History Museum of Paris (France). Her main areas of interest have been carnivore and mammal conservation, population restoration (through reintroduction and population reinforcement) and relationships between people and their environment. In the future, she wishes to pursue the field of research in mammal in situ conservation.

Oksana Grente obtained her Bachelor's degree in Biology of Organisms from the Department of Life and Environmental Sciences, University of Rennes 1, France, in 2015. She is currently a Master's student in Tropical Ecology, Biodiversity and Evolution at the University of the French West Indies in Guadeloupe and is planning to pursue her training in French Guyana. Her primary areas of interest have been feline and social insect's biology. She is now interested in pursuing research on conservation biology.

Variation in Sociologists' Perspectives About Sustainability in Higher Education: Outcomes from a Phenomenographic Study

Patrick Baughan

Abstract

Sustainability issues are increasingly being adopted in higher education, in areas such campus initiatives, student-led schemes, in teaching, and through growing research activity. Much of the sustainability research focuses on discussions and debates about its inclusion in teaching and curricula in different disciplinary contexts. However, relatively little attention has been given to sustainability in relation to sociology, which is surprising since both have an interest in society and social change, and sociological research addresses areas including the environment and consumption. Sociologists' understandings and experiences of sustainability might have particular value and provide new lessons and ideas relevant to those interested in promoting sustainability in the higher education sector. Consequently, and using the phenomenographic approach, this study sought to cast light on sociologists' conceptions of sustainability and its relevance to sociology as a discipline. The project addressed the following research question: *What variations exist in sociology academic staff and students in their accounts about and experiences of sustainability in higher education?* The intention of the work was to collect broad-based perspectives from a diverse range of sociology staff and students about sustainability. The study comprised 24 semi-structured interviews with academic staff and students based in sociology departments at three UK-based universities. This chapter reports on the study and presents the findings in the form of two outcome spaces entitled *Sustainability and me* and *Sustainability, sociology, and sociology curricula*. In addition, the chapter reflects upon the phenomenographic research approach, and suggests that it offers considerable value for the research of sustainability in higher education.

P. Baughan (✉)
Senior Lecturer in Educational Development, City,
University of London, London EC1V 0HB, UK
e-mail: p.baughan@city.ac.uk

W. Leal Filho (ed.), *Sustainable Development Research at Universities in the United Kingdom*, World Sustainability Series,
DOI 10.1007/978-3-319-47883-8_18

Keywords
Sustainability · Sociology · Phenomenography · Higher education · Curriculum

1 Introduction and Background

What can sociology and sociologists tell us about sustainability? This project sought to address this question, and examine sociologists' conceptions of sustainability and its relevance to sociology. The rationale for the project lies in the point that whilst an important and popular area of sustainability research is its role in different disciplines (e.g. Barlett and Chase 2013; Johnston 2013; Jones et al. 2010), relatively little of this discipline-based interest considers *sociological* perspectives, even though sociology and sustainability share interests in society and social change. Thus, a central aim was to capture sociological perspectives about sustainability, on the basis that such perspectives might offer particular insights for the advancement of sustainability in higher education.

Sustainability is a topic, issue, and set of practices that has attracted positive attention in higher education, but has also tended to confuse and polarise opinion (Baughan 2015). It has been influential in informing aspects of higher education activity, policy and research, but it is often treated with scepticism because there appears to be no overall agreement as to what it encompasses, and because it is sometimes interpreted as an imposed agenda through which staff and students are expected to enact certain 'sustainable' behaviours. Nevertheless, many published articles and books—in Europe, America and elsewhere—have examined different aspects of sustainability in higher education, often in the context of individual disciplines (Barlett and Chase 2013; Reid and Petocz 2006; Jones et al. 2010; Johnston 2013). These disciplinary perspectives have been valuable, providing insights into how sustainability is perceived and can be addressed positively within disciplinary research and teaching. In spite of this, relatively few studies have focused on *sociological* interpretations, even though sociology and sustainability share an interest in society and social change, and sociological research and teaching have addressed related areas such as the environment and consumption (Soron 2010). Indeed, some sociologists have advocated closer links between their discipline and sustainability, Passerini (1998) opining "…sociology is uniquely equipped with the theoretical and methodological background to contribute scientifically accurate understandings of this phenomenon to a world much in need of such guidance" (p. 59). In addition, a number of sociology departments already include coverage of sustainability in their curricula.

The purpose of this chapter is to discuss a phenomenographic research project recently undertaken by the author which sought to understand, from sociologists' accounts, what sustainability is, its relevance to sociology, and whether it should be addressed in sociology curricula at higher education institutions.

It will be argued that sociologists have varied understandings and views about sustainability in relation to their own discipline, but this variation in accounts provides important insights for our understanding of the complexities, challenges and debates associated with sustainability. It will also suggested that, with its emphasis on variation in experience, phenomenography provides an appropriate research approach for researching sustainability and for highlighting those aspects that need more clarification and development in a higher education context.

2 The Research Question

This is a phenomenographic research project, in which the task of the researcher is to identify and focus on *variation* in interpretations and accounts amongst participants of a given phenomenon, so the wording of the research question reflects this approach. The central research question is: *What variation exists in sociology academic staff and students in their accounts about and experiences of sustainability in higher education?* In addition, the project addresses two subsidiary questions, these being: (a) *What do sociology staff and students understand by sustainability?* (b) *Should sustainability be included in sociology curricula?* In using phenomenography, an intention of the project was to gain an understanding of diverse understandings and experiences that sociologists have of sustainability. By appreciating a full range of sociologists' views, the study is potentially more valuable in that its outcomes can be used to cast light on the different conceptions about, and complexities associated with sustainability.

3 Additional Literature Which Informed the Project

The development of the project was informed by two substantive areas of literature: sustainability in higher education and phenomenography as an approach for researching education. However, before further considering these areas, some further comment should be offered about sustainability itself. Various definitions and interpretations of sustainability and sustainable development have been offered, some broader, others more specific. Williams and Millington (2004) suggest sustainable development to be a '…notoriously difficult, slippery and elusive concept' (p. 99). The following provide example definitions, particularly for readers new to the area, though there are many others:

> [Sustainability is about] development that meets the needs of the present without compromising the ability of future generations to meet their own needs (United Nations 1987).

> [Sustainability] 'represents a condition, or set of conditions, whereby human and natural systems can continue indefinitely in a state of mutual well-being, security and survival (Blake et al. 2013).

> Sustainability efforts are defined broadly to include changes in campus operations, financial and administrative planning and/or policy, and/or academic curricula and research that facilitate positive environmental changes (Brinkhurst et al. 2011, p. 340).

Whilst often considered a novel or niche area, there actually exists a fast growing corpus of research and scholarly activity in the field, with dedicated books, case studies, journals and (many) conferences in Europe, America and around the world. Amongst other issues, this research has considered the views of higher education staff (e.g. Reid and Petocz 2006) and students (e.g. Kagawa 2007) about sustainability. It has been suggested that universities should have an important role in promoting sustainability (Orr 2002). Further, a number favour its integration into the learning and teaching process, including Sterling (2001) who discussed the importance of 're-orienting' learning approaches. Examples of innovative sustainable curricula initiatives have been published in sources including Barlett and Chase (2013); Cotton et al. (2012); Drayson et al. (2013); and Johnston (2013). Other published works offer models or ideas for infusion of sustainability in curricula (for example, De La Harpe and Thomas 2009). For Weller (2016) sustainability, or to use its more specific relation, education for sustainable development (ESD) offers an example of the benefits to be gained from adopting interdisciplinarity in the curriculum—utilising a holistic approach to uncover connections between different methodological and conceptual positions.

Nevertheless, the matter of sustainability in the curriculum appears to be a divisive one, with some commentators providing a persuasive case in favour (Orr 2002) and others suggesting that there are barriers and disadvantages to integrating sustainability in curricula (Chase 2010; Reid and Petocz 2006). Even so, more recent research by Drayson et al. (2013) found that 80 % of (UK) students believe that sustainable development should be encouraged at their institutions, with two thirds believing that sustainability issues should be in some way integrated into their curricula.

If these challenges are to be overcome, and sustainability is to be integrated in more curricula, there is also a need to capture views and perspectives from specific staff and student groups. As mentioned above, whilst there is already sociological work about sustainability (Passerini 1998; Soron 2010; Warde 2005), there remains a need to develop this further and more directly examine sociologists' conceptions —which is what this study attempted to do.

4 Phenomenography as an Approach for Researching Education Issues

This project adopted a phenomenographic approach in examining variation in the way a particular phenomenon (sustainability in higher education) is experienced amongst a group of participants (sociology staff and students). Phenomenography has its basis in studying how people experience, understand and ascribe meaning to a particular phenomenon (Marton and Booth 1997). It assumes that experiences may be depicted by a limited number of qualitatively distinct categories of description (Marton 1981), the researcher seeking to develop an understanding of the meanings of those categories and the way they relate to one another (Entwistle 1997). Findings are mapped through the presentation of *outcome spaces* and constituent *categories of description*. It is suggested here that phenomenography is well suited to researching sustainability, which is also characterised by differences, in definitions, understandings, and views of its role in higher education.

Phenomenography has been adopted for researching a range of higher education issues including learning and teaching (Shreeve et al. 2010), curriculum (Fraser 2006), academic development (Åkerlind 2007), and study support (Hallett 2010). It has also been used in other sustainability-focused projects (Baughan 2015; Carew and Mitchell 2006; Corney and Reid 2007).

5 Research Design

Phenomenography can be undertaken using various methods, with semi-structured interviews being the most frequently selected—and used in this study. When using this approach, it is important to aim for variation in the sample base, so as to maximise variation in the later outcomes. This study comprised a total sample of 24 sociologists—12 academic staff and 12 students—based at three (UK based) sociology departments. The sample also incorporated variation in gender, course; level of experience, and 'types' of department (research foci and curricula). Students included first, second and final year undergraduates undertaking various sociology degree courses.

The project gained ethical approval from both the author's employing university and the three other universities involved. As a means to check for variation, pilot interviews were undertaken, revealing clear examples of variation, thus providing confidence that phenomenography offered an appropriate approach for the project. The inclusion of a student perspective is noteworthy as, until recently, fewer studies have addressed the 'student voice' in sustainability.

6 Implementation and Data Analysis

Interviews were conducted with staff and students at their own institutions, the researcher spending two days in each department Most interviews took 30–40 minutes, the staff interviews tending to take slightly longer, reflecting the fact that staff often responded to questions in more detail. During the interviews, each participant was invited to comment on their experiences of sustainability, whether and how higher education institutions should be involved in it, and on their views and experiences of sustainability in the sociology curriculum. Participants were also asked to provide examples to illustrate their points.

Phenomenography is interested in variation, and this emphasis is reflected in the way data collected through this approach should be analysed. The analysis necessitated a multi-stage, iterative process, organising and reducing the data into outcome spaces and categories of description. The intention was to produce categories depicting variation as opposed to themes showing commonality. During analysis, the researcher strives to exclude (or bracket out) their own experience (Ashwin et al. 2013), focusing on the relationship between participants and phenomenon. Data analysis comprised detailed reading and re-reading of the transcripts, noting down patterns and ideas for categories, all the time looking for evidence of variation in accounts. Analysis was also organised under a series of phases, as a means of enabling checks for variation. For example, the first analysis phase involved reading and note-taking from the first ten transcripts, identifying relevant quotations and gradually building a sense of preliminary outcome spaces and categories. Once a preliminary analysis was complete, a further check of outcome spaces and categories was undertaken, to check that the latter were genuinely qualitatively distinct. Towards the end of the process, a colleague was asked to review the categories, following which further refinements were made. The process yielded two outcome spaces, each depicting participant experiences of different components of sustainability as raised in the original research questions. The outcome spaces are entitled "Sustainability and me" and "Sustainability, sociology and sociology curricula". These are subdivided into the aforementioned categories which each show one way the phenomenon was experienced, in relation to the other categories within the same outcome space.

7 Findings

The analysis of interviews yielded two outcome spaces, each of which is presented with its constituent categories of description, below.

7.1 Outcome Space A: "Sustainability and Me"

The first outcome space depicts five related but qualitatively distinct categories of description, based on participant understandings of sustainability, and their accounts of how they enact and apply sustainability. Illustrative quotations are provided under each category:

Category 1: Sustainability is about sustaining and protecting in higher education This category provides a contained account of sustainability as being based within higher education. Sustainability is about sustaining, protecting, or 'keeping things going' in higher education or within an aspect of higher education.

> Making sure the campus stays nice and pleasant and that it stays functioning. And making sure everything stays up to date and making sure it stays within health and safety regulations.

> I suppose if you say sustainability in higher education, maybe I would think that you are talking about… the sustainability of the higher education endeavor and what you see as important within higher education, so perhaps the teaching-research things.

> I think the first thing that comes to mind would be about the financial sustainability for them. And also maybe about how to keep on improving the quality of their education.

Category 2: Sustainability is about managing and controlling in higher education In this category, sustainability is interpreted to be a tool or device used for managing, controlling, or cost-cutting. Participants were also more critical in their accounts, relating sustainability to terms (and practices) including 'neo-liberal discourses', 'control mechanisms', as a 'management tool' or for 'behaviour modification'.

> I think a lot of what is passing for sustainability is really about the struggle for intellectual control over universities. Not on ideological lines… I think it is more in particular the growth of managerial power.

> It's painful to think about, and people are frightened about it, so it's much easier to not think in those macro-terms. But I think some of that kind of individualising, it sort of fits very well with the kind of neo-liberal ideology of, well, it's all your fault.

Category 3: Sustainability is about environmental issues and looking after the environment In this category, accounts of sustainability itself broaden beyond sustaining and protecting, to take in environmental issues. This conception offers an environment-based account of sustainability, and includes more sophisticated examples of practice

> I think immediately it's an environmental side for me. It's recycling and, erm, sort of the environmental, protecting the world… I think the immediate impression is more, is an environmental base or side of it.

> [It's] to do with the sustainability of the natural environment as a resource for human living, well-being, economic productivity, you know, renewable energies and safe clean drinking water supplies under context of population growth and all those other sorts of things.

Category 4: Sustainability is about things I do for the environment This category is also about the environment, but in this case, how participants *relate* to the environment and seek to be involved in environmental issues and enact environmental and sustainable behaviours.

> It's like, living in our house, we're very strict with our recycling, we've got one girl who's very on it, and she'll check what you're putting in the recycling.

Category 5: Sustainability is about my identity and lifestyle This is the most complete category in which accounts of sustainability are based around a diverse range of issues. For participants adopting this category, sustainability is also a lifestyle, or forms an important part of personal identity, or both. Participants actively enact various sustainable behaviours in their day-to-day lives and sustainability is seen as central to lifestyle.

> The kind of way I like to live, eat… travel… and the kind of society I would like to live in, definitely sustainability is right at the heart of it.

7.2 Outcome Space B: "Sustainability, Sociology and Sociology Curricula"

This outcome space depicts five related but qualitatively distinct categories of participant accounts about the nature of the relationship between sustainability and sociology, and the relationship between sustainability and sociology curricula. To an extent, these categories in turn reflect participant accounts about sustainability itself.

Category 1: Sustainability and sociology are different In this category, sociology and sustainability are understood to be mainly or completely distinct. Sustainability is not interpreted to be part of the discipline of sociology; it is not usually of sociological relevance. Participants did not view sustainability to form any major part of their role or identity as a sociologist.

> I think it's a thing that's sort of there on the sidelines as it were but, yeah, I mean it's not really sort of a core part of my research, you know it doesn't really come into the stuff that I teach.

> If you integrate it then you are sort of imposing something onto the curriculum which means other things have to come out, and I mean I guess practically speaking you've got to take people with you as well, and I suspect that it would come to be regarded, even by people who are relatively sympathetic to the goals of sustainability, it could come to be regarded as, you know, tokenistic nonsense to satisfy bureaucrats.

Category 2: Sustainability and sociology are connected through the institution Participants adopting this account cited links between sociology and sustainability, but via the role of the institution (their university). Sustainability does not usually hold any special, discipline based link to sociology, but maybe linked or included via institutional policies or curriculum requirements. In this category, sustainability is related to institutional activity.

> It seems to me that when people talk about sustainability in this context, it has now become part of the whole strategic plan, one of the key pillars of what the university is supposed to be doing.

> Sustainability should be about investing in kind of, local parks, or investing in studies which look at what people do, who own allotments, things like that. Actually doing something cultural and local which basically engages with sustainability, but also a localist kind of agenda.

> When we come to university they could also implement policies to do with sustainability in terms of the environment… and I'm sure university students would also get excited about it… so like we would feel a connection to our uni so we would want to do it as well.

Category 3: Sustainability is something that we do in sociology Participants adopting this category cited links between sociology and sustainability through the institution and through the discipline. In particular, participants cited a relationship between sociology and sustainability: the focus of sociology affords it a particular relationship with sustainability, although that relationship may be variable or unspecified.

> I have found the sort of discourse that has come about from environmental activism… a willingness to talk about capitalism not being sustainable as a positive step. So I guess I feel a bit hopeful about that in some ways. And I think sociology can bring a lens on that, you know in terms of talking about capitalism and the relationship between neo-liberalism, capitalism, climate change and sustainability.

> I would imagine that it would probably be very well received in sociology, certainly… I think it would probably be a very receptive environment for that.

Category 4: Sustainability is something that we learn and teach in sociology This builds on the previous category in that participant accounts add reference to a relationship between sustainability and the sociology curriculum; that sustainability is relevant and appropriate for inclusion in sociology degree courses.

> There's a developing field of the sociology of the environment and I think that would be the ideal place for it. I think it would fit in well there… One of my colleagues offers a module on consumer society and we could look at how the making and buying of all these goods is actually depleting the environment. I think it would fit into quite a lot of different places.

> I think that sustainability should be taught. I mean this is one world. Everybody lives in one world. What's the point in destroying it?

> I know that on our curriculum we teach things about how consumption of too many material goods can be bad for the environment, for example. And the students seem quite open to that as well. Whether they actually take it on board I don't know but I think, yeah, it should be a key part of the curriculum.

Category 5: Sustainability is integral to sociology In this category, participants cite an integral link between sustainability and sociology. Sociology has a key role in addressing sustainability due to strong commonalities, and the focus of sociology on group and societal issues. As a consequence, participants also referred to sustainability as forming an important part of their professional identity, of their *being* a sociologist.

> Sociology is like actually studying human behavior and like seeing how people think about their future even though it's not their personal own future, but seeing how they think about future generations and how are they concerned about it, [sustainability] is a huge part of sociology.
>
> I see it [sustainability] as part of my professional responsibility… to me that was part of my professional function to do this.

8 Discussion

The outcomes reveal variation in sociologists' accounts of sustainability in two key dimensions: in their accounts of and relationship to sustainability, and in their conceptions of the relationship between sustainability and sociology. In the case of the first of these (the first outcome space), interpretations of sustainability were very broad based: definitions range from those which are narrow and higher education based, to much broader interpretations in which participants perceived themselves as active agents in sustainability. Of particular interest is the second category, in which accounts refer to sustainability as a management tool, or part of a particular ideology. This might point to some misuses of sustainability, or at least the sustainability 'term' in higher education. For the second outcome space, the latter three categories cite a relationship between sociology and sustainability which may provide insights for our understanding of the complexities and challenges associated with sustainability. The findings appear to corroborate previous research in favour of exploring and furthering the relationship between sociology and sustainability (Passerini 1998).

Sociology offers a range of valuable perspectives about sustainability and many sociologists interviewed in this study suggested that it can inform sustainability practices in higher education—albeit in various different ways. On this basis, and in view of the relative interest amongst sociologists in sustainability, sociology curricula might provide a good 'place' to model teaching, from which lessons may be learned, relevant to infusion of sustainability in other disciplines. This should of course only be on a voluntary basis. However, pro-sustainability in the curriculum views are a characteristic of several of the categories, with various ideas proffered by which this might be done, and with interest shared amongst both staff and students.

In addition, some further issues and discussion points of note emanated from the results:

- In discussing sustainability in the curriculum, participants warned against using 'preaching' approaches (proselytism) as these approaches, often associated with other higher education agendas (for example, employability), maybe interpreted as threats.
- As mentioned above, 'misuses' of sustainability were also cited, including as an instrument to justify neo-liberal policy and as justification for cost-cutting within the sector. These negative associations may deflect from more positive attributes of sustainability, and derail well intended efforts.
- Institutional sustainability policies and initiatives should be clearly explained and justified. Sustainability should involve engaging staff and students critically, in relation to their everyday activities, and also link to local communities. There should be meaningful social and community links.
- Sustainability remains a contested term. Outcomes from this work point to an on-going need to 'find the right language' to capture imaginations. Intuitively, many people subscribe to sustainability but not to its current discourses, so more meaningful approaches are needed.

Finally—some comments on phenomenography, the research approach used for the study. As an approach which foregrounds variation, there is potential for phenomenography to cast further light on areas of ambiguity and debate in sustainability, in different disciplines and amongst different staff and student groups. This could help inform fuller and more inclusive engagement, taking account of the varied understandings and views about sustainability in different disciplines and sections of universities. This could, for example, help shape future policy and provide guidance for teaching staff about whether and how to infuse sustainability into their curricula.

9 Limitations of the Study

There are, of course, a number of limitations of the study. It was undertaken within three universities but it would be interesting to expand the research and add an international perspective. The question might also be raised as to whether the outcomes really represent those of sociologists; might, for example, similar points have been made by staff and students based in other disciplines? In fact, this might be the case in certain aspects of participant accounts, particularly those relating to the first outcome space. But for much of the interview, participants were talking specifically about their own discipline and curricula, so this would seem only a partially accurate criticism to levy. As with any research approach, phenomenography itself has limitations, discussed by authors including Cousin (2009). Finally, this chapter reports on a study which is not fully completed, as some of the categories are being further refined. It is expected that more discussion points and recommendations will be drawn from the research as it reaches completion.

10 Closing Comments

To conclude, sociologists' perspectives offer deep and varied insights about sustainability which could be of genuine value for sustainability researchers and teachers, as well as to policy makers. Sociologists are generally favourable, if varied, in their views about the role of higher education institutions in promoting sustainability activity, whilst interest in sustainability in the curriculum points to the possibility that certain sociology curricula might provide a suitable home for modelling teaching in sustainability, even if on a trial basis. Caution is needed here, however, since sociologists had very different ideas about *how* sustainability might be included in curricula. Follow up research would surely be of value.

As an approach which foregrounds variation, there is potential for phenomenography to cast further light on areas of ambiguity and debate in sustainability, particularly in the context of different disciplines and amongst different stakeholders. This too could help shape future policy and provide guidance for teaching staff about whether and how to infuse sustainability into their curricula.

References

Åkerlind, G. (2007). Constraints on academics' potential for developing as a teacher. *Studies in Higher Education, 32*(1), 21–37.

Ashwin, P., Abbas, A., & McLean, M. (2013). How do students' accounts of sociology change over the course of their undergraduate degrees? *Higher Education, 67*, 219–234.

Barlett, P., & Chase, G. (2013). *Sustainability in higher education; stories and strategies for transformation (urban and industrial environments)*. Massachusetts, MS: MIT Press.

Baughan, P. (2015). Sustainability policy and sustainability in higher education curricula: the educational developer perspective. *International Journal for Academic Development, 20*, 3. doi:10.1080/1360144X.2015.1070351

Blake, J., Sterling, S., & Kagawa, F. (2013). Getting it together: Interdisciplinarity and sustainability in the higher education institution. Plymouth: Pedagogic Research Institute and Observatory. PedRIO Occasional Paper 4 (https://www1.plymouth.ac.uk/research/pedrio/Documents/PedRIO%20Paper%204.pdf)

Brinkhurst, M., Rose, P., Maurice, G., & Ackerman, J. D. (2011). Achieving campus sustainability: Top-down, bottom-up, or neither? *International Journal of Sustainability in Higher Education, 12*, 338–354.

Carew, A. L., & Mitchell, C. A. (2006). Metaphors used by some engineering academics in Australia for understanding and explaining sustainability. *Environmental Education Research, 12*(2), 217–231.

Chase, G. (2010). Large scale university curriculum change: campus stories and strategies for change (and back again). Paper presented at *Tomorrow's Sustainable Universities Conference*, University of Bradford, Bradford, UK, 15–16 July.

Corney, G., & Reid, A. (2007). Student teachers' learning about subject matter and pedagogy in education for sustainable development. *Environmental Education Research, 13*(1), 33–54.

Cotton, D., Sterling, S., Neal, V., & Winter, J. (Eds.). (2012). *Putting the 'S' into ED—Education for sustainable development in educational development*. SEDA Special 31. London: Staff and Educational Development Association.

Cousin, G. (2009). *Researching learning in higher education: An introduction of contemporary methods and approaches*. London: Routledge.

De La Harpe, B., & Thomas, I. (2009). Curriculum change in universities: Conditions that facilitate education for sustainable development. *Journal of Education for Sustainable Development, 3*(1), 75–85.

Drayson, R., Bone, E., Agombar, J., & Kemp, S. (2013). *Student attitudes towards and skills for sustainable development*. New York: Higher Education Academy/National Union of Students.

Entwistle, N. (1997). Introduction: Phenomenography in higher education. *Higher Education Research and Development, 16*, 127–134.

Fraser, S. (2006). Shaping the university curriculum through partnerships and critical conversations. *International Journal for Academic Development, 11*, 5–17.

Hallett, F. (2010). The postgraduate student experience of study support: A phenomenographic analysis. *Studies in Higher Education, 35*(2), 225–238.

Johnston, L. (2013). *Higher education for sustainability: Cases*. Oxon, Routledge: Challenges and Opportunities from Across the Curriculum.

Jones, P., Selby, D., & Sterling, S. (2010). *Sustainability education: Perspectives and practice across higher education*. London: Earthscan.

Kagawa, F. (2007). Dissonance in students' perceptions of sustainable development and sustainability: Implications for curriculum change. *International Journal of Sustainability in Higher Education, 8*(3), 317–338.

Marton, F. (1981). Phenomenography—describing conceptions of the world around us. *Instructional Science, 10*, 177–200.

Marton, F., & Booth, S. (1997). *Learning and awareness*. New Jersey: Lawrence Erlbaum Associates.

Orr, D. (2002). *The nature of design: Ecology*. Culture and Human Intention: New York, Oxford University Press.

Passerini, E. (1998). Sustainability and sociology. *The American Sociologist, 29*(3), 59–70.

Reid, A., & Petocz, P. (2006). University lecturers' understanding of sustainability. *Higher Education, 51*, 105–123.

Shreeve, A., Sims, E., & Trowler, P. (2010). A kind of exchange: Learning from art and design teaching. *Higher Education Research and Development, 29*(2), 125–138.

Soron, D. (2010). Sustainability, self-identity and the sociology of consumption. *Sustainable Development, 18*, 172–181.

Sterling, S. (2001). *Sustainable education: Re-visioning learning and change*. Schumacher Briefings: Bristol.

United Nations. (1987). *Report of the world commission on environment and development: Our common future*. Retrieved from http://www.un-documents.net/ocf-02.htm

Warde, A. (2005). Consumption and theories of practice. *Journal of Consumer Culture, 5*(2), 131–153.

Weller, S. (2016). *Academic practice: Developing as a professional in higher education*. London: Sage.

Williams, C., & Millington, A. (2004). The diverse and contested meanings of sustainable development. *The Geographical Journal, 170*, 99–104.

Evaluating Strategy in Practice: Implementing Carbon Reduction and Sustainability Strategy in University of Westminster

Lez Rayman-Bacchus and Jandi Pearman

Abstract

Within the UN Framework Convention on Climate Change, The European Union's commitment to reducing carbon emissions, and the British Government's introduction of appropriate legislation, universities across the UK have been tasked with developing and implementing strategies for reducing their own emissions. Universities and colleges generate large carbon footprints, due to many factors (especially large and aging infrastructure) so that, as one sector of the economy, higher education provides an obvious target for carbon reduction policy initiatives. However the British Government's carbon reduction targets for the higher education sector are looking increasingly unattainable. The aim of this chapter is to examine the University of Westminster's journey toward sustainability through the lens of 'strategy as everyday practice' set in the context of new (environmental) regulation and the wider development of a sustainability agenda and corporate responsibility projects. The approach of using strategy practice helps shine light on implementation by looking at change at the micro-social level of organisational life. The paper examines how the implementation of carbon management strategy was organised, while at the same time trying to introduce sustainability thinking within the organisation. The study reveals challenges around plural interpretations of sustainability, a flowering of both informal and formal sustainability initiatives, and an evolving perspective on the university's appropriate response, from compliance to ethical.

L. Rayman-Bacchus (✉)
University of Winchester, Winchester SO22 5HT, UK
e-mail: lezmichael@gmail.com

J. Pearman
Cavendish House, University of Westminster, 101 New Cavendish Street, London W1W 6XH, UK
e-mail: J.Pearman@westminster.ac.uk

W. Leal Filho (ed.), *Sustainable Development Research at Universities in the United Kingdom*, World Sustainability Series,
DOI 10.1007/978-3-319-47883-8_19

Lessons learnt from this study are put forward, including the need to build shared understanding, and encouraging shared responsibility for success.

Keywords
Strategy practice · Sustainability · Implementation

1 Introduction

Climate Change is one of the most significant challenges facing our planet and its inhabitants, though governments around the world continue to argue about whether it is real or man-made. Whether natural or man-made, there is broad consensus that reducing our carbon emissions is necessary. While some governments resist taking meaningful action, others are already introducing legislation aimed at reducing carbon emissions. Indeed member states of the European Union have committed to forcing organisations of all types to reducing their carbon footprint. Emerging roughly in parallel with concerns about climate change, are separate and broader calls for sustainable development (World Business Council for Sustainable Development) and greater corporate (social) responsibility (CR). Indeed the need to tackle climate change is one of a spectrum of themes and social movements that highlight the need for action on sustainable development and/or CR. The aim of this chapter is to examine strategy practice and discourse in one British organisation, as it implements a compulsory carbon reduction strategy overlaid with the emergence of a wider demand for a voluntary sustainability strategy.

Everyone should have an interest in sustainability, since it has implications for our way of life today and in the future. This requires organisational leaders and employees to change their way of thinking and working, which in turn demands a re-examination of organisational strategy. The field of 'strategy as practice' is a branch of research focusing on the micro-social level, and tries to understand how strategy is made, by whom, when, and where. This understanding could help shine light on the challenges and potential lessons for organisational practitioners, regardless of their functional specialisation, as they engage with compulsory environmental legislation while embracing internally driven socially responsible practices. Such insights could contribute to more effective organisational adaptation while contributing to the alleviation of climate change. The approach taken in this study is to employ the notion of everyday strategy practice as a theoretical lens. As an idea with sociological roots, this approach is useful for shining light on the micro-practices of practitioners doing strategy, rather than strategy as a formal corporate practice.

This is an empirical study, involving the collection and analysis of qualitative data about activities within the University of Westminster taking place over nine years (2007–2016). This is also a reflective study, drawing on the experiences of both authors as practitioners. To some extent this work could also be seen as practitioner research, and as a mode of knowledge generation. One author is a practitioner specialising in the sustainability field, and the other was for many years a practitioner specialising in strategy development and implementation. The work reported here is therefore informed by practitioner enquiry and knowledge. This study is offered as a basis for further discussion and enquiry within several communities of practice, especially among researchers interested in strategy as everyday practice, and among organisational practitioners charged with integrating compliance with external regulatory priorities while satisfying internal socially conscious demands. First, it adds to the body of knowledge on strategy as everyday practice, exploring its potential for explaining strategy change in organisations. Using concepts of practice and discourse, and practitioner work in context, this study offers insight to the dynamics of organisational strategy change in the context of a globally significant event unfolding over a number of years. This essay also offers insight to the work of organisational practitioners implementing new environmental regulation against the background of a socially conscious community within and beyond the organisational boundary.

This chapter is organised as follows. First, *Context* outlines the wider political agreement underpinning recognition that climate change is a serious global problem, and one that requires individual nations to address by introducing policies to mitigate, if not roll back, the harmful consequences of growing industrialisation and consumption. Taking the UK as an example, this section then sketches out the British government's approach to targeting the need for carbon reduction, focusing on the Higher Education sector. It notes that this sector is unlikely to achieve the carbon reduction targets set. Following this outline of the global and British Higher Education context, the second section presents the *Theoretical Background.* This section introduces three elements. The concept of strategy as everyday practice and discourse is laid out; the importance of (external) context as a site that provides individual organisations with a frame of reference and purpose; and the important role of practitioners as carriers of practice. The third section outlines the research design behind this study, and the fourth section *University of Westminster's Journey*, presents the case study of one organisation's progress between 2007 and 2016 as it engages with the sustainability agenda. The journey is divided into roughly three parts: awakening, forming, and embedding. These give a sense of movement and development of strategy practice at the University, rather than describing distinct episodes of its history. In the fifth section *Discussion* uses the ideas about strategy practice and discourse in order to analyse and offer insight to strategy practice at the University. This section highlights that the meaning of sustainability is plural, ambiguous and negotiable, and that strategy practice is both ordered and equivocal, shaped on one hand by regulatory demands, while on the other hand pervious to the influence of a socially conscious staff and wider community. The final section presents *Conclusions*.

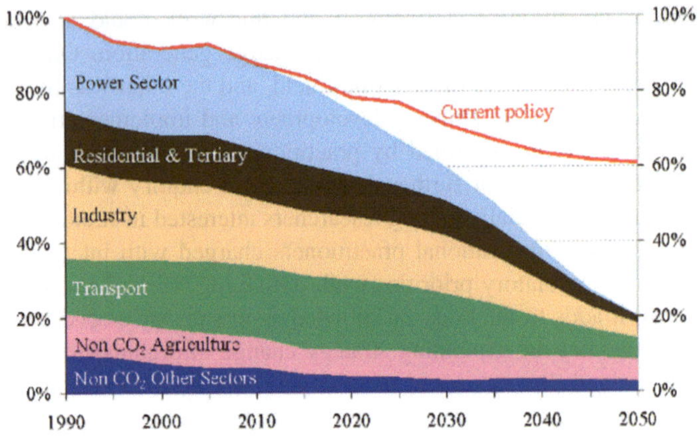

Fig. 1 EU emissions by sector.
Source http://ec.europa.eu/clima/policies/strategies/2050/index_en.htm

2 Context

A global commitment to reducing carbon emissions through the UN Framework Convention on Climate Change (UNFCCC) has been under construction for some time, starting with the Kyoto Protocol adopted in 1997 and coming into force in 2005. However progress on adoption has been hesitant, as reflected in the lack of agreement at the Copenhagen talks of 2009, and the poor level of adoption some twenty years on at the 2016 Paris Agreement.[1] Global agreements are always difficult to achieve and maintain, but there seems to be greater consensus at a regional level: the European Union's (EU) commitment to a Low-Carbon Economy with a goal of achieving an 80 % reduction by 2050 is in line with these Agreements (Fig. 1). Encouragingly, the EU is on track to meet the 20 % reduction target for 2020; in 2014, EU emissions were 24 % below 1990 levels. According to national projections, emissions will further decrease until 2020, but additional policies will need to be implemented to achieve the 2030 target of a 40 % reduction.

2.1 UK Policies

The UK strategy is currently in line with European commitments and has also set an 80 % reduction target by 2050 through the Climate Change Act and various regulations (Fig. 2). UK emissions were 35 % below 1990 levels in 2014 and

[1]As of June 2016 just 18 out of 197 parties have ratified the convention, representing 0.18 % of Green House Gas emissions [http://unfccc.int/paris_agreement/items/9485.php; accessed June 20, 2016].

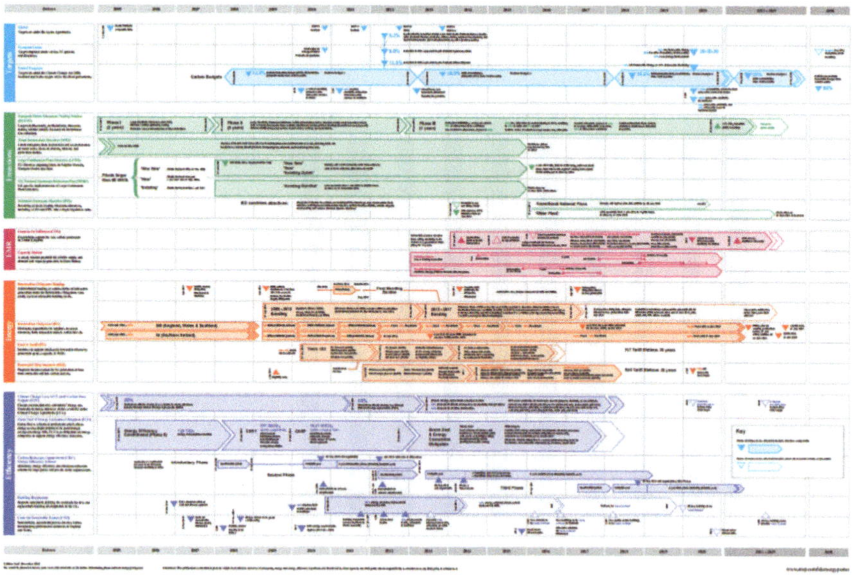

Fig. 2 UK legislation timeline.
Source http://publications.arup.com/publications/u/uk_energy_legislation_timeline

provisional figures show emissions fell a further 3 % in 2015. Figure 2, shows the different mechanisms deployed by the UK government to realise the targets. Blue at the top shows the different targets set internationally, at EU level and nationally. It illustrates the timeline on how the targets and agreements have developed on all three levels, and shows their relationship with the different mechanisms used to achieve carbon reduction targets. The different mechanisms are colour coded green, pink, orange and purple. The green row shows the progress of two directives for reducing European emissions: the European Union Emissions Trading Scheme (EUETS), and the Industrial Emissions Directive. Both European directives are intended to phase out greenhouse gas emissions from large energy consumers and industrial processes. EUETS is gradually capping the amount of carbon available to purchase for heavy users and IED is phasing out old industrial equipment, indicating that emissions from heavy users are intended to gradually decrease. The pink row details the Electricity Market Reform (EMR) and relates to the Power Sector's carbon emissions illustrated in Fig. 1. EMR is the UK government's programme for directing investment towards securing a low carbon supply of electricity. The orange row shows the different policies for supporting renewable energy, such as the Renewables Obligation, which requires energy suppliers to source electricity from renewable sources. The last row in purple shows legal requirements intended to encourage reduction of energy consumption by the end-user, such as the Climate Change Levy, an additional energy charge to businesses.

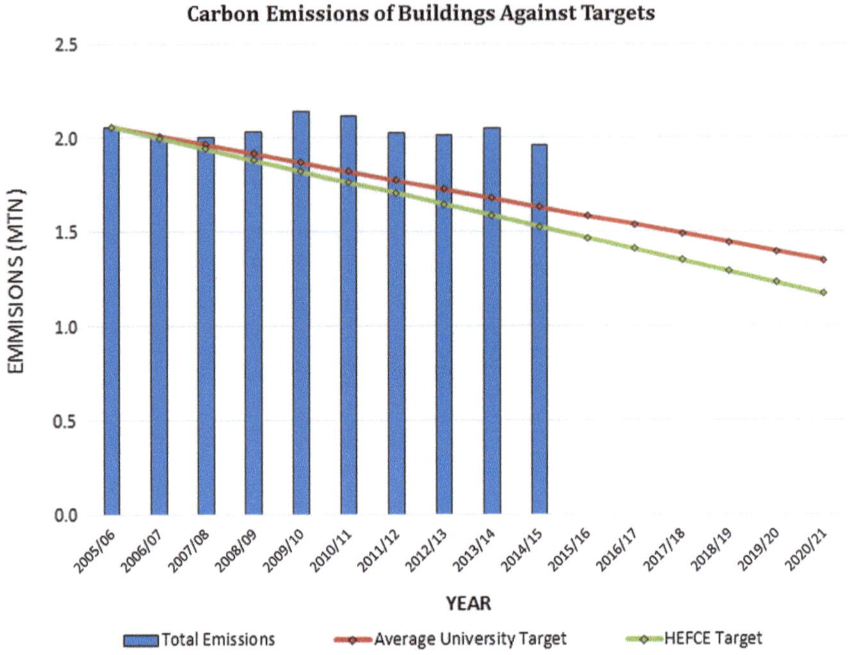

Fig. 3 UK universities carbon emissions against targets

2.2 UK Higher Education Carbon Reduction Policy

The Higher Education Funding Council for England (HEFCEE)—the UK government agency tasked with implementing its carbon reduction policy within Higher Education Institutions (HEIs) across England—developed and introduced in 2005 their approach for the sector under 'Sustainable Development in Higher Education'. Five years later (in 2010) HEFCEE set a sector target of 43 % carbon reduction by 2020. Figure 3 shows this sector target (green), the consolidated target set by individual universities (red), and their actual emissions (blue). Some 76 % of Universities are failing to meet their carbon reduction targets, while 25 % have reduced their targets. So far there has been a weak 7 % reduction of carbon emissions over the last six years, and with only 4 years left to 2020 there is little prospect of achieving much further reduction toward what looks like a distant 43 % target.

This poor performance hides that there has nonetheless been important progress in thinking. For example, there has been a shift in thinking by the University of Westminster about the scope of the Estates Departments' responsibilities, from regarding carbon reduction as being about emissions from University buildings to a growing focus on the wider notion of sustainability.[2] No doubt this shift in thinking

[2]The term sustainability has become a familiar qualifier within organisational strategy, no longer exclusively belonging to environmental concerns. For example, one of the University's aims is to maintain its long term financial sustainability.

Fig. 4 The growing sustainability agenda used for assessing Universities (from People & Planet University League)

is due to both a search for ways of reaching emissions targets, and at the same time a growing realisation that emissions are manifest in almost all university activities. In addition various independent organisations have emerged, concerned to promote and monitor sustainable behaviour across the HE sector, spurring debate, awareness, and competition through league tables. This can be seen from some of the reporting required within the sector, such as the UI Green Metric World University Ranking, the People & Planet University League and the Higher Education Statistics Agency's Estates Management Records. All these reporting mechanisms have in the past 5 years expanded the criteria used to measure sustainability within HEIs. For example, The People & Planet's criteria grew between 2007 and 2016 (Fig. 4).

Similarly, back in 2010, the Estates Management Records, managed by the Higher Education Statistics Agency, requested HEIs to provide information on traditional Estates based activities, such as energy consumption and waste disposal. Since then the range of factors of interest has grown to include Supply Chain issues and Staff & Student Commuting data.

3 Theoretical Background

An understanding of the practice of strategy is central to making sense of how organisational strategy develops and evolves. Mainstream literature on strategic management presents strategy as a tool used by organisation leaders in order to achieve goals. Another stream of thought posits there is much to learn about strategy by focusing on the everyday practice and discourse of practitioners at work, (Rayman-Bacchus 1996; Whittington 2006). From the mainstream view, strategy is an object belonging to the organisation that may be modified at will, while from an everyday practice perspective, strategy is an integral part of organisational daily life and practitioners' identities are bound up with that everyday activity. Here everyday practices are not simply embedded routines and rules carried out unthinkingly. Rather, organisational actors need to be creative since the rules, procedures and structures of the organisational social systems of their workplace are not simply given and unambiguously readable, but unavoidably are plural, requiring interpretation according to the needs of the situation. Through such practice, organisational actors are not slavishly bound by existing ways of working and thinking like automata, but must interpret, refine or redefine, elaborate or diverge from the (often invisible) hand of practice (formal and informal).

Both management teaching and strategy research tend to distinguish levels of strategy (corporate, business, functional), and to delineate intra-organisational strategy from extra-organisational practices. Clearly organisational strategy practice and discourse do not operate in a vacuum, but to a significant extent reflect internal hierarchical priorities, as well as currents flowing in a wider sectoral and societal context. The interrelationship between intra-organisational strategy practice and wider societal currents need to be recognised as part of a whole.

Context comprises all manner of institutions (political, cultural, economic, social norms, shared discourse and practice) operating at global, national, and industry sector levels. Context here is not a homogenous backdrop to social action, or detached and somehow passive and unresponsive to social action. Rather context is created and maintained by those with a shared frame of reference. Adapting Russell (2002: 68), context 'is a web of sociocultural interactions and meanings that are integral' to everyday practice. It is a shared frame of reference; a common resource of knowledge, discourse and practice that develops over time and to which practitioners look for guidance (Edwards and Mercer 1987). It is persistent, enabling communication and the intelligibility of that communication.

This discourse and practice provides practitioners some level of mutual understanding, based on their shared experience. However, this does not mean that practice is the same everywhere. Practitioners the world over operating in particular communities of practice contribute to the reproduction and interpretation of strategy, sensitive to differences between some universally accepted discourse and local variations. The interpretive flexibility of regulation, and individual jurisdictions with their differing regulatory heritages and customs, means that local discourse and practice does differ and matter to the particular organisation and responsible jurisdiction. Where organisations operate in a regulatory environment where new policy is overt yet somewhat ambiguous and developmental, practitioners need to be more creative, exercising much more judgement about what approach to take as there are few precedents to guide them. Building on Whittington (2006: 625), there is also need to acknowledge the role of practitioners as 'carriers of practice', including not only strategists, 'establishment elites' and other influential actors, but also those possessing specialist knowledge and skills in areas such as finance, governance, operational management, environmental stewardship, health and safety, and employment rights. We also need to recognise the influence of practitioner networks and communities of practice, where practice and discourse is shared across organisational boundaries.

4 Research Design

This exploratory study seeks to better understand, and draw lessons from, the everyday strategy challenges to implementing new regulation (on reducing carbon emissions), while at the same time incorporating wider demands for sustainability thinking. Consequently this study is grounded in examining elements of the everyday work of practitioners within an organisation as they make sense of regulatory demands, and try to read the evolving mood within the HE community toward sustainability. The approach to this study is both interpretive and pragmatic, guided by grounded theory (Strauss and Corbin 1990) and abductive reasoning (Reichertz 2009), and case study (Yin 1989). Interpretive here means trying to understand how practitioners make sense of their emerging regulatory reality, not to find some truth from the data, and to generate knowledge that - while remaining tentative - is useful for the purpose of drawing lessons on implementing sustainability strategy in context. The case study is appropriate for studying a contemporary social phenomenon in context, where the phenomenon—context boundary is shifting and symbiotic (Yin 1989: 23), as is the case of implementing environmental regulation within the University of Westminster. The case study presents reflections and observations of everyday strategy within the University, from which we offer likely explanations and lessons.

As Strauss and Corbin (1990) and Eisenhardt (1989) note there is potential value in exploring existing literature in order to stimulate ideas. Reviewing existing ideas and knowledge in this way allows us to interpret the data in potentially novel ways. Following reflection on the theoretical perspective of strategy as everyday practice, and the role of context, we collected a variety of data: documented commitments by the UN, EU and UK government, to tackling climate change through carbon management; the University of Westminster's documented strategies and policies, and interviews with practitioners from various function around the University (Estates, Finance, HR, Procurement, Academic Faculty). Following Schutt (2006) we take an interpretive historical approach to analysis, continually comparing similarities and differences over time, and continually comparing theory with data. This also accommodates that we, the authors, may have prior knowledge (of strategy as everyday practice, and of the organisation's practical engagement with environmental regulation). Since one of the authors works in sustainability at the University of Westminster this opens the research design to a charge of bias, for example in the account presented as the case study. On the other hand her subjectivity is also valuable, enabling a more nuanced and reflexive understanding of the environment under study (Olesen 1994). The reliability and validity of the account given here has been confirmed by two independent readers familiar with the issues presented.

5 University of Westminster's Journey

The regulatory drivers mentioned in the previous section focused on how carbon emission was addressed by UK universities, including the University of Westminster. This case study demonstrates how the University of Westminster's approach to carbon management slowly transformed into an elaborate and structured approach to sustainability thinking, with an attendant and continual shift in the discourse, from a focus around carbon emissions to one on a broader interpretation of sustainability. This approach involved the University adopting an overarching institutional framework, the recruitment of a team of sustainability specialists, and the growth of an increasingly sophisticated communication strategy. The account highlights how the alignment between top down and bottom up pressures combined to shape the journey: regulatory pressure from HEFCE; the University's proclaimed corporate sustainability values; and the commitment and broader aspirations of employees to working and living sustainably, and (importantly) to be seen as doing so.

5.1 Awakening

In 2007 the University created a new post of Energy and Environment Manager, based within the Estates & Facilities Department, to develop and implement a 5 year Carbon Management Plan. Working with the Carbon Trust, a government-funded body at the time, a lengthy, technical plan was developed to evaluate and reduce the University's carbon footprint based on its building usage. By 2010, it was evident that the University's Carbon Management Plan, developed with good intentions as a response to the HEFCE strategy, was not being effectively implemented. There was no clear accountability. As part of a wider reorganisation, the post of Energy and Environment Manager was replaced by a new post of Asset Performance Manager. Under this new post, another Carbon Management Plan was developed in 2011, but again few actions in this plan were being realised, mainly due to very few staff being familiar with the plan, or were even considering how the idea of carbon reduction might affect their projects or actions. This was true of both the core carbon management implementation team[3] and the rest of the University community. Indeed responsibility for carbon reduction was limited to one department (Estates) and then ultimately to one person, whose job title and scope of

[3]The Carbon Management implementation team was a group of people identified in the plan as core stakeholders, and included among others Capital Projects Manager and Maintenance Manager.

responsibilities was redefined three times between 2007 and 2012. First there was the Energy and Environment Manager, superseded by the Asset Performance Manager, then in 2012 a new post of Sustainability Manager was created, replacing that of Asset Performance Manager. This reflected a sense of fuzziness about what the task entailed. Indeed, the incoming postholder, seeing carbon reduction management as part of a larger sustainability challenge, suggested the title of Sustainability Manager as more appropriate than Asset Performance Manager.

Alongside the (revised) University 2011 carbon management plan, and consistent with the University leadership's commitment to becoming 'sustainable', there were localised initiatives around the wider sustainability agenda in various departments, such as Human Resources, Central Procurement. These departments recognised the need to operate sustainably and appointed a member of existing staff to incorporate sustainability within their normal duties. However, these tasks were being treated as additional to the existing roles of staff, and little training or support was given to help these individuals understand and manage their additional sustainability/CSR work load. On reflection, there appears to have been an assumption among university leadership that 'sustainability' could be straightforwardly understood and applied as part of existing work. Further, inadequate attention was being paid to maintaining project momentum, resulting in carbon-related and wider sustainability projects being abandoned when responsible personnel moved on. This reliance on a few people with additional, delegated and vague responsibility for sustainability and/or carbon management generated 'more heat than light', i.e. little progress. In addition to the lack of clear responsibilities and poor project continuity, there was also no shared understanding about the meaning of key ideas including: sustainability, carbon footprint, green, corporate social responsibility. Indeed they were used interchangeably across the institution. This ambiguity in the interpretation of key ideas existed not only within the institution but also within the wider Higher Education community and beyond.

Despite these difficulties, there was strong interest among all levels of staff in engaging with the wider sustainability agenda. Localised grassroots initiatives were being developed, often without the knowledge of senior University staff, and with limited wider engagement by staff. For example, a group of staff started a small-scale local food growing project on campus, which very few people knew about. Staff was frequently requesting financial support to explore or develop sustainability related ideas that had organisation-wide relevance, such as additional waste bins with clear information on recycling. Where centralised University support was not provided, individuals sought support from within their work group, or abandoned their idea. This mismatch between espoused corporate commitment to sustainability and lack of tangible support or action led to frustration amongst staff and reinforced their perception that the University was not genuinely committed.

As the result of a sector wide consultation exercise, the University adopted the Learning in Future Environments Framework (LiFE) (Fig. 5) to help guide its sustainability strategy. LiFE categorises sustainability into four main groups:

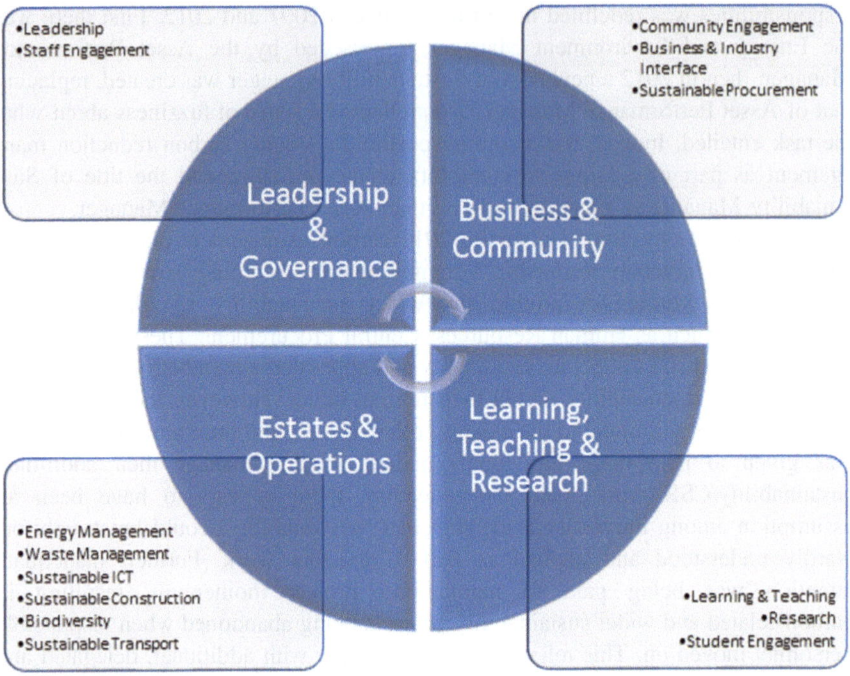

Fig. 5 Learning in future environments framework (LiFE)

- Leadership & Governance,
- Business & Community,
- Learning, Teaching & Research and
- Estates & Operations.

The Estates & Operations section of this framework was approached by adopting ISO 14001 (Environmental Management System). Also, the Business section of the framework was approached by adopting the Flexible Framework. These strategic initiatives, plus ongoing internal discussions have helped discipline the internal discourse on sustainability, so that there was increasing agreement on defining sustainability task areas and responsibilities, and on priorities against resources. The LiFE framework was instrumental in bridging the gap between the University's vision to be 'sustainable' and the local sustainability projects developing organically. This also helped guide the subsequent development of a sustainability governance structure, key performance indicators, and where to draw the boundary of sustainability within an organisation.

At this stage, apart from the above mentioned Carbon Management Plan and an Environmental Policy, there was little in the way of any formalised or published information regarding the University's commitment to sustainability, despite the University having 'Sustainable' as one of its five core values, and the introduction of the LiFE Framework. The few formal commitments that existed (i.e. the two noted above), seemed to have been used as a tick-box exercise to satisfy auditors, attract funding or build reputational credit. With an eye on funding, the University had these plans prepared and made available to external organisations through various reporting requirements. That the plans existed could be interpreted as 'the plans are being implemented'. This was to satisfy HEFCE as future capital allocations from HEFCE were linked to carbon performance and the establishment of a Carbon Management Plan. Universities that had developed baselines, set targets and achieved them would potentially be eligible for future funding. Another external pressure group, People and Planet managed the UK Green League (subsequently renamed University League in 2013/14) whereby having an Environmental Policy earned HEIs further points and potentially increased the Institution's rankings in this league table. However, these University plans, did not have clear ownership, targets, accountability for poor performance, or formal monitoring of progress.

Critically, the regulatory driver behind the University's carbon reduction plan was weakening. While the University was grappling with how best to organise for, and interpret emissions regulation, and apply carbon reduction strategies, the focus of HEFCE seemed to be shifting, with the original push on carbon reduction slackening as capital allocation from HEFCE was diminishing. Legislation was constantly evolving, accompanied by a reducing sense of urgency. It was also leading to some confusion in the HE sector. There was myriad different energy related legislation whose applicability to HEIs was not clear. For example, the government announced in 2014 that the Energy Saving Opportunities Scheme would apply to privately funded Universities, including Universities funded mainly from student tuition. However, within 18 months, the government announced that this would not apply to such universities. Also, the government's approach to enforcement was changing. For example, within the Carbon Reduction Commitment Energy Efficiency Scheme (CRCEES) the regulator, (the Environment Agency) stated that all participants will eventually be audited within each Phase. However, this has not been the case for the University which at the time of writing is well into Phase 2. Still, there was a new sentiment brewing within the University and across HE institutions in general. The University was experiencing a growing sense of being responsible for its own stewardship of increasingly scarce and expensive natural resources, plus the notion that doing good is also good for business. Staff Engagement Surveys increasingly showed interest amongst staff and their personal commitment to sustainability. This sentiment was further strengthened as external reporting requirements were asking for the University's ethical stance.

5.2 Forming

From around 2012 there were increasing calls from inside the University, and from the wider HE sustainability community, to see the broader sustainability agenda incorporated within the University's strategy, and to include topics such as carbon management, social responsibility, and ethical investment. The focus was no longer limited to ways of reducing energy consumption of buildings, but whether wider sustainability initiatives were being embedded across the University at all levels. For example, within the HE sector, the UI Green Metric was not only asking Universities for their environmental impacts from estates activities, but also wanted to know about areas such as student engagement, and funding for sustainability research projects. So the internal discourse evolved from a focus on carbon management to one on sustainability strategy with carbon management being one element within a more complex debate about the University's relationship with the natural environment. University of Westminster staff was increasingly regarding sustainability not as an additional (external) regulatory burden, but as an intrinsic and collective social responsibility. All levels of management were looking for ways of making sustainability an integral part of daily life within the University. For example, a number of grassroots sustainability projects were formed, such as the Bicycle User Group led by one of the University's IT Technicians. This group not only developed a network of bicycle users of staff, but also started Bicycle Maintenance Workshops during lunch breaks to fix bikes and transfer skills.

Government too has been playing a part in the University's transformation of thinking around what counted in measuring carbon emission reduction. Whether by intention or accident, government policy required the University to carry out audits, thereby forcing the institution to confront the consequences of its own strategies on a range of issues, including energy use, waste management, staff awareness, and supply chain influence. This gradual transformation can be seen in the University instituting a variety of formal mechanisms for implementing sustainability across all areas of the organisation: Health, Safety & Environment Committee, Energy Committee, Value for Money Group, Carbon Emissions Group, and Sustainability Group. These separate committees and groups were subsequently set within a comprehensive management and reporting framework, in the form of the sustainability governance structure (Fig. 6). Each of the groups identified in this latest structure have specific objectives, key performance indicators and detailed action plans of how they will achieve the University's sustainability vision.

As noted, being a sustainable organisation was now more than about regulatory compliance. There were now additional pressures that, while less tangible, were

Fig. 6 University of Westminster sustainability governance framework

perhaps more significant to university staff. These include reputational risk, HE sustainability league tables and performance comparisons from peer institutions, and rising expectations about the University's social responsibility. University staff at all levels and functional areas saw in the sustainability agenda an idea they valued at a personal level, and a source of inspiration. The agenda also came with a reasonably clear framework (the trilogy of economic, environment, social) allowing individual initiatives towards addressing social and environmental concerns.

Against this more facilitative background, ongoing discussion between the Sustainability Manager and key senior managers led to agreement that sustainability should and could be more effectively instilled in all departments. The strategy was that each department would be responsible for incorporating sustainability, but to achieve this, the University needed an Implementation Team of specialists. It was further agreed this team of specialist practitioners would report to the Sustainability Manager. This approach would overcome the problem of additional seemingly distracting responsibilities being heaped on existing staff, and would support departments that were already enthusiastic, but had no means of implementation. For example, departments frequently accumulated unwanted items they wanted to give away rather than throw away, but there was no enabling mechanism. Still, the Sustainability Manager had to show the business case for recruiting these specialists as this would require a significant investment and a leap in the dark as there was no certainty this would deliver the intended organisation wide cultural change. A more tentative approach in the same direction was adopted. The Sustainability Manager was given the authority to recruit temporary roles, so by 2013 started recruiting specialist practitioners on a temporary basis. Working within the LiFE Framework, this implementation team began catalysing the take up of sustainability thinking across all departments, helping to initiate and organise action, and thereafter facilitate ongoing development. For example, one specialist from the implementation team began helping departments select items for donation and appointing the preferred charity. Another specialist began working with the University's Procurement team in order to integrate sustainability thinking, using a recognised protocol.[4]

What constituted the University's carbon emissions now encompassed a wider scope, defined by the Greenhouse Gas Protocol, and included areas such as staff and student commuting and the supply chain. One key role of the sustainability implementation team was to bridge the gap between the University's corporate ambition to be sustainable, and the localised ad hoc sustainability initiatives operating across the University, some receiving little or no formal corporate or financial support. Another role of the team was to identify potential new sustainability risks to the University and mitigate such risks by addressing them early. Individual departments started paying attention to the growing momentum of the University's sustainability initiatives, and either linked their existing projects to the University's vision and/or developed their own projects. Various staff engagement programmes were launched or emphasised to get more people understanding their own footprint and applying that awareness within working practices. This

[4]*Sustainable Procurement in Government: Guidance to the Flexible Framework.*

strengthening of commitment translated into material innovations, which would ensure the University's vision would not be lost in a document, but be taken up in real and tangible practices.

5.3 Embedding

Over the last five years (from around 2011) the University's espoused position on (environmental) sustainability has evolved and broadened, becoming more closely linked with the University's social responsibility agenda, as evidenced by the new Social Responsibility Team located within the Human Resources Department. Now environmental and social concerns together form the core of the organisation's corporate social responsibility: 'The University of Westminster's Corporate Social Responsibility (CSR) programme reflects our commitment to operate in an environmentally sustainable and socially responsible manner'.[5] As noted earlier, sustainability is identified within its vision, and is one of five core values underpinning the *University's 2020 Strategy*.[6] The University's understanding of sustainability compared with five years earlier can be seen in the way sustainability now is more comprehensively regarded and featured in various other University strategic plans:

- The *People Strategy*, delivered through a variety of mechanisms, including the Key Competency Framework. Here sustainability thinking is a required competency, included in individual development plans, and supported by appropriate sustainability training.[7]
- The New Sustainability Training Plan. This aims to inform individuals of their environmental responsibility within the workplace, and also raise awareness of wider sustainability issues such as ethical purchasing, organic food growing, healthy eating, and sustainable travel.
- The *2015 Financial Statement*, with a commitment to ethical investment.[8]
- The *Sustainable Procurement Strategy*[9] and the *Sustainability Charter*[10] stating what the University expects of its suppliers.
- The new *Carbon Management Plan*, encompassing a wider remit than the original focus on the University's buildings, now including supply chain, staff and student commuting.[11]
- The *Employability Strategy*, recognising the need for graduates to acquire knowledge of, and skills in developing, sustainability initiatives.[12]

[5]University of Westminster About Us: vision, mission, and values.

[6]University of Westminster 2020 Strategy.

[7]Westminster 2020 People Strategy.

[8]University of Westminster Financial Statements 2015.

[9]University of Westminster Sustainable Procurement Policy & Strategy.

[10]University of Westminster Sustainability Charter.

[11]University of Westminster Carbon Management Plan 2014.

[12]University of Westminster Employability Strategy.

Results from the sustainability implementation team until now made up of temporary staff, showed there was a case for building a permanent team. For example, the Energy Coordinator role was justified on potential financial savings, and the Waste Coordinator role was justified on regulatory compliance grounds. In 2014 a permanent sustainability team was established, still located within the Estates & Facilities Department, and still mainly focussing on environmental issues, but with a broader lens than the traditional estates focus on building emissions. Since then this team has grown to include: Energy Coordinator, Waste Coordinator, Data Analyst, Environmental Management System Advisor, Sustainable Food Assistant, Waste Assistant, and Sustainable Travel Assistant.

New, unprompted staff campaigns and projects are mushrooming across the University. Staff feel inspired to take initiatives, leading to unexpected positive departures from the original sustainability vision, and which help meet or surpass the University's key performance targets. For example, individuals are creating more localised flora and fauna habitats, beyond the communal areas initially identified in the University's Biodiversity Action Plan, and donations to charities from departmental clear-outs reduce the amount of waste being sent to landfill or incinerators. Indeed there is a snowball effect as events are being organised without support from the sustainability implementation team, and more effective communication is enabling further unexpected initiatives.

Further, individuals and teams throughout the University are organising projects that are not directly linked to sustainability key performance indicators, such as food/clothes bank collections from other members of staff, but which nevertheless support the University's wider corporate social responsibility programme. These initiatives show a clear overlap between the University's sustainability and corporate social responsibility agenda and that of individual staff. This overlap seems to be self-reinforcing as rising expectation from the University's internal community push the University's corporate responsibility agenda, which in turn is inspiring more individual initiatives.

While there has been increasing understanding and consensus, new challenges have emerged. For example, internal discussion about the meaning of key terms and attendant managerial responsibilities continue, but now these are conducted from a position of a more common understanding relative to that of earlier phases. Also, from the early days there have been informal 'sustainability champions', who share a commitment to sustainability values, and whose passion has been contagious enough to influence anyone coming within their sphere of contact. However, more recently, as the university leadership reasserts top-down direction and feel once again in control of the sustainability agenda, proposals have emerged to create (formal) sustainability champions, few of whom are the existing informal champions. Unlike their informal peers, the (new) champions have formal responsibilities and key performance indicators. There is potential difficulty in the creation of formal champions through the possible disenfranchisement of the established informal champions.

6 Discussion

In this study we see strategy practice and discourse as a collective process shaped by the exigencies of context, in which practitioners organise, interpret, and negotiate differentiated assessments of meaning in pursuit of some shared but tentative understanding of an ambition (to become sustainability); yet the goal is not fixed but equivocal and developmental. Strategy practice involves making sense of the immediate past, of regulatory conditions and organizational commitments (historical, cultural, structural), while at the same time engaging with the urgency of the immediate future and exploring longer term ambitions. Here strategy in practice is routinely guided by, and contributes to, a more or less coherent shared reality; one that is at the same time routinised and a creative process, ordered and disordered, reinforcing and elaborating, intuitive and rational. It is a process full of unstated assumptions, ambiguity and interpretive flexibility, individuals seeking identification with organizational goals, political bargaining and symbolism.

As an HE institution, the University of Westminster's obligations towards addressing climate change were initially driven by both HEFCE, the HE sector regulator, and subsequently by the Environment Agency, the UK government body directly responsible for regulating the reduction of carbon emissions across all UK industrial sectors. The UK's obligations towards the climate change challenge are in turn shaped by its commitment to the UNCCC, a global convention. It is a context that is multi-layered and complex, comprising overlapping regulatory demands. It is also in a state of flux, and subject to shifting political priorities. At a global level, there is the ongoing attempt to engage more nations in the UNCCC; an engagement that seems hesitant at best and subject to individual national priorities. There is also the initiation of new environmental regulations, some focused on reducing carbon emissions, and subsequent uncertainty about the scope and significance of one or more regulation. This fluidity and uncertainty translates into uncertainty within individual HE institutions about what has to be done.

Within the UK context, the push from HEFCE requiring HE institutions to meet carbon reduction targets, led to the creation of a new position within the University of Westminster, initially focused on energy consumption, then through two further reorganisations to a broader focus on sustainability. The University leadership was primed having already debated and concluded a need to include 'sustainability' as one of its core values. However, this readiness does not mean there was a well worked out strategy for delivering sustainability. The discourse within this and many Universities, for a time focused mainly on energy consumption and emissions from existing buildings. However, there was a growing awareness of the wider sustainability agenda and external pressure groups also expanded their requirements: Indicators they used were no longer limited to environmental concerns within estates operations. Hence the People & Planet's Green League widening its focus and changed its name to People & Planet's University League to reflect the wider scope of interest (in sustainability). Environmental impacts were only a part of sustainability indicators which also included universities' social responsibility.

Ambiguity in the meaning of key terms (sustainability and others) and the scope of their applicability to the everyday work of the University, reflects a lack of shared understanding not only within the University but also within the wider Higher Education community and beyond. Even the job title of Sustainability Manager rather than Asset Performance Manager could be negotiated, reflecting the uncertainty around the scope and detail of the job to be done. Practitioners operating all levels and functions, from strategists to functional specialists, were busy trying to make sense of this new policy on carbon management and more. Strategy became increasingly elaborate through a combination of influences, as practitioners worked, singly and collectively, at developing an appropriate organisation, interpreting and translating regulatory policy into University procedures and guidelines, and managing its coordination with, and integration into, ongoing University operations. This element of interpretation is not simply the straightforward reading and unambiguous understanding of how the policy is to be implemented. For example the sustainability manager was negotiating meaning while seeking clarification, with the regulator on one side, while engaging with other practitioners within the University on how to make sense of the policy in the particular circumstance within the University. In the process there were three successive reorganisations partly driven by weaknesses in existing arrangements and by a desire to better integrate environmental concerns, and subsequently sustainability/CSR thinking; the introduction of successive specialist practitioners, and their engagement with other specialists (e.g. professional staff from Estates Management, Finance, Human Resource Management, Procurement, and Faculty staff) influenced the sharing of ideas and changes in practices; and the various networks of sustainability practitioners working in other institutions also contributed to the exchange of ideas and practices. These elements of implementation (organisation, interpretation and integration) unfolded under the gaze and support of interested leadership, and senior University influencers paying attention to the execution of the University's regulatory obligation.

Witness the increasing integration of organisation and coordination around sustainability through an elaboration of procedures and structures, the latest manifestation being the Sustainability Governance group. Note also a shift in the attitude among staff, away from seeing their everyday practice as being burdened with regulatory compliance and separate from their main tasks, and toward everyday strategy practice as being infused with a discourse of intrinsic responsibility, shared values and ethical drivers. University staff find within the sustainability agenda the potential and mechanisms for making a personally rewarding social investment, through how they think about sustainability and do everyday work, and by becoming involved in one or other sustainability initiative. The range of practitioners with a formal sustainability responsibility continues to expand. The opportunity to engage with socially rewarding work, through everyday strategy practice and discourse around the University inspires innovation and commitment. Government demand for audits has been helping that shift of mindset, as has an ongoing debate about, and greater shared understanding of, key terms, plus the ongoing creation and support of sustainability initiatives. While staff find intrinsic value in working and living sustainably, there is at the same time a need among staff to be recognised for their commitment and industry, whether through reputational credit from external agencies or by individual acknowledgement from within.

7 Conclusions

The aim of the research presented in this chapter is to offer useful insights to the challenges one UK HEI, the University of Westminster, faced in implementing government regulation on carbon emissions, and how it became part of the institution's development of a wider sustainability strategy. This better understanding provides a basis for highlighting lessons on implementing new regulation where there is no clear consensus on the meaning of key ideas (such as sustainability), or on how to translate these into strategy practice.

The chapter first outlines the wider political context of an international agreement, the UN Framework Convention on Climate Change, to address climate change. In line with this convention, the European Union member states set ambitious goals to reduce their carbon emissions. The UK introduced legislation requiring all sectors of the economy, including Higher Education, to take measures to reduce their carbon emissions. Set within this context, the chapter presents the case of the University of Westminster's journey towards sustainability in its practice and discourse. This journey is explained as developmental, passing through three stages: awakening, forming, and embedding. In order to analyse this journey, an analytical framework was adopted, based on theoretical ideas around strategy as everyday practice, ideas about context, and about the work of practitioners as carriers of strategy.

In this journey, sustainability thinking at the University of Westminster started from a narrow focus on reducing carbon emissions of its buildings, taking place alongside other ad hoc initiatives, but without any overarching framework. Over a period of about nine years, strategy practice has evolved and more structured, guided by a sector-wide framework (LiFE), the creation of a team of specialists to support implementation, and the creation of an overarching sustainability governance framework encompassing a wider spectrum of environmental and social issues. This elaboration of perspective reflects the engagement of an increasingly wide range of staff, as they gradually see sustainability as the responsibility of all staff and not just a few. One manifestation of this development is that sustainability has become an established agenda item for various university committees and groups apart from the traditional environmental groups, to include the Audit Committee, the Efficiency Task Group, Senior Management Team Meetings, and Faculty Executive Groups.

Critically, there were, and continue to be, plural interpretations of the meaning of key terms, which unavoidably influences the direction of future development and the form of new procedures and guidelines. The collegiate environment of the University means that there are few hierarchical barriers to the development of a wide variety of initiatives, many of which are local, informal, though often wider university staff remain unaware of them. The introduction of regulation on reducing carbon emissions catalysed the university into rethinking the meaning of sustainability, instituting organizational changes, and triggering changes in everyday strategy practice and discourse around sustainability.

8 Lessons Learnt

1. Importance of developing shared meaning and responsibilities. Incorporating the wider University community helped build momentum and the development of new norms and create more meaning for staff. Delegating responsibility to a small number of people for the University's carbon management led to project failure. In contrast, a broader spectrum of staff could relate to sustainability goals, and were therefore able to volunteer their commitment and share responsibility. This led to a higher rate of success with sustainability projects, as staff were able to initiate and link their particular projects to the University's vision of sustainability.
2. Consulting and cooperating widely helps bond staff. Strategic decisions based on consulting widely resulted in shared responsibility and higher levels of ownership. Staff from different departments came together through sustainability projects of common interest, for example ethical purchasing, renewable energy, and organic food.
3. Linking organizational goals with personal values. Approaching sustainability as an organizational goal without also tapping into individual personal interest is a wasted opportunity. Sustainability could be used as a motivating tool that relates to individuals' lifestyles and values. Removing the boundary between work objectives and personal objectives with regards to sustainability encouraged self-motivation and brought like-minded people together. These groups were then able to share spaces to discuss and inspire new activities.
4. More robust accountability across various departments led to higher rate of success.
5. Reliance on one driver. As regulations and regulators changed, the relationship between the University and the regulators changed. Compliance was a poor driver of the University's sustainability agenda. A balance between reputational risk, financial stability and the University's sense of responsibility has provided a far greater foundation for addressing the sustainability agenda.
6. The importance of frameworks and standards, collectively agreed, in defining how sustainability is approached within the University. Prior to frameworks or standards being available, sustainability was approached randomly, without structure. As the meaning of sustainability was not clear internally or more widely,

sustainability was interpreted differently by many if not most staff. Frameworks provided boundaries, key indicators, benchmarking tools to use off-the-shelf.

9 Limitations

Unavoidably, there are limitations in our study, and these invite further study. The case study presented here represents the view from those most closely involved with implementing the sustainability strategy. Given the aim of this paper is to examine the evolution of strategy in practice, such a focus is justifiable on the basis that these actors/practitioners have intimate knowledge and experience of the process. Nevertheless, the view of the University executive is likely to offer a contrasting perspective, and may offer interesting insights for example on the challenges of leading an institution through culture change. Although this study is exploratory, and aims to contribute to theory (of strategy practice), the lessons offered are drawn from the particular circumstances of one large British university. It would be instructive to learn what kinds of lessons would emerge from other settings, from within education and other public sector organisations, and from private enterprise in different sectors.

Acknowledgments We would like to thank Mitch Dalgleish, Head of Procurement, and David Haylor, Space Manager, both at the University of Westminster, for reading and commenting on the case study 'University of Westminster's Journey', as part of the drafting of this paper.

References

Edwards, D., & Mercer, N. (1987). *Common knowledge: The development of understanding in the classroom*. London: Methuen/Routledge.

Eisenhardt, K. M. (1989). Building theories from case study research. *The Academy of Management Review, 14*(4), 532–550.

Olesen, V. (1994). Feminisms and models of qualitative research. In N. K. Denzin & Y. S. Lincoln (Eds.), *Handbook of qualitative research*. London: Sage.

Rayman-Bacchus, L (1996), The practice of strategy, University of Edinburgh, unpublished thesis.

Reichertz, J (2009). Abduction: The logic of discovery of grounded theory. *Forum: Qualitative Social Research, 11*(1), Art. 13, http://nbn-resolving.de/urn:nbn:de:0114-fqs1001135. Accessed 12 July 2016.

Russell, D. (2002). Looking beyond the interface: Activity theory and distributed learning. In M. Lea & K. Nicoll (Eds.), *Distributed learning*. London: Routledge.

Schutt, R. K. (2006). *Investigating the social world: The process and practice of research*. London: SAGE.

Strauss, A. L., & Corbin, J. (1990). *Basics of qualitative research—grounded theory procedures and techniques*. London: Sage.

Whittington, R. (2006). Completing the practice turn in strategy. *Organization Studies, 27*(5), 613–634.

Yin R. (1989). *Case study research: Design and methods* (revised. ed.). London: Sage.

Websites

carbontrust.com/home/
eauc.org.uk/life/home
greenmetric.ui.ac.id/
HEFCEe.ac.uk/workprovide/carbon/
hesa.ac.uk
iso.org/iso/iso14000
gov.uk/guidance/energy-savings-opportunity-scheme-esos
gov.uk/government/collections/crc-energy-efficiency-scheme
peopleandplanet.org/university-league
blog.westminster.ac.uk/sustainability/2015/03/cycle-skill-sharing-session-3/
ghgprotocol.org/
University of Westminster Forward Thinking. (2015). *Westminster 2020 People strategy: Valuing our staff experience*, July 2015. Available at: http://westminster.ac.uk/file/50291/download. Accessed 30 June 2016.
University of Westminster Report and Financial Statement. (2015). Available at: https://westminster.ac.uk/sites/default/files/Report-and-financial-statements-2015.pdf. Accessed 30 June 2016.
University of Westminster Sustainable Procurement Policy & Strategy, March 2016. Available at: https://westminster.ac.uk/file/33016/download. Accessed 30 June 2016.
University of Westminster Carbon Management Plan, March 2014. Available at https://westminster.ac.uk/file/13896/download?token=F1mQeSJM. Accessed 30 June 2016.
University of Westminster Employability Strategy 2015–2020, September 2015. Available at: https://westminster.ac.uk/file/16051/download?token=jgJz3rW8. Accessed 30 June 2016.
University of Westminster About Us: vision, mission, and values. Available at https://westminster.ac.uk/about-us/our-university/vision-mission-and-values. Accessed 30 June 2016
University of Westminster About Us: environment policy. Available at: https://westminster.ac.uk/about-us/our-university/corporate-information/policies-and-documents-a-z/environment-policy. Accessed 30 June 2016.
University of Westminster 2020 Strategy. Available at: https://www.westminster.ac.uk/sites/default/files/University-of-Westminster-2020-strategy.pdf. Accessed 30 June 2016.

Authors Biography

Lez M Rayman-Bacchus PhD Founding Director of the Centre for Corporate Responsibility and Sustainable Development and an international bi-annual symposium, and a visiting Research Fellow at Winchester University, UK. Previously a business practitioner his focus is consulting, teaching and researching strategy, corporate responsibility, ethics, sustainability and related subjects.

Jandi Perman Masters in Environmental Assessment & Management. Sustainability Manager at University of Westminster, UK, helping embed sustainability within the University as well as influencing external stakeholders in doing the same. Trained as an environmental engineer and in humanitarian work, now focusing on implementing environmental management systems and sustainability strategies, and in driving environmental and social change.

The Role of Education for Sustainability in the Sustainable Development Goals—Changing Policy and Practice?

Ros Wade and Hugh Atkinson

Abstract

The key focus of this paper is on the crucial role of education and learning in achieving the global shift of policy and practice which is needed in order to implement the sustainable development goals (SDGs). As highlighted in 2015 by the International Council for Science (ICSU) and the International Council for Social Science (ISSC) 'Education has been recognized for many years as a critical factor in addressing environmental and sustainability issues and ensuring human well-being'. This paper will argue that in order to achieve the SDGs it is an imperative to embed education for sustainable development (ESD) within all future policy and practice. It will examine the relationship between Education for All (the focus on basic and primary education in the Millennium Development Goals) and ESD. It will refer to the work of UNESCO in trying to develop synergy between these two policy approaches. It will argue that neo liberal agendas of marketisation and privatisation of education have impeded progress in re orienting systems towards sustainability. However, it will then indicate that there are some small signs that the global paradigm of neo liberalism is beginning to shift and that this will present opportunities to open up space and to highlight the need for ESD. The main focus of this paper will of necessity be on formal education though some reference will be made to the role of informal social learning and non-formal education. This research has been carried out with the support of colleagues in the disciplines of political science and education and has been subject to peer review at a number of fora including a social science seminar series at LSBU, a panel of the specialist group on

R. Wade (✉) · H. Atkinson
Division of Law and Social Sciences (LSBU), London South Bank University, 103 Borough Road, London SE1 0AA, UK
e-mail: wader@lsbu.ac.uk

H. Atkinson
e-mail: atkinsh@lsbu.ac.uk

W. Leal Filho (ed.), *Sustainable Development Research at Universities in the United Kingdom*, World Sustainability Series,
DOI 10.1007/978-3-319-47883-8_20

Environmental Politics at the UK Political Science Association Conference, a research panel at the Uganda EFS Conference and at the Mid West American Political Studies Association Conference. The research process has drawn heavily on the authors' experience over 21 years of working in the area of education, politics and policy change, it has involved desk based reviews of a large number of UN and UNESCO policy documents and discussions with a number of policy makers and practitioners. The paper also draws on research from a range of countries and makes reference to masters research dissertations by scholars of the MSc Education for Sustainability at LSBU.

Keywords
Sustainability · Education · Policy and practice · Paradigm shift

1 Introduction

The Paris Conference Of the Parties (COP) in December 2015 represented a clear acknowledgment by all member states that humanity is at a pivotal moment when future decisions on policy and practice will effectively determine our future survival as a species. The changes needed to chart a course towards sustainability and to prevent the worst effects of climate change are immense and require a major change of behaviours and practice in the social, economic and the environment realms. This of necessity will entail education, innovation, creativity and learning new ways of relating to the natural world of which we are a part.

Since the Millennium Development Goals (MDGs) come to an end in 2015 the new global policy framework has coalesced around the 'Sustainable Development Goals' (SDGs). The need for a more sophisticated and integrated set of future goals has long been identified (Sachs 2012), to bring more policy cohesion and to link environmental concerns with economic and social concerns and moves towards a process of sustainable development. The Stiglitz Report (Stiglitz et al. 2009) and others have also highlighted the need to move from a narrow use of GDP to measure economic success to one which is more related to human well being and flourishing. An emphasis on economic growth at any price has led us into an unsustainable impasse, with recession, low wages and a noticeable lack of attention to environmental concerns. At the same time, man made climate change, as well as posing an existential threat to the animal species and the whole ecology of the world, is the biggest threat to the continued existence of humankind (Stern 2006; IPCC 2013; Klein 2014). A consensus is now developing that we are in a new geological era of the anthropocene with human activity becoming the key influence and threat to the future of our planet.

Some noticeable changes of policy and practice have been taking place over the last 20 years, such as the 1997 Kyoto Protocol, the 2008 UK Climate Change Act and the emerging Green Economy agenda. Most recently the agreement reached at the Paris COP 2015 offers some more promising ways forward. But so far this has only just touched the surface and not led to substantive change. A number of the so called 'developing' nations (such as low lying Bangladesh and the Maldives) are under the gravest immediate threats but the global changes in the climate mean unprecedented challenges for every nation (IPCC 2013). These threats cannot be confined to national boundaries and will require global policy responses so the international community needs to start to address these at the global level. The achievements of the MDGs (2000–2015) have illustrated the potential of global policy making to make inroads into a range of global issues such as the education and poverty gap but there is a danger that without radical future commitments these achievements may not advance. This was highlighted in the UNDP Report of 2007 which drew particular attention to the impact of climate change on the achievements of the MDGs, 'Looking to the future, the danger is that it will stall and then reverse progress built up over generations not just in cutting extreme poverty, but in health, nutrition, education and other areas' (UNDP 2013: 1). The development of the SDGs offers the potential to move policy and practice in a new direction and put the world on a course which promotes human well being and development within the ecological and planetary boundaries of our Earth.

2 The Sustainable Development Goals and the Role of Education and Learning

According to a report by the International Council for Science (ICSU and the International Social Science Council (ISSC) (2015) 'the proposed Sustainable Development Goals offer major improvements on the rather fragmented Millennium Development Goals (MDGs). The SDG framework addresses key systemic barriers to sustainable development such as inequality, unsustainable consumption patterns, weak institutional capacity, and environmental degradation that the MDGs neglected.' However, this report also highlights key weaknesses, in particular it notes that

- The SDG framework would benefit from an overall narrative articulating how the goals will lead to broader outcomes for people and the planet. An overarching goal could be formulated, for instance in the political declaration framing the Post-2015 Development Agenda, binding together the 17 goals, thus providing a clearer means-to-end continuum.
- The current SDG framework does not identify the wide range of social groups that will need to be mobilized to deliver on the goals as agents of change alongside governments (CSU, ISSC 2015).

Educators form one of these key social groups but unfortunately (as with Agenda 21 1992) they are not formally acknowledged. There is a single Goal 4 on education which does, nonetheless, offer a much broader perspective than the education goals of the MDGs, with the emphasis on quality as well as quantity in the commitment to 'ensure inclusive and equitable quality education and promote lifelong learning opportunities for all'. The focus on lifelong learning is to be greatly welcomed, however, there is no overview of the key role that education and learning will have to play in order to achieve all the 17 goals. Urgent changes in policy and practice at the global as well as national levels are needed. However, in order to change policy and to change attitudes and behaviours then it is clearly an imperative (in democracies especially) to develop public understanding and support for these agendas. Politicians find it very difficult to enact the 'brave decisions' needed to make radical change without the support of voters. They are also frequently surrounded by powerful lobby groups (such as the oil lobby) who have only their sectional interests in mind. The dilemma over the Keystone XL pipeline is an example of this where the environmental lobby was pitted against the powerful energy lobby in the USA.

However, the SDGs do represent a much more integrated, holistic view of world development than the previous MDGs which failed to integrate environmental goals with development goals. Unlike the MDGs they emerged after a huge process of consultation and engagement with policy makers, business and communities across the world, the UN Open Working Group then identified 17 SDGs. These consultations generated inputs into global policy making from individuals and groups in 88 countries through meetings and conferences, on-line discussions, and larger public debates in the participating 88 countries. This stakeholder involvement offers much more potential for ownership and participation of relevant groups and communities. This is essential to make them happen, in particular, the involvement of the education community.

If we just look at two of the SDGs, it is quite clear that these cannot be achieved without education. Goal 14 is a commitment to 'Conserve and sustainably use the oceans, seas and marine resources for sustainable development' and goal 15

'Protect, restore and promote sustainable use of terrestrial ecosystems, sustainably manage forests, combat desertification, and halt and reverse land degradation and halt biodiversity loss'. These goals require a combined effort between policy makers, public and business leaders to change behaviours and practice. Without considerable public awareness and support, politicians (especially in democracies) will find it very difficult to enact policies which involve a necessary paradigmatic shift in change to lifestyles and behaviours. Hence the importance of education and public awareness raising, but also the necessity for *knowledge* of biodiversity and sustainable land use which will enable informed policy making.

If an informed public is essential to achieving the SDGs then according to the 2013 UNESCO Global Monitoring report on the education goals of the MDGs 'Education helps people understand democracy and promotes the tolerance and trust that underpin it, and motivates people to participate in politics. Education also has a vital role in preventing environmental degradation and limiting the causes and effects of climate change. And it can empower women to overcome discrimination and assert their rights. Education improves people's understanding of politics and how to participate in it.' In terms of political understanding UNESCO highlight the fact that 'across 12 sub-Saharan African countries, 63 % of individuals without formal schooling had an understanding of democracy, compared with 71 % of those with primary education and 85 % of those with secondary. People with higher levels of education are more interested in politics and so more likely to seek information.' Furthermore, 'By improving knowledge, instilling values, fostering beliefs and shifting attitudes, education has considerable potential to change environmentally harmful lifestyles and behaviour' (UNESCO 2013).

Education is also important to enable local communities to protect land and ecosystems. Research by Vicent Muhumaza in the Albertine region of Uganda for his ESD masters' dissertation identified that a lack of even basic education among many local communities left them vulnerable to the destruction of ecosystems and land grabs and unable to claim their rights (Wade R with Muhumaza V: 2015, 157). Local and indigenous communities are often best placed to understand the complex biodiversity interactions of environments and may have tried and tested ways of living sustainably with the natural world. This knowledge can be equally as valuable as western scientific knowledge but is rarely recognised as such because within the politics of knowledge the latter is prioritised. Robin Wall Kimmerman (a native America botanical scientist) draws on western scientific research as well as traditional indigenous knowledge to demonstrate the value of the synergy between the two. She highlights the anthropocentrism of our educational and political system and argues that we have learnt much and indeed have much more to learn from the natural world if we would only pay more attention to all livings things as co creators of the world (Kimmerman 2016). This point of view could be challenged as overly ecocentric, yet she is not arguing that it take the place of scientific thought, but rather that indigenous knowledge and ways of being are also relevant and important. In relating this to global policy making, one might conclude that until we see the need to give the natural world a place at the negotiating table, humans will continue to view nature as theirs to dominate and over use. In the form of climate change nature is speaking out but unfortunately, nature continues to be seen in global policy terms mostly as a resource for humans than of intrinsic value in itself. ESD aims to find a balance between the environmental and the social, to promote ecological and social justice and as such involves different ways of living and being and relating with the natural world. This is counter hegemonic to the current neo liberal global paradigm and it has implications for the role of ESD and for how ESD relates to mainstream educational agendas.

It should be noted of course that education policy and practice does not occur in a vacuum but is set within the wider social, political, economic context. As such it is often a 'site for struggle …. and should be seen as dialectically linked to broader national and global contradictory dynamics' (Ginsburg et al. 1991: 29). Hence, the context of policy and practice in each member state will influence the kind of education that is promoted (Blenkin et al. 1992). As educators we need to be aware of the policy context as we have an opportunity and a responsibility to influence the policy agenda by 'building alliances locally and globally with other groups and social movements' (Ginsburg 1991: 29).

3 The Role of Education—Quality or Quantity?

The proposed outcomes for Goal 4 on education include the following commitment:

> 'By 2030, ensure that all learners acquire the knowledge and skills needed to promote sustainable development, including, among others, through education for sustainable development and sustainable lifestyles, human rights, gender equality, promotion of a culture of peace and nonviolence, global citizenship and appreciation of cultural diversity and of culture's contribution to sustainable development' (UNESCO Open Working Group 2014a, b). This rightly underlines the crucial role of education in achieving sustainable development and represents an acknowledgement of this at a global policy level. Sterling (CSU, ISSC 2015) offers a note of caution 'This goal (4) is both an end and a cross-cutting means.... What is much weaker in the current articulation of the goal and its targets is education as a vehicle or instrument for change'. In other words, how exactly is education to perform this transformative role? What kind of education is needed? It is clear that there is a crucial role for education and learning in achieving this global shift of policy and practice in order to implement all the SDGs. As highlighted by the CSU and ISSC (2015) 'Education has been recognized for many years as a critical factor in addressing environmental and sustainability issues and ensuring human well-being. The importance of education and learning in supporting change is justified by research evidence.'

The focus of the MDGs on basic education and education for all (EFA) enabled some very positive achievements in raising global literacy rates and access to schooling but the disconnect between quantity and quality of education also presented a number challenges which were demonstrated in the lack of synergy between the two key global education programmes of EFA and ESD, both led by the work of UNESCO (Wade and Parker 2008). Access to education (quantity) is important but the kind of education (quality) on offer is also crucial. Educational achievements so far have continued to lead us into living unsustainably, in fact the countries with the highest levels of education 'also have the biggest footprints presenting the biggest challenges to sustainable development on the planet' (Wade and Parker 2008: 5).

As David Orr reminds us: 'Education is no guarantee of decency, prudence or wisdom. Much of the same kind of education will only compound our problems. This is not an argument for ignorance but rather a statement that the worth of education must now be measured against the standards of decency and human survival—the issues now looming so large before us in the twenty-first century. It is not education but education of a certain kind that will save us' (Orr 2004 p. 8).

As a result of the Education for All (EFA) targets of the MDGs, there have been some substantial achievements in terms of universal primary education and educational access, however, more generally there remain a number of concerns about the relevance, appropriateness and above all the quality of the education on offer.

According to Irina Bokova, Director-General of UNESCO, the 2013 EFA Global Monitoring Report 'makes a powerful case for placing education at the heart of the global development agenda after 2015…. Fifty-seven million children are still failing to learn, simply because they are not in school. Access is not the only crisis—*poor quality* (my emphasis) is holding back learning even for those who make it to school' (UNESCO 2013: i) Bokova goes on to emphasise the importance of education within the SDGs' We must learn from the evidence as we shape a new global sustainable development agenda after 2015. As this report shows, 'equality in access and learning must stand at the heart of future education goals. We must ensure that all children and young people are learning the basics and that they have the opportunity to acquire the transferable skills needed to become global citizens' (UNESCO 2013: ii) The key question here is what is meant by the term 'global citizens?' Does it just mean fitting into the current unsustainable global system or does it mean being able to question and challenge unsustainable practices and become empowered to make change. In other words, is it just more of the same education or is it ESD? The development of a new set of sustainable development goals presents us with the opportunity to embed ESD within the educational goals and so enshrine ESD within the human right of education for all. Unfortunately, ESD was not included in the education goals of the MDGs and the strong emphasis on access to primary schooling led to some unintended consequences in terms of quality. For example, the need for a whole systems approach to education which included the development of secondary education and teacher training was neglected. As a result a huge shortage of teachers and of school classrooms led to class sizes of over 70 as the norm in many countries in sub Saharan Africa.

4 Education as Empowerment?

UNESCO's Global Action Plan for ESD highlights the important role of education in empowerment and 'societal transformation: Empowering learners of any age, in any education setting, to transform themselves and the society they live in'.

This should include:

'Enabling a transition to greener economies and societies.

- Equipping learners with skills for 'green jobs'.
- Motivating people to adopt sustainable lifestyles.

Empowering people to be global citizens who engage and assume active roles, both locally and globally, to face and to resolve global challenges and ultimately to become proactive contributors to creating a more just, peaceful, tolerant, inclusive, secure and sustainable world' (UNESCO 2014a, b).

The key role of education in empowerment is also highlighted by UNESCO's report 'Sustainable Development 2015 begins with education' which provides research evidence of how education will contribute to each and every one of the SDGs. For example, for Goal One on poverty eradication: 'Education enables those in paid formal employment to earn higher wages. Better-educated individuals in wage employment are paid more to reward them for their higher productivity. On average, one year of education is associated with a 10 % increase in wage earnings. Returns to schooling are highest in sub-Saharan Africa, highlighting the need to invest in education in the region' (UNESCO 2014a, b).

In relation to Goal 5 on gender equality, 'Around 2.9 million girls are married by the age of 15 in sub-Saharan Africa and South and West Asia, equivalent to one in eight girls in each region. If all girls had secondary education in these two regions, child marriage would fall by 64 %' (UNESCO 2014a, b). Furthermore, for Goal 16 on developing peaceful and inclusive societies, 'Education is a key mechanism promoting tolerance to diversity. In Latin America, people with secondary education were less likely than those with primary education to express intolerance for people of different race (by 47 %). In the Arab States, people with secondary education were 14 % less likely than those with only primary education to express intolerance' (UNESCO 2014a, b).

When the SDGs were drafted there was also an opportunity to embed education and learning within all the SDGs which would have highlighted its role in empowerment, but the report identifies only one specific goal on education. This is disappointing but it perhaps reflects the lack of status of education within the policy process generally. UNESCO's report (2014a, b) demonstrates how education and learning is an essential building block to the achievement of all the goals and it is essential that future processes of policy making take this into account. As Sterling (CSU, ISSC 2015) points out in the CSU report 'Education is a key part of working to reduce vulnerability to economic, social and environmental dislocation and building more resilient systems. In developed countries, research indicates that education enables people to perform better economically, enhances health and extends life span, promotes civic engagement, and improves sense of wellbeing'.

Nonetheless, there is undoubtedly a strong argument for a specific goal on education to ensure accountability and policy action. Goal 4 certainly promotes a wider, holistic, more overarching view of education and does represent a step forward from the limited, rather instrumental education goal of the MDGs. This in no small part owing to the influence of the education lobbies of both EFA and ESD, which are increasingly coalescing around the concept of quality education (Pigozzi 2003).

The key question here is what kind of education is required if we wish to live sustainably? Current educational practices have led to some important innovations, for example in relation to sustainability practice in the built environment where the requirements of professional bodies have gone hand in hand with new undergraduate and master's courses in sustainable engineering and renewable energy. However, these are small steps on the road to achieving sustainable lifestyles and many of these innovations are not yet common practice. Current educational practice has failed to address substantially our unsustainable lifestyles driven by the continued focus on economic growth and the resulting over consumption. The negative impact on personal health of the focus on continuing economic growth and consumerism has been described by various psychologists as affluenza which is like a 'painful, contagious, socially transmitted condition of overload, debt, anxiety and waste resulting from the dogged pursuit of more' (Graaf et al. 2001: 122). Developing the concept of affluenza Oliver James has linked rising consumption and the influence of advertising with high levels of anxiety and depression (James 2007: 142). Wilkinson and Pickett note how available evidence shows 'that further economic growth in the developed world no longer improves health, happiness or measures of well being' (Wilkinson and Pickett 2010: 217). Indeed as Jackson argues there is 'yet no credible, socially just, ecologically sustainable scenario of continually growing incomes for a world of 9 billion people' (Jackson 2011: 85).

In the UK, the government's response to addressing future societal, economic and environmental needs has resulted in an on-going emphasis on the STEM subjects at the expense of the social sciences. This can lead to extensive scientific and technological innovation but without an understanding of human behaviours and social change then this is unlikely to lead to adoption and support for change. ESD's focus on systems thinking and interdisciplinarity has the potential to maximise the effectiveness of innovation and to change behaviours. ESD also recognises the importance of local and indigenous knowledge which is key to achieving understanding of and commitment to appropriate sustainability practice.

5 The Transformative Role of Education for Sustainable Development

The notion of sustainable development and that of education for sustainable development are closely interlinked, and ESD can be viewed as the learning (formal, non formal and informal) that is necessary to achieve sustainable development (UNESCO 2007). UNESCO as the lead UN agency for ESD has succeeded in achieving a broad global consensus about ESD:

- ESD is facilitated through participatory and reflective approaches and is characterised by the following:
- is based on the principles of intergenerational equity, social justice, fair distribution of resources and community participation, that underlie sustainable development;
- promotes a shift in mental models which inform our environmental, social and economic decisions;
- is locally relevant and culturally appropriate;
- is based on local needs, perceptions and conditions, but acknowledges that fulfilling local needs often has international effects and consequences;
- engages formal, non-formal and informal education;

- accommodates the evolving nature of the concept of sustainability;
- promotes life-long learning;
- addresses content, taking into account context, global issues and local priorities;
- builds civil capacity for community-based decision-making, social tolerance, environmental stewardship, adaptable workforce and quality of life;
- is cross disciplinary. No one discipline can claim ESD as its own, but all disciplines can contribute to ESD; (UNESCO 2007).

Among educational practitioners there is considerable agreement around the pedagogy and approaches, which underlie an effective and empowering curriculum (Wade and Parker 2008). This growing consensus could be expanded and developed to bring in the additional dimensions of ESD, which are currently missing. The most obvious of these include the futures' dimension as well as a linked understanding of ecological and social processes, together with commitments to social and ecological justice. This presents an opportunity for the shared development of a framework (which might be called Education for Sustainable Development for All or ESDFA), which could bring the social and environmental dimensions together more effectively in the context of learning.

UNESCO recognised that this synergy could facilitate the achievement of both EFA and ESD together (Parker and Wade 2007; Bangay and Blum 2010). However this is where policy has fallen short and despite numerous commitments, limited progress has been made as demonstrated by the lack of reference to ESD within the 2013 Global Monitoring Report which addressed the EFA goals. Without this synergy the danger is that the same education systems and curricula which have led us to unsustainable development will be perpetuated and as a human race we may be left without the skills, competences, values and knowledge to tackle the major challenges which we are facing. Disappointingly the importance of this synergy was not reflected in the earlier discussions around the SDGs and in two key reports linked to discussions on the SDGs (Report of the High Level Panel May 2013 and the Report of the Sustainable Development Solutions Network (SDSN) June 2013) there was no mention at all of Education for Sustainable Development (ESD). There is clearly ongoing work for educationalists to do to keep ESD on the agenda in taking forward the SDGs. Numerous educational fora have endorsed the need for the transformative role of ESD (Bonn Declaration 2009, Tokyo Declaration 2009, GAP 2014) but policy makers seem slow to engage with the educational community on this.

Although the UNESCO overview in 2012 of ESD policy and practice across a range of countries indicated that national policy commitments have increased in the last 20 years and ESD practice has developed considerably, this is obviously very variable from nation to nation. Concerns about climate change and related threats have clearly helped to put this on the political agenda. The global fiscal crisis has also presented an opportunity for global leaders to review current unsustainable economic and social practices but so far there is little evidence of this in actual policy or practice. To many working in ESD this illustrates the urgent need for ESD for politicians and policy makers!

The UNESCO report was of course constrained by the complexity of capturing the full range of ESD activity and by limitations in the data, however, UNESCO found that progress in re orienting education systems towards sustainability has been very uneven. According to UNESCO's report to the Rio plus 20 Conference (UNESCO 2012: 12),

> '...in 2008, the proportion of countries evoking ESD or related fields in their development education programs (was) about 50 %. In some cases, ESD (was) evoked or included as a theoretical frame without the evidence of inclusion on the curricula or project development. Education by itself (was) sometimes described as a tool for sustainable development, without really including ESD. From a 50 country sample 26 countries reported no evidence of ESD in 2008, but by 2012 after the boost of the Bonn Conference in 2009, 16 of them fall no longer in that category. We can perceive an estimate increase of 34 % from 2008 to 2012. This allows us to have an approximation of the rate of adoption of ESD.'

Nonetheless it would seem that policy and practice in ESD have certainly developed from very small beginnings over the last 10 years and in many countries there is at least some government policy in place in all areas of the formal education sector, from schools to higher education (UNESCO 2013). In addition, national legal requirements on sustainable development in relation to other sectors, such as the built environment, have created space and demand for training at a range of levels. Additionally, in 2005 the UN acknowledged the Decade of ESD 2005 to 2014 and an implementation plan was produced and agreed. In this plan, education was viewed as a prime lever for social change, described by UNESCO in the implementation plan for the Decade in the following way: 'It means education that enables people to foresee, face up to and solve the problems that threaten life on our planet' (UNESCO 2005). Furthermore, at the international level, ESD was again strongly endorsed at the Rio+20 World Summit on Sustainable Development (WSSD) in 2012. UNESCO is, of course, dependent on funding and resources from member state governments and the response of governments to this challenge reflected their response to ESD generally. In other words, it was rather limited with the exception of a few countries. We will now seek to examine some of the issues which impede progress on policy and practice for ESD.

6 What Are the Blocks and Obstacles to ESD?

Since the 1980s, we have seen neo liberal perspectives form the overarching framework for policy making and this has shaped educational policy trends. Education represents one of the largest resource commitments of the public sector so it is not surprising that governments take a close interest in it and that it reflects certain ideological perspectives. In the last two decades many (Selby and Kagawa 2011; Blewitt 2013) would argue that it has reflected the rise of neo-liberal ideas both in terms of the purpose as well as the delivery of education. In many countries, such as the USA, UK, Australia, this has led to what is often called a 'compliance culture' within education, with a focus on targets, tests and tick boxes, something that leaves little space for the critical thinking and questioning required by ESD. This view of education fits most closely with the 'job slots' view of education' (Kemmis 1983) which 'aims the whole of the school system at the job market and the structured set of inequalities that constitutes society as we know it' (Kemmis 1983: 1).

At the same time education has come to be seen more and more as a commodity rather than a process and this is illustrated nowhere more clearly than within the international agreement on General Agreement on Trade in Services (GATS) (World Development Movement (WDM) now Global Justice Now 2006) and the Trade in Services Agreement (TiSA) (WDM 2016). According to Nick Dearden, Director of Global Justice Now, 'This deal (TiSA) is a threat to the very concept of public services. It is a turbo-charged privatisation pact, based on the idea that, rather than serving the public interest, governments must step out of the way and allow corporations to 'get on with it' '. This could open the door for private corporations to take over the running of education systems, with all the possible implications for democracy and accountability. 'The dangers of such deregulation have already been highlighted, for example, by disastrous water privatisations in countries such as Tanzania (Rice 2007). Access to education like access to clean water is a basic human right and therefore needs to be seen as a public good, not a commodity which is subject to the vagaries of the market. Unfortunately this trend is continuing with the proposal for the EU to sign up to Transatlantic trade and Investment Partnership (TTIP) which would effectively allow corporations to set up unaccountable private courts to sue national governments which refused to open up key sectors to them (GJN 2016a, b).

Marketisation and privatisation trends have frequently skewed educational practice towards unsustainable development rather than helping to address the huge challenges which the world is facing in the 21st century. Politicians and policy makers have increasingly involved themselves in the detail of educational curricula and delivery and increasingly marginalised the expertise of practitioners and educational experts and theorists. Increased centralised control of formal educational agendas has resulted and this is illustrated by the work of Ball, Blewitt, Apple in their investigations of formal sector curricula. Increasingly, educators have been marginalised in the policy process—in 2014 Faul identified the key actors in global educational policy making as 'donor countries, multilateral agencies/regional banks, civil society and the private sector' (Faul 2014). Faul goes on to demonstrate that 'the global education policy space and implementation mechanisms are being constructed to prioritise literacy and numeracy, and gender parity alone' (Faul 2014). The global policy focus on targets and indicators has also arguably hampered progress towards ESD. As Disterheft et al. (2015) point out 'Shifts of perception are at the core of transformative learning, however it is difficult to assess these shifts'.

In the UK within the English National curriculum under the previous Labour administration there was a strong focus on numeracy and literacy, although there were some steps to introduce concepts of sustainability (for example, through the Doorways programme for primary schools). Nonetheless, it was very difficult to find much evidence of a commitment **to re orient** educational systems towards sustainable development (a key commitment of Agenda 21). And one of the first actions of the UK Conservative led coalition was to withdraw funding from the Sustainable Schools network and to abolish the Sustainable Development Commission.

Nonetheless, there is an increasing demand from UK students to embed sustainability knowledge and skills within the taught curriculum (Bone and Agombar 2011). A survey in 2011 of over 5000 first year UK students also found that 'overwhelmingly, skills in sustainable development are viewed as significant for employability and over 80 % of respondents believe these skills are going to be important to their future employers;—respondents placed high value on many of the aspects of sustainable development for use in HE in relation to increasing their ability to perform well in their course;—sustainability concerns are significant in students' university choices;—the vast majority felt that sustainable development is something universities should actively incorporate and promote (Drayson et al. 2011: 6).

These initiatives are supported by demands from the business sector which is increasingly highlighting the need for employees to bring skills and understanding of sustainable development. While recognising the importance of enhancing employability skills there is a danger that too great a focus on the skills agenda alone will ignore the challenge of a changing world and a changing global economy which cannot rely for ever on unlimited energy supplies at a time when global warming is changing the very planet we live on. As Drayson et al. (2011: 12) go on to caution 'the EfS agenda advocates the need for a broader range of skills that can challenge societal norms, and transform educational practice'.

Porrit and many others point out in his book, Capitalism as if the world really mattered (Porritt 2005), business as usual is no longer an option. Poritt demonstrated how things could be different with a more people centred and planet centred attention to capital, and his work as chair of the UK Sustainable Development Commission highlighted education a key element of this.

Unfortunately, there is a limited focus at the national and global level on transformative education for sustainable development. For example, a recent paper written for the United Nations Development Programme (UNDP) on the post 2015 agenda for development contains only one very brief reference to education (Pettinato and Vasquez 2013).

7 Catalysts for Change?

Nonetheless, there are at least some small positive signs that the discourse of on-going, energy consuming economic growth is starting to change. A number of initiatives may offer the potential to break through these blocks and obstacles. The so called 'Green Economy', for example, was central to discussions and debates at the 2012 United Nations Conference on Sustainable Development (commonly known as Rio+20). It is based on the assumption that green or sustainable growth can be achieved by utilising the latest science and technology. A 2011 UNESCO policy paper stated that 'Science holds many of the answers to the complex questions we face (UNESCO 2011a, b: 5). It talks of the need for 'resolute science and technology based solutions' to combat the many social and environmental challenges (UNESCO 2011a, b: 29). These ideas sit within the ecological modernisation school of thought and is part of the mainstream thinking within the UN and international development circles. But the concept of the Green Economy 'and strategies to promote a green economy are highly contested' (UNRISD 2011). Jones has argued about the important link between creating green jobs and protecting the environment (Jones 2012: 187). But for Jackson nobody has yet come up with an honest and clear definition of what is actually meant by sustainable growth (Jackson 2011). Cable goes further arguing that 'Sustainable growth is nonsensical: growth is not sustainable because resources are not infinite' (Cable 2012: 12). And yet the idea of sustainable growth has gained significant leverage in policy circles. No one should deny the important role that science and technology can play in shaping a more sustainable world but the Green Economy approach is in danger of perpetuating the myth that science and technology are all that is needed. Indeed as Bowen has argued 'it is not clear whether this new emphasis on green growth represents a paradigm shift or just spin to cover up inconsistencies between economic and environmental objectives of government' (Bowen 2012: 7). The challenges that the world faces today are multi-faceted and require a variety of social, environmental and economic policy responses, of which science and technology is but a part. Indeed there is recognition in international circles that 'Green economies on their own are not enough' There is also a need to build 'green societies' which 'must be fair, equitable and inclusive societies' (UNESCO 2011a, b: 8). The concept of green societies offers us a potentially important way forward. But we must careful to avoid prioritising the green economy as the driver for social change over the green society.

A school of thought that challenges the dominant paradigm of neo liberal economics is New Economics (Simms and Boyle 2009). For the UK based New Economics Foundation 'The UK and many of the world's economies are increasingly unsustainable, unfair and unstable'. What is needed, argues the Foundation is a 'Great Transition—to transform the economy so that it works for people and planet' (New Economics Foundation 2014). In similar vein the USA based New Economy Coalition talks of 'an economy that is restorative to people, place and planet' (New Economy Coalition 2014). In essence, New Economics challenges neo liberal assumptions about the value of traditional measures of economic growth such as GDP. It aims to place the well being of people and planet at the heart of the economic policy agenda.

The 2008 global financial crisis presented world leaders with the opportunity to address some of the problems of resource consuming, poorly regulated global capitalism and to deliver a New Economics. The failure to take this opportunity may be looked back on as one of the greatest betrayals of the 21st century. In the UK context, Porritt's 'resignation [from the UK Sustainable Development Commission] and the failure of the Green Deal to become counter hegemonic at the moment when neo liberalism had failed showed just how resilient neo liberal capitalism is' (Blewitt 2013: 54). If ESD had been strongly embedded within the education of the public and of politicians, perhaps some more forward thinking would have helped politicians to take the 'brave decisions' needed to make the changes needed. Blewitt argues that to a great extent education has been captured by the neo liberal mainstream agenda and that a critical pedagogy is needed in order to challenge and change this' (Blewitt 2013). He is critical of educationalists and indeed also of some ESD practitioners in making too much accommodation with the mainstream agendas which will only perpetuate more of the same social, economic and environmental relations. He maintains that 'It has worked within the paradigm it wants to shift and in so doing helped to sustain it' (Blewitt 2013: 53).

Nelson Mandela is often quoted as saying that 'education is the most powerful weapon we can use to change the world' but what kind of education? Paulo Freire's literacy work in Brazil among peasant farmers did much to enable them to claim their rights—so much so that he was seen as a major threat to the dictatorship at the time and was forced into exile! Freire espoused a critical view of education as liberation, not the individualised 'banking' style of education of which he was highly critical (Freire 1972). ESD can be seen as a challenge to current neo liberal hegemony as indeed it does encourage critical questioning and involve a more holistic approach to learning and to addressing global challenges. Its emphasis on equality, supported by increasing evidence that human well being and healthy societies depend on social and ecological balance, (Picket et al. etc.) does not sit easily with the current market driven economy, unsustainable consumption and uncritical economic growth. ESD represents both a challenge to current unsustainable practices but also a process and a framework to move forward. In progressing this within higher education, Leal Filho proposes the concept of applied sustainability as 'An action-oriented and project-based approach, which uses principles of sustainable development and applies them to real contexts and to real situations, yielding the benefits which can be expected when methods, approaches, processes and principles of sustainable development are put into practice' (Leal Filho 2015: 15). This approach presents a challenge for current models of global policy making as it implies a high level of participation form all key actors, including the least powerful.

8 Seeds of Change?

In seeking to achieve the SDGs, it is crucial to acknowledge that it is not more of the same education but the **kind** of education that is essential and it is important to note that 'the concepts of ESD and indeed, sustainable development have relatively recent origins and are both seen as 'emerging' and contested. How they are interpreted will depend very much on the ideological, philosophical, cultural and ethical perspectives of those using them' (Wade 2015). We have argued that the

SDGs cannot be achieved without ESD but how then can educators take this forward in policy and practice terms?

There are a number of initiatives in progress which offer opportunities for taking forward the ESD agenda. At a practitioner level, the UNESCO International Network (INTEI) of Teacher Education Institutions is comprised of teacher education institutions from about 60 nations around the world and the member institutions work to incorporate sustainability into their programmes, practices and policies. 'Each member institution addresses environmental, social, and economic contexts to create locally relevant and culturally appropriate teacher education programmes for both pre-service and in-service teachers' (UNESCO 2013). This network has produced a number of very useful resources, meets bi annually and offers mutual support and the potential for collaborative engagement. UNESCO has viewed it as flagship project for the UN Decade of ESD and the influence of the network can be seen in the, the development of EFA ESD synergy and even in the wording of Goal 4 of the SDGs.

In addition, the RCE initiative (Regional Centres of Expertise in ESD), co-ordinated up by the United Nations University-Institute of Advanced Studies is also another potential driver for change. In 2016 this involved over 138 regional centres of expertise in ESD operating in a wide range of countries and global regions including Europe, Asia, Africa and the Americas. Their purpose is to mobilize individuals and communities towards sustainable development, using the most appropriate expertise, knowledge and skills and they are founded on the principles and values of ESD in relation to social and ecological rights and justice, locally and globally. An RCE is a network of formal, informal and non-formal organisations mobilised to act as a catalyst for the delivery of education for sustainable development (ESD) to local and regional communities. The network is made up of schools, community and voluntary groups, the business sector, universities and non-governmental organisations, local authorities and other interested individuals. Most (though not all) RCEs are founded and co-ordinated by Higher Education institutions (HEIs).

Examples of the work of a number of different RCEs which illustrate the seeds of a new more engaged model of education can be found in the 2015 book 'The Challenges of Sustainability: linking politics, education and learning (Atkinson and Wade 2015). RCE Saskatchewan in Canada, for example, was able to mobilise a consortium of universities, educators and local communities to examine and win the argument against the proposal to build a nuclear power station in the region (Petry and Benko 2015: 192). In Japan RCE Greater Sendai is based in the region of the terrible East Japan earthquake and tsunami of 2011 and has been greatly involved with helping to address issues relating to 'the disaster the area experienced. Each region worked on restoration and redevelopment and each is now working on their redevelopment program. Greater Sendai RCE now includes the promotion of education for disaster risk reduction (EDRR) and redevelopment in its agenda' (Koganezawa and Ichinose 2015: 197). Schools which had been working with the RCE for some time already had strong networks with the community and were more prepared for the disaster, hence evacuation plans were carried out and the students not only survived but were able to work to build up their communities again. The RCE has organised regional seminars to share experience and produced textbooks for schools and 'The concept of sustainable development provides an important framework for relationship building between local communities and schools. We learned through the Great East Japan Earthquake how effective it is when we have to cope with a natural disaster to have had a cooperative relationship between the two and have deepened such liaisons and communications' (Koganezawa and Ichinose 2015: 197).

Of course education is but one facet of wider social agendas and without wider social change it is unlikely that educational change will result. At the same time education can also influence social change as highlighted earlier in this paper by the 2013 Global Monitoring Report. Governments have often had a tendency to try to use education in an instrumental way as a tool for their policy. Understanding the causes of change in policy and practice is of course a complex area of study and is like trying to unravel a complicated, interconnected and tangled web of relationships and conflicts with a vast array of actors and influences. It is an iterative, not linear process where policies are 'the operational statements of values, statements of 'prescriptive intent' which are then 'contested in and between the arenas of formation and implementation' (Bowe et al. 1992: 20). Educational change and social change are closely interlinked in a symbiotic, mutually dependant relationship. In order to change education policy there must be understanding and commitment from policy makers and to do this we also need an informed electorate and general public. The discussions around the development of the SDGs offer a real opportunity to put our planet on a more sustainable trajectory and it is an opportunity which global policy makers cannot afford to miss. ESD offers both a framework and a process to take this forward.

9 Conclusion

The Paris COP agreement of 2015 has been heralded as a global turning point and a breakthrough moment in addressing climate change yet it has been heavily criticised by civil society organisations, especially for putting the main burden onto developing countries (CSO 2016). Nonetheless, together with the SDGs this global agreement offers our best hope yet for making the deep and substantive changes needed for the survival of humankind on our fragile 'pale blue dot' (Sagan 1996). We have seen that there are some small signs that the global paradigm of neo liberalism is beginning to shift and that this will present opportunities to open up space and to highlight the need for ESD. The challenges we face are immense, not least the more immediate concerns and humanitarian crises caused by conflicts in the middle east and by changes in the global balance of power.

This paper has taken a big picture approach to policy and practice change and drawn from a wide range of sources. This of necessity provides some limitations and constraints as it seeks to be both reflective and normative. Its main focus has been on educators and the role of education and it could be developed further by a review of attitudes of decision makers and politicians to the role of education in the SDGs. We feel that this would probably highlight the inherent contradictions described earlier and possibly highlight the generally low status of education within policy agendas. However, this is for future exploration.

The 1992 commitments of Agenda 21 recognised the imperative of integrating development and environmental issues in order to address poverty and the aspirations of a 'developing' world while also tackling the environmental degradation and depletion caused by the unsustainable development of the past decades. Agenda 21 recognised that 'development' takes place within the finite limits of the earth's resources and that we all have a responsibility to respect these both for current but also for future generations. It was a huge achievement for the world's governments to sign up to such commitments to achieving sustainable development for all and the summit recognised the key role of a number of major groups, including Trades Unions, Indigenous People, NGOs, Local Authorities and the Business sector. Education, awareness raising, informal and non-formal learning were all seen as key to these commitments but one major group that was not mentioned was the Education community. As educators we believe that we have a particular role and responsibility in taking forward these agendas, as academics and as educational activists as well as being members of a global community which has signed up to the commitments of the SDGs. In order to do this, we can build on and develop current initiatives and strengthen learning communities of practice in ESD, such as the RCEs and INTEI. We can also ensure that our work has real wold impact by engaging with progressive social movements such as the Transition Town movement and New Economics.

The SDGs offer a real opportunity to set the world on a course to a more sustainable future. ESD can provide the framework, the learning and the process for this and this is a great opportunity for the education community to make an impact. The seeds of change have been scattered but are starting to grow, through solidarity and mobilisation networks and communities of practice both locally and globally. It is perhaps not surprising that many of them are operating outside and beyond mainstream structures, across sectors and different communities, across regions and countries. New patterns of living, working and being are undoubtedly needed and ESD can provide the link which connects us all with our common humanity and our relationship with the natural world.

References

Atkinson, A., & Wade, R. (Eds.). (2015). *The challange of sustainability: Linking politics, education and learning*. Bristol: Policy Press.

Bangay, C., & Blum, N. (2010). Education responses to climate change and quality: Two parts of the same agenda? *International Journal of Educational Development, 30*, 359–368. Elsevier.

Blenkin, G., Edward, G., & Kelly, A. (1992). *Change and the curriculum*. London: Paul Chapman publishing.

Blewitt, J. (2013). EfS: contesting the market model of higher education. In S. Sterling, L. Maxey, & H. Luna (Eds.), *The sustainable university*. London and New York: Routledge.

Bone, E., & Agombar, J. (2011). *First year attitudes towards, and skills in, sustainable development*. New York: Higher Education Academy.

Bowe, R., Ball. S., & Gold, A. (1992). *The policy process and the processes of policy: Reforming education and changing schools*. London: Routledge.

Bowen, A. (2012, December). Green growth, what does it mean? *Environmental Scientist*. http://www.lse.ac.uk/GranthamInstitute/wp-content/uploads/2014/02/green-growth-environmental-scientist-dec12.pdf. Accessed June 27, 2016.

Cable, S. (2012). Sustainable failures: environmental policy and democracy in a petro dependent world. Philadelphia: Temple University Press.

CSO, Civil Society Review. (2016). Fair shares: An equity review of the INDCs. http://civilsocietyreview.org/wp-content/uploads/2015/11/CSO_FullReport.pdf. Last accessed March 20, 2016.

CSU, ISSC. (2015). *Review of the sustainable development goals: The science perspective*. Paris: International Council for Science (ICSU). http://www.icsu.org/publications/reports-and-reviews/review-of-targets-for-the-sustainable-development-goals-the-science-perspective-2015. Accessed June 27, 2016.

Disterheft, A., Caero, S., Leal Filho, W., Azeiteiro, U. M. (2015). Ecological indicators. *63*(2016), 172–186. Elsevier.

Drayson, R., Bone, E., & Agombar, J. (2011). *Student attitudes towards and skills for sustainable development*. Higher Education Academy. https://www.heacademy.ac.uk/sites/default/files/student_attitudes_towards_and_skills_for_sustainable_development.pdf. Last accessed June 27, 2016.

Faul, M. (2014). Future-perfect/present imperfect: contemporary global constraints in the implementation of a post-2015 education agenda. *International journal of educational development*. http://dx.doi.org/10.1016/j.i.jedudev.2014.07.011. Accessed June 27, 2016.

Ginsburg, M., Cooper, S., Raghu, R., & Zerrara, H. (1991). Education reform: Social struggle, the state and the world economic system. In M. Ginsburg (Ed.), *Understanding educational reform in a global context: economy, ideology and the state*. NewYork and London: Garland publishing.

Global Justice Now (GJN). (2016a). http://www.globaljustice.org.uk/sites/default/files/files/resources/ttip_briefing_update_feb_16_2016.pdf. Accessed March 20, 16.

Global Justice Now (GJN). (2016b). http://www.globaljustice.org.uk/news/2016/jan/30/controversial-trade-services-agreement-be-voted-european-parliament-week. Last accessed March 20, 16.

Graaf, J., Wann, D., & Naylor, T. (2001). *Affluenza: The all consuming epidemic*. San Francisco: Berret-Koehler publishers.

ICSU, ISSC (2015). *Review of the sustainable development goals: The science perspective*. Paris: International Council for Science (ICSU)

IPCC Climate Change. (2013). *synthesis report: Contribution of working groups I, ii and iii to the fifth assessment report*. Cambridge: Cambridge University Press.

Jackson, T. (2011). *Prosperity without growth: Economics for a finite planet*. London: Earthscan.

James, O. (2007). *Affluenza*. London: Random House.

Jones (2012) *Rebuild the dream*. New York: Nation Books.

Kemmis, S. (1983). Getting our thinking straight: three views of education. In A. Greenhall-Gough (Ed.), *Foundations of environmental education*. Geelong, Australia: Deakin University.

Kimmerman, R.W. (2016). On being podcast and ted talk.https://itunes.apple.com/gb/podcast/on-being/id150892556?mt=2&i=363675986 TED talk https://www.youtube.com/watch?v=Lz1vgfZ3etE&feature=youtu.be.

Klein, N. (2014). *This changes everything*. London: Allen Lane.

Leal Filho, W. (2015). *Chapter one Applied Sustainable Development: A Way Forward in Promoting Sustainable Development in Higher Education Institutions. World trends in education for Sustainable Development*. Hamburg: Peter Lang publishers.

New Economy Coalition. (2014). Homepage. Available at www.neweconomics.net.

New Economics Foundation. (2014). Homepage. Available at www.neweconomics.org.

Orr, D (2004) *Education, the environment and the human prospect*. Washington DC: Island Press.

Parker, J., & Wade, R. (2007). Journeys around education for sustainability.

Pettinato & Vasquez. (2013). From the MDGs to the multi dimensional measurement of poverty as part of the post 2015 Agenda for (human) development, working Paper no 6 UNDP.

Petry, T., Benko, L., Koganezawa, T., & Ichinose, T. (2015). chapter 8, Regional Centres of expertise as mobilising mechanisms for education for sustainable development in Atkinson, A. & Wade, R (eds.). *The challenge of sustainability: linking politics, education and learning*. Bristol: Policy Press.

Pigozzi, M. J. (2003). Reorienting education in support of sustainable development through a focus on quality education for all. In *Paper presented at GEA Conference*, Tokyo, October 25, 2003, UNESCO.

Porrit, J. (2005). *Capitalism as if the world matters*. London: Earthscan.

Report of the High Level Panel (HLP) of Eminent Persons on the Post-2015 Development Agenda. (2013). A New Global Partnership: Eradicate Poverty and Transform Economies through Sustainable Development. http://www.post2015hlp.org/the-report/. Last accessed May 30, 2013.

Rice, X. (2007). *The water margin. The guardian*. www.theguardian.com/business/2007/aug/16/imf.internationalaidanddevelopment. Last accessed April 14, 2015.

Sachs, J. (2012). Rio+ 20: Jeffrey Sachs on how business destroyed democracy and virtuous life. http://www.guardian.co.uk/sustainable-business/rio-20-jeffrey-sachs-business-democracy. Last accessed October 29, 2012.

Sagan, C. (1996). *Pale blue dot: Vision of the human future in space*. Ballantine books.

Simms, A., & Boyle, D. (2009). *The new economics: A bigger picture*. London: Earthscan.

Stern, N. (2006). Stern Review on economics of climate change. http://webarchive.nationalarchives.gov.ukhttp:/www.hm-treasury.gov.uk/sternreview_index.htm. Last accessed June 27, 2016.

Stiglitz, J., Sen, A., & Fitoussi, J. P. (2009). *Report of the commission on the measurement of economic performance and social progress*. London: HMSO. www.stiglitz-sen-fitoussi.fr/en/index/. Last accessed March 19, 2016.

UNDP. (2013). Assessing progress in Africa towards the MDGs. New York: UNDP.

UNESCO. (2005). UN Decade for sustainable development 2005–2014. http://unesdoc.unesco.org/images/0014/001403/140372e.pdf. Last accessed March 29, 2016.

UNESCO. (2007). Introductory note on ESD—DESD. In *Monitoring & Evaluation Framework*. Paris: UNESCO.

UNESCO. (2011a). *EFA global monitoring report the hidden crisis: Armed conflict and education*. Paris: UNESCO.

UNESCO. (2011b). From green economies to green societies. Paris. UNESCO (2013). UNESCO Chair on Reorienting Teacher Education to Address Sustainability: International Network. http://www.unesco.org/new/en/education/themes/leading-the-international-agenda/education-for-sustainable-development/partners/educators/teacher-education/. Last accessed March 29, 2016.

UNESCO. (2012). Shaping the education of tomorrow: 2012 Report on the UN Decade of Education for Sustainable Development authored by Arjen E. J. Wals, Wageningen University, The Netherlands. http://unesdoc.unesco.org/images/0021/002166/216606e.pdf. Last accessed March 26, 2016.

UNESCO. (2013). EFA Global monitoring Report 2013/14: Teaching and Learning—achieving quality for all. UNESCO Paris. 2014.

UNESCO. (2014a). *UNESCO Roadmap for the implementation of the Global Action Plan for ESD*. Paris: UNESCO.
UNESCO. (2014b). Global action plan. http://en.unesco.org/gap. Last accessed March 29, 2016.
UNRISD. (2011). *(Research institute for Social Development) Social dimensions of the green economy, Research and Policy Brief 12*. Geneva, Switzerland: UNRISD.
Wade, R. (2015). In Atkinson, H. and Wade, R. (Eds.), *The challenge of sustainability: linking politics, education and learning* (Chapter 7 and 8). Bristol: Policy Press.
Wade, R., & Parker, J. (2008). *EFA/ESD Dialogue: Educating for a sustainable world*. Paris: UNESCO.
Wilkinson, R., & Pickett, K. (2010). *The spirit level: Why equality is better for everyone*. London: Penguin books.
World Development Movement. (2016). http://www.wdm.org.uk/campaigns/GATS.htm. http://www.wdm.org.uk/news/presrel/current/biwatersues.htm. Last accessed March 14, 2016.

Authors Biography

Professor Ros Wade chairs the London Regional Centre for Education for Sustainability (EFS) and she is also a Distinguished Research Fellow of the Schumacher Institute for Sustainable Solutions. Her main teaching responsibility is directing the PG Education for Sustainability programme which entered its 21st year in 2015 (www.lsbu.ac.uk/efs). This is an international distance learning programme which covers all global regions and which is producing cutting edge research on education for sustainability. Her prime concern in research and scholarship is to explore the ways that issues of social justice, poverty eradication and development can play a vital role in the field of EFS and to examine how these concerns can be addressed within the parameters of the ecological limits of our planet. Recent publications include (with Hugh Atkinson) 'The challenge of sustainability: linking politics, education and learning', Policy Press 2015.

Dr. Hugh Atkinson is a Distinguished Fellow of the Schumacher Institute of Sustainable Solutions and a member of the steering group of the London RCE. He is a visiting fellow in politics at London Southbank University and is currently writing a book on the politics of climate change in the USA. He is the author of 'Local democracy, civic engagement and community: from New Labour to Big Society' (Manchester University Press, 2012). He is a founder member of the Political Studies Association specialist group on environmental politics. Recent publications include (with Ros Wade) 'The challenge of sustainability: linking politics, education and learning', Policy Press 2015.

Sustainable Development Research at Universities in the United Kingdom: Moving Forward

Walter Leal Filho

Abstract
This short paper presents some international perspectives on sustainable development research, describes the contribution of the Inter-University Sustainable Development Research Programme, and outlines some areas where a stronger emphasis may assist universities in the United Kingdom to reach excellence in this fast growing field.

Keywords
Sustainability · Higher education · Research · Impacts

1 Introduction

Over twenty year ago, in the summer of 1996, a Symposium was organised at the University of Bradford, in cooperation with the Association of Universities Rectors (CRE) which, at the time, spearheaded the movement towards fostering sustainable development in a higher education context in Europe.

Bradford University was one of the pioneers in the debate then, since the discussion on matters related to the integration of sustainable development in higher education was at a rather embryonic stage in the UK. The publication produced as part of that meeting, titled "Implementing Sustainable Development at University Level" (Leal Filho et al. 1996) opened the way for a set of further works,

W. Leal Filho (✉)
School of Science and the Environment, Manchester Metropolitan University,
Chester Street, Manchester M1 5GD, UK
e-mail: w.leal@mmu.ac.uk

W. Leal Filho (ed.), *Sustainable Development Research at Universities in the United Kingdom*, World Sustainability Series,
DOI 10.1007/978-3-319-47883-8_21

encouraging many universities across the UK to engage in what is now a rapidly growing field.

In 1999, a publication called "Sustainability and University Life" (Leal Filho 1999), aimed at documenting the variety of works in this field in the 1990s, only had a handful of inputs from UK universities. The next publication, "Communicating Sustainability" (Leal Filho 2000) published a year later, was also characterized by a small number of research projects on the issue of sustainability communication.

It was not until 2002 that many UK universities started to engage more systematically on sustainability efforts in higher education, and actively pursued the documentation of their experiences. The book "Teaching Sustainability—towards curriculum greening" (Leal Filho 2002) captured some of these initiatives, and the volume "Handbook of Sustainability Research" (Leal Filho 2005) congregated a variety of examples of sustainability research efforts being pursued by universities in the United Kingdom.

If one considers what this short overview of developments says, the trend is rather clear: even though a variety of research efforts have taken place at higher education institutions in the UK since the 1990s, not of many of these efforts were documented in the literature, and even fewer have been widely disseminated. Unfortunately, this trends continues today.

Even though it is no longer appropriate to suggest that sustainable development research in the UK is at an embryonic stage, it is a fact that it does not as yet yield the impacts it may be expected to. This is so for three main reasons:

i. despite substantial efforts and concrete moves towards curriculum greening and campus operations at HEIs across the country especially over the past 15 years, sustainability research is still the weakest link. There are seldom examples of institutions where sustainable development research is a field centrally funded, and supplied with the funding and resources needed to allow them to operate in a continuous way. As a result, most research efforts are ad hoc, and many research teams suffer from lack of funding and/or institutional support;
ii. provisions for research funding from UK research councils do not always cater for the interdisciplinary nature of sustainable development issues, and tend to focus on better established fields instead (e.g. ecology or pure education);
iii. in the periodical Higher Education Research Assessments (REF) it is difficult for researchers to allocate their sustainable development research efforts to one of the given areas. Apart from this, many journals in the field to not have the high impact factors many universities desire, and many researchers feel tempted not to report their research there, preferring other outlets to promote their works.

In addition, there is a lack of specific statistical data on funding to sustainability-related projects.

This state of affairs illustrates the need to identify ways to move forward with sustainable development research in the UK in a more systematic way, and overcome some of the barriers which have prevented progress in this field.

2 Opportunities and Trends Elsewhere

A comparison across Europe shows that things can run differently. As far as funding for sustainability research is concerned, Germany offers a clear example of what may be achieved. The programme "Research for Sustainability" (Forschung für Nachhaltigkeit) or FONA as it is known, is the world's largest funding programme for sustainable development research. Now on its 3rd Phase, FONA has spent several hundred million Euros on sustainable development research in Germany, across areas such as:

- biodiversity
- energy
- land use
- social sciences
- climate change
- green economy and
- resource efficiency

among others. One specific scheme funded by FONA, called sustainability research in the social sciences, has provided substantial funding to projects with a social orientation.

Thanks to its multi-thematic focus and the preference for Consortia-led research, FONA has allowed various teams of researchers to work together on a common issue or problem, further consolidating a community of sustainability researchers.

In Austria, the programme "Technologies for a Sustainable Development" provides funding for a variety of sustainability research efforts, primarily focusing on areas such as energy, resource efficiency and social science projects. In addition, the Ministry of Education provides a variety of grants for sustainable development initiatives. Austrian universities are encouraged to describe their efforts in the field of sustainable development, and an arrangement is in place via which universities commit to particular sustainability goals and indicators, with specific funding being allocated to it.

In the Netherlands, "NWO-WOTRO Science for Global Development" programmes, funds and monitors innovative research on global issues, with a focus on sustainable development and poverty reduction. NWO-WOTRO's research projects are realised by interdisciplinary teams of researchers from the Netherlands (but also from other European countries) and countries in the southern hemisphere, and in close collaboration with non-academic stakeholders. These partnerships should yield solutions for development challenges and strengthen the bridge between research, policy and practice.

In Finland, the Academy of Finland and the Finnish Ministry for Foreign Affairs fund problem-oriented and multidisciplinary development research (on developing countries). In sustainable development research, the problems to be studied may derive from the local as well as the global level, or from a search for and analysis of the connections between development phenomena at different levels. Projects are also funded on how international, national and cultural systems work and about the limitations that these systems present to and the opportunities they offer for sustainable development.

The European Commission has also a wide range of funding mechanisms to support sustainable development research. Whereas it is unclear as to the extent to which UK based universities may benefit from such funding in the future, it is a fact that their participation right now is open, and they are fully eligible in all current funding streams.

Many opportunities for UK-led sustainability research efforts at a global level are also seen. For instance, the momentum created after the UN Decade of Education for Sustainable Development (UNESCO 2015), with the launching of the Global Action Programme (GAP) (Leal Filho et al. 2015) acknowledged by UN General Assembly Resolution A/RES/69/211, means that new opportunities to develop sustainability research at an international level can be pursued. Thematically, GAP focuses on five priority action areas, considered as key leverage points to advance ESD:

- advancing policy;
- transforming learning and training environments;
- building capacities of educators and trainers;
- empowering and mobilizing youth;
- and accelerating sustainable solutions at local level.

In addition, in areas as varied as corporate social responsibility (UN Global Compact 2013) where sustainability efforts at enterprises are pursued, there is still much to be done, so a fertile ground for research is available.

Moreover, the "Sustainable Development Goals" provide another window of opportunity to foster the cause of sustainable development research in the UK (United Nations 2015). The 17 Goals are:

(1) End poverty in all its forms everywhere
(2) End hunger, achieve food security and improved nutrition, and promote sustainable agriculture
(3) Ensure healthy lives and promote wellbeing for all at all ages
(4) Ensure inclusive and equitable quality education and promote lifelong learning opportunities for all
(5) Achieve gender equality and empower all women and girls

(6) Ensure availability and sustainable management of water and sanitation for all
(7) Ensure access to affordable, reliable, sustainable and modern energy for all
(8) Promote sustained, inclusive and sustainable economic growth, full and productive employment, and decent work for all
(9) Build resilient infrastructure, promote inclusive and sustainable industrialisation, and foster innovation
(10) Reduce inequality within and among countries
(11) Make cities and human settlements inclusive, safe, resilient and sustainable
(12) Ensure sustainable consumption and production patterns
(13) Take urgent action to combat climate change and its impacts (taking note of agreements made by the UNFCCC forum)
(14) Conserve and sustainably use the oceans, seas and marine resources for sustainable development
(15) Protect, restore and promote sustainable use of terrestrial ecosystems, sustainably manage forests, combat desertification and halt and reverse land degradation, and halt biodiversity loss
(16) Promote peaceful and inclusive societies for sustainable development, provide access to justice for all and build effective, accountable and inclusive institutions at all levels
(17) Strengthen the means of implementation and revitalise the global partnership for sustainable development

Within the goals there are 169 targets, which also offer a variety of opportunities for sustainability research. Apart from the fact that there is a wide diversity of research themes which can be pursued by researchers from different backgrounds and settings, many opportunities for national and international cooperation projects exist, which may lead to interesting projects, PhD training and publications.

3 Supporting Sustainable Development in the UK: The Contribution of the Inter-University Sustainable Development Research Programme

In order to support sustainable development research in the UK and strengthen links between UK-based and international organisations, and with a view to develop the abilities of universities in the UK and elsewhere to perform high quality sustainability research, the **Inter-University Sustainable Development Research Programme** (IUSDRP) has been created. Launched in 2015, IUSDRP will provide a contribution towards the efforts of member universities to consolidate their initiatives in the field of sustainable development in many ways. For instance:

(a) by increasing their research income: income from research projects may be increased by means of sustainable development research. Statistics show that sustainable development expertise clustered around research centres or programmes are more successful, than ad hoc efforts and research bids by individuals alone;
(b) by an enhancement of institutional research profiles in the field of sustainable development: at present, only a few universities already have an international and authoritative research profile in this field, even though they have a great potential to develop it. This potential can be further optimized and increased, by means of more coordinated approaches to bid and secure national and international funding, for interdisciplinary and cross-faculty projects on matters related to sustainable development, and by increasing their visibility;
(c) by an increase in the intake of PhD students: many universities have the potential to be attractive places for PhD students focusing on sustainability research, but at present—for various reasons—do not fully use it. The PhD programme to be led by the Inter-University Sustainable Development Research Programme will help to realize this potential, and offer the critical mass needed to train a new generation of sustainability experts, across the whole spectrum of themes and topics associated with it;
(d) by an increase in publication output: apart from income and profile-building, the Inter-University Sustainable Development Research Programme will help its members to achieve more substantial research outputs, which may be translated in the publication of more papers in indexed journals, as well as innovative books and book chapters. The International Journal of Sustainability in Higher Education and the newly created "World Sustainability Series", the leading book series on the topic, are some of the instruments to be used, as well a wider access to the world's leading journals on environment and sustainable development matters.

The Inter-University Sustainable Development Research Programme will consolidate and further develop the available know-how and profile on sustainable development among its member universities across the world, helping them to take full advantage of the many possibilities for institutional consolidation and individual career development which sustainability research offers.

A working programme (2015–2020) for the Inter-University Sustainable Development Research Programme has been prepared and is now under implementation. It is coordinated by Manchester Metropolitan University (UK), but each member university is autonomous in respect of their individual activities, funding and operations.

4 Conclusions

As this paper has tried to show, the level of excellence reached in the UK in respect of sustainability research can only be up kept, if UK based higher education institutions strive more intensively towards enlarging the scope and the remit of their research efforts. Also, as far as "applied sustainable development" is concerned, there is a perceived need to focus more on the practicalities of sustainable development themes, since much has already been written on the theory of sustainable development, as part of the debate held in the 1990s and up to 2014. Among other things, there is a need for greater collaboration between higher education institutions, and government offices, enterprises and NGOs active or interested at in the field of sustainable development. These exist at present, but are not as widely spread as they should be.

By doing so, a new momentum may be created and UK higher education institutions as a whole and researchers in particular, can take better advantage of the many opportunities sustainable development research offers to them.

The creation of the Inter-University Sustainable Development Research Programme is meant to support these efforts, and participation is open to any interested university in the UK and beyond.

References

Leal Filho, W., MacDermott, F., & Padham, J. (Eds.). (1996). *Implementing sustainable development at university level*. Geneva: CRE.

Leal Filho, W. (Ed.). (1999). *Sustainability and university life*. Frankfurt: Peter Lang Scientific Publishers.

Leal, Filho W. (Ed.). (2000). *Communicating sustainability*. Frankfurt: Peter Lang Scientific Publishers.

Leal Filho, W. (Ed.). (2002). *Teaching sustainability—towards curriculum greening*. Frankfurt: Peter Lang Scientific Publishers.

Leal Filho, W. (Ed.). (2005). *Handbook of sustainability research*. Frankfurt: Peter Lang Scientific Publishers.

Leal Filho, W., Manolas, E., & Pace, P. (2015). The future we want: Key issues on sustainable development in higher education after Rio and the UN decade of education for sustainable development. *International Journal of Sustainability in Higher Education, 16*(1), 112–129.

UN Global Compact. (2013). *Corporate sustainability and the United Nations Post-2015 development agenda*. New York: UN Global Compact.

UNESCO. (2015). *Shaping the future we want—a report of the UN Decade of education for sustainable development 2005–2014*. Paris: UNESCO.

United Nations. (2015). *Draft resolution submitted by the president of the general assembly-draft outcome document of the United Nations summit for the adoption of the post 2015 development agenda*. New York: United Nations.

Meeting Workforce Needs?: Developing and Delivering Education for 'Sustainable Communities'

Philip Brown, Helen Sharman and Graeme Sherriff

Abstract

Embedding notions of sustainability within both higher education and practice occasionally faces resistance. This chapter details one such experience of resistance by drawing on attempts in the last decade to develop and embed the concept of 'sustainable communities' in higher education and professional practice within the United Kingdom (UK). The Foundation Degree in Sustainable Communities (FdSc) was developed by the Homes and Communities Agency (HCA) in partnership with a select number of Higher Education Institutions (HEIs). The development of the FdSc was spurred by, what was perceived as, the significant lack of skills, within the various sectors, required to deliver New Labour's 'Sustainable Communities' agenda within a framework of regeneration. By drawing upon research with the HCA, HEIs and students this chapter explores the development of the FdSc and reflects upon the experience of the various stakeholders who have played a part in the delivery of the programme. A positive unanticipated outcome of this process; the collaborative working, provides ideas as to how to increase the effectiveness of collaboration across HEIs generally. The chapter also highlights various challenges and dilemmas' facing the FdSc as it was delivered within a very different political and public milieu to that of the 2000s. The chapter focuses on the difficulties that can be faced by HEIs when they become the delivery agents of political discourse.

P. Brown (✉)
Sustainable Housing & Urban Studies Unit (SHUSU), School of Nursing, Midwifery, Social Work & Social Sciences, University of Salford, Salford M6 6PU, UK
e-mail: p.brown@salford.ac.uk

H. Sharman · G. Sherriff
University of Salford, Salford, UK
e-mail: H.K.Sharman@salford.ac.uk

G. Sherriff
e-mail: G.Sherriff@salford.ac.uk

W. Leal Filho (ed.), *Sustainable Development Research at Universities in the United Kingdom*, World Sustainability Series,
DOI 10.1007/978-3-319-47883-8_22

Keywords
Skills · Sustainable communities · Sustainable development · Higher education

1 Introduction

The concept of Sustainable Development and its rise to prominence within policy making can be tracked back to the 1987 'Our Common Future' report (World Commission on Environment and Development 1987) and subsequently the prominent 1992 Rio de Janiero Earth Summit (Bell and Morse 2008, p. 3), which endorsed Local Agenda 21 (Dryzek 1997) which set out plans for action from national governments, and was followed by increased pressure on governments to formulate comprehensive policies for sustainable development (Burke 1995). The term itself has proved to be almost 'chameleon-like' (Raco 2005, p. 329), finding itself reinterpreted by a range of interest groups to 'justify a range of often conflicting and divergent agendas' (Raco 2005, p. 329). Not all attempts at embedding notions of sustainability in policy, professional practice and education though have been successful. Some attempts at doing so have experienced resistance (see Corcoran and Wals 2004) this chapter details one way in which this resistance has occurred.

Notions of sustainability have to varying extents informed UK policy development in the areas of transport, energy, biodiversity and overseas development to name a few. One particular area of interest here is urban regeneration, as developed by the New Labour Government, which became closely linked to notions of sustainability (Tallon 2010, p. 163) and, within this, the idea that professionals and communities should work together to create "sustainable communities". The language and terminology used in related reports reveal the influence of two closely related ideas: sustainable development and "new urbanism" (ODPM 2003, 2005). The various definitions of sustainable development are all linked by the "notion of 'Equity'" the view that resources are to be "used fairly to meet the needs of both current and future generations" (Jones and Evans 2008, p. 83).

The rhetoric of "new urbanism" can also be found in Labour's sustainable communities documentation. Cochrane (2007) suggests that that the new urbanism movement's influence on the concept of sustainable communities is manifested in the support for the idea that "better 'communities' can be developed by professionals rather than focusing on the "social processes of segregation and exclusion" (p. 54). Raco agrees with this criticism but attributes it to the related neoliberal principles underpinning sustainable communities (Raco 2005, p. 331). Thus the seemingly uncontentious ambition of working towards a "sustainable community" has been vulnerable to criticism by academics concerned about its underpinning philosophies. Raco (2005, p. 342) associates New Labour's version of 'sustainable communities' with a '"light green" manifestation of [Sustainable Development]' in that it does not question fundamental environmental and social issues associated with development. Indeed, there are concurrent conceptualisations of 'sustainable communities' that can be seen to provide a deeper shade of green that resonates more strongly with the issues raised in Rio, such as Agyeman's (2005) focus on environmental and social justice and Barton's (2000) reinventing of the neighbourhood as a site of more ecological living.

These voices of concern were arguably relatively marginalised by the dominant "mantra" that the main challenges enshrined in the Sustainable Communities Plan (ODPM 2005)—the lack of "key worker" housing in the South East and the decline of communities in the North and Midlands—could be resolved through, amongst other things, the activities of professionals. With such a key role, the skills of these professionals involved in developing and managing "sustainable communities" attracted specific government attention (Rogerson et al. 2010, p. 505) and led to the Egan Review (2004). This Review is the main starting point in understanding the perceived need for and development of a specific course which became the Foundation Degree in Sustainable Communities (FdSc).

2 The Egan Review

In his review John Egan focused on analysing the specific skills that were necessary in order to deliver 'sustainable communities'. He defined sustainable communities as places that:

> …meet the diverse needs of existing and future residents, their children and other users, contribute to a high quality of life and provide opportunity and choice. They achieve this in ways that make effective use of natural resources, enhance the environment, promote social cohesion and inclusion and strengthen economic prosperity. (Egan 2004, p. 7)

This statement was supported by a more detailed explanation of the key characteristics of sustainable communities that were collated in a diagram that become known as the "Egan Wheel". However, whilst some professionals such at the Chartered Institute for Housing and Royal Town Planning Institute broadly welcomed the review (Chartered Institute for Housing and Royal Town Planning Institute 2003, p. 10) Rogerson et al. (2010, p. 505) argue that it was a "poorly defined" concept.

The Egan Review concluded that a range of technical skills as well as more generic skills, were needed to ensure the best chance of success in implementing housing and regeneration projects. The targeted 'core occupations' included those working as town planners, architects, urban designers, developers as well as staff from local, regional and central government and workers within voluntary and community organisations. The Review asserted the desirability of encouraging people to enter such core occupations and 'upskill' in order to ensure the creation and maintenance of sustainable communities. Following the Review the Labour Government of the time supported the creation of what was to become the Academy for Sustainable Communities (ASC) the role of which, after further permutations, was, for the later stages of the 2000s, embedded within a new quasi-autonomous non-governmental organisation the Homes and Communities Agency (HCA).

The remainder of this chapter discusses the Foundation Degree in Sustainable Communities (FdSc) developed by the ASC/HCA in partnership with a select number of Higher Education Institutions (HEIs). It describes the rollout of the FdSc. and summarises the findings of an evaluation of the new qualification and the associated partnership working in HEIs. The issues that arise when HEIs become delivery agents of an activity centred around a particular political discourse are considered. This chapter also presents some of the unanticipated positive consequences of development and delivery of the FdSc experienced by HEIs. These consequences have implications for those involved or contemplating collaborative working in higher education in general.

3 The Foundation Degree in Sustainable Communities

Following the Egan Review a report by the Homes and Communities Academy 'Mind the Skills Gap: The skills we need for sustainable communities' (Academy for Sustainable Communities 2007), forecast a shortfall in the supply of suitably qualified professionals that could work in, what was perceived as, a growing sector. In taking steps to address the identified skills gaps it was decided, following a gap analysis and market testing, that a new work-based qualification should be established: the Foundation Degree in Sustainable Communities (FdSc).

A report was commissioned to support the HCA in designing and developing the FdSc. The report that followed the development of the FdSc, observed that there was 'a clear and growing need for an entry level qualification in Sustainable Communities' (Sheffield Hallam University 2007). This research highlighted growing skills gaps and difficulties with recruitment across the sector. The report recommended that the FdSc should:

- allow and encourage progression to further qualifications to allow specialisation;
- extend and enhance generic skills in a professional context;
- introduce technical skills (with a view to further specialisation at higher levels) with an emphasis on cross-disciplinary working;
- adapt to evolving issues e.g. climate change; quality of life (including health); green issues; and,
- provide for flexible learning approaches.

In addition, the research highlighted a number of issues worthy of further consideration such as:

- concern over the terminology of 'sustainable communities' which was thought to be possibly ambiguous, confusing and fragmented. More definition was recommended;

- the funding available for students and employers was seen as limited. A sliding scale and bursaries were suggested as strategies to overcome barriers posed by finite individual or public sector capacity to fund enrolment on the programme; and,
- the need for close partnership working in order to attract non-traditional students.

The FdSc that was subsequently developed aimed to:

- engage students in a challenging, critical and interdisciplinary education in sustainable communities' policy and practice;
- stimulate the students' awareness of the links and tensions between theory, policy and practice and to support the development of their professional community management skills though activities that have strong links with practice;
- enable students to develop their academic and professional key skills and competencies in an interdisciplinary and inter-professional educational environment;
- enable students to develop the qualities of reflective, professional and empathetic sustainable communities practitioners;
- offer 'pathways' that will enable students to meet the requirements of a range of 'core' sustainable communities' professional bodies, for professional accreditation by including assessment of work and voluntary experience thus providing a route to professional membership; and
- provide students with transferable, as well as specific vocational skills, which can be used to provide a foundation to enable and empower students to make choices in work, training and education throughout their life.

Homes and Communities Agency (2008)

Sheffield Hallam University became the first HEI to launch the FdSc and the then Academy for Sustainable Communities (2008), now Homes and Communities Agency (HCA), subsequently embarked on a three year Higher Education Strategy where the new FdSc was the centrepiece. A core component of this strategy involved the rollout of the FdSc across England with the aim of identifying HEIs in each of the English regions which had the reputation, capacity and capabilities to deliver the degree. The Strategy outlined a number of characteristics that the ASC expected of the FdSc namely:

- the focus upon generic skills;
- multi-disciplinary learning;
- knowledge and understanding of sustainable communities policy and practice;
- pathways to further study; and,
- pathways to progression into sustainable communities professions e.g. housing, planning and environmental studies.

At the same time there was an expectation that the HEIs delivering the FdSc would adapt the content and add modules as is relevant to their local/regional and employer needs.

4 The Rollout of the Foundation Degree in Sustainable Communities

The rollout of the FdSc was supported to a significant extent by the ASC and its successor, the HCA. This support included the provision of a modest bursary for a small number of students at each HEI to assist in meeting tuition costs for their first year of study. The ASC/HCA also provided specialist consultancy support to aid the development of the FdSc, resources for marketing and secretariat support to assist in the formation and maintenance of a network of HEIs involved in the delivery of the programme.

Within the Higher Education Strategy it was perceived that the successful rollout of the FdSc relied, to a significant extent, on the regional distribution of HEIs providing the FdSc. However, at the height of the delivery of the FdSc in 2011 it had not achieved total coverage across regions of England. As of 2011 (at the height of the rollout) the programme was validated in the following regions:

- **North East**: Northumbria University.
- **North West**: The University of Salford.
- **Yorkshire and the Humber**: Sheffield Hallam University.
- **West Midlands**: Staffordshire University with Stafford College (delivered jointly) and Birmingham City University.
- **East Midlands**: De Montfort University; University of Northampton with University Centre Milton Keynes (delivered jointly).
- **London**: London Metropolitan University.

The regional 'gaps' in the distribution of providers compared to that originally envisaged were:

- East of England;
- South West; and,
- South East.

The HCAs Higher Education Strategy was designed for the 2008–2011 period therefore prior to the 2010 General Election the HCA decided to wind down its involvement in the Foundation Degree in Sustainable Communities with a view to encouraging the network of providers to continue to work together to market and develop the qualification. However, the withdrawal of HCA support for the programme was accelerated as part of the reduction in HCA budget that followed the 2010 General Election.

The remainder of this chapter details the findings from a research study that was commissioned by the HCA and undertaken by the authors, to analyse and reflect on the FdSc programme and delivery between the 2008–2011 period. This research study had a number of more specific objectives which included undertaking an investigation into the impact the FdSc had had; ascertaining any lessons that could be learned from this process; and developing any recommendations for taking the FdSc forward. In order to undertake the study this involved bringing together various sources as well as consultations with key stakeholders involved in the design, delivery and receipt of the FdSc over this period. A total of ten face-to-face interviews were held with staff in all HEIs delivering the FdSc as well as interviews with twelve students who were enrolled on the programme at the time. Discussions were also held with a number of the national organisations centrally involved in the development of the FdSc. These interviews were recorded and subsequently translated for thematic analysis.

4.1 Findings and Discussion

Overall, there was a mixed picture with regards to the delivery of the FdSc. As of January 2011, recruitment for the FdSc across the various HEIs was at very low levels. Sheffield Hallam University, which was the first HEI to commence delivery of the FdSc in 2007–08 had not recruited for the last two years. Salford, Northumbria, London Metropolitan and the partnership between Staffordshire and Stafford were all running the programme with relatively small student cohorts (circa 10 students). The remaining HEIs of Birmingham City, De Montfort and the partnership between Northampton and Milton Keynes had been unable to recruit sufficient numbers to commence the delivery of the programme within their respective institutions. In total, there were currently 71 students enrolled on the FdSc across the various HEIs as of 2011. Recruitment onto the FdSc had been a real barrier for all HEIs. Discussions with staff revealed that there were very few enquiries about the courses despite significant attempts by most if not all HEIs to market the programme as widely as possible. At the same time HEIs have also experienced problems with retention of students which were largely attributable to students finding themselves in an insecure position both in terms of potential staff redundancies and reductions in funding available for staff training in light of the austerity policies of the then incoming Coalition Government in the UK. This, to a large extent, is a result of the 'perfect storm' of factors involving: anxiety about HE funding, public sector funding cuts, the lack of regeneration activity and a move away a 'sustainable communities' political discourse. Those HEIs who had ran the programme the longest were either being forced to close the programme or merge with other programmes. Similarly, those staff within HEIs with the FdSc validated and ready to commence delivery were seemingly becoming under pressure to justify retaining the programme.

A number of HEIs saw the resolution to recruitment problems as either engaging more effectively with public sector employers, particularly social housing providers who were, at the time, becoming more and more involved in development initiatives, or a need to engage more widely and articulately with a broader potential student base. There was also a suggestion that in order to ensure the content of the FdSc reaches as many people as possible the FdSc might be discontinued and the content embedded within more popular subjects such as Housing, Planning, Regeneration, etc. Alternatively, other HEIs suggested that there needed to be a more concerted effort to more clearly articulate what the FdSc was and what the benefits of it were.

5 Impact on Addressing Skills Shortages

The key reason for developing the FdSc was the intention to address some of the key perceived skills shortages in the housing and regeneration sectors. As the FdSc was arguably in its infancy, compared to more established programmes, it is difficult to ascertain how the programme had been meeting these needs. However, there was some evidence gathered during the consultation as to ways in which such skills gaps were being addressed.

In the Milton Keynes and Northampton partnership it was perceived that the planning and development of the FdSc had helped to create a dialogue with, and between, a wide range of public/private/community sector players within the Milton Keynes/South Midlands area. The result of this dialogue was that the differing needs and expectations of the various local stakeholders were reportedly more clearly understood. It was thought that this probably would not have happened without the HCA and the FdSc acting as catalyst. The University of Salford though talked about the challenge faced by delivering learning in sustainability skills. Here the experience was that students requested 'knowledge' about sustainable communities as opposed to the development of 'skills'. It was therefore difficult to know how the principles learned on the programme took shape 'on the ground' within the fabric of community settings. It was also thought that the concept of 'skills' was something that employee organisations struggled with too, but that some head-way had been made around issues such as 'collaborative' skills with some students.

Another HEI commented that although the FdSc tends to be designed for people who are in work in the related sectors already, a good number of their students were either unemployed or currently working outside of the sectors. Here it was thought that part of the role of the FdSc should be to develop the skills necessary to enter employment in the housing and regeneration sectors. As a result an employability element (i.e. interview protocols, application completion) was being built into this programme to help these students secure employment in the future.

One HEI was however sceptical about the potential for the FdSc to fill the gaps in skills. It was perceived that the programme entered an already crowded market where there was existing provision that met similar objectives such as courses in Community Development, Regeneration, Planning and Housing.

6 The Impact of the FdSc upon Individuals

In terms of the views of current students on the FdSc there appeared to be an even split between the students on the programme who saw the FdSc as a route to assisting them in their career or work, with those who saw the programme as a way of increasing their knowledge of sustainability and/or sustainable communities for personal development.

Similarly, students' expectations of the FdSc were varied which might perhaps be best explained by the diversity of the sustainable communities area and the diversity offered by the FdSc programme. Some students talked about their expectations that the programme would help them in their work around community involvement, sustainability and environmental issues. Other students simply hoped the programme would provide them with more skills, knowledge and experience that would help them at some unspecified point in the future. These latter statements were particularly common amongst students who had been away from formal education for a period of time. For many students this was the preferred route to learning about sustainable communities and meeting their learning needs. A number of people had engaged in some prior reading around the general area—with one student reviewing the related material online via the HCA website—but who preferred the more traditional classroom learning environment instead.

Although students were mostly positive with regards to how the course had been delivered and 'new' methods for delivery, trying to adequately satisfy the expectations of all students was clearly impossible. Some people liked the timings of the programme in one institution whilst for others this did not suit their other commitments. Similarly, whilst some view online delivery positively others perceived this as an occasional barrier and preferred more traditional (i.e. classroom) modes of delivery.

7 Impact on Their Work

For a number of the students consulted it was too early in the course to detail specific ways in which the course had impacted on their working lives. Some people talked about specific projects they were involved in and how the content of the course had helped them:

> I have applied both knowledge and practical skills learnt so far to my current job role. I have also been able to understand more in meetings with other agencies and colleagues. As previously stated because the course is up to date with the current Government's legislation and guidance it means I am able to bring this knowledge into work.

> I am now facilitating communities to engage in their own planning and we use the knowledge gained through my course to help structure community plans. Including making sure that consideration is given to all factors of a Sustainable Community. I now assess projects for their environmental, economic and social impacts.

However, the single most pervasive impact mentioned by students was the confidence gained by taking part

> I am more confident to work in partnership with other organisations such as police, health, schools and councillors as I have a better understanding that a holistic approach is the only way.

From the people who worked in the housing or regeneration sectors all thought they were either slightly or significantly better equipped as a result of the FdSc. Students' statements and employers suggest that they were managing to apply their learning from the Foundation Degree to their workplace/community. This needs to be tested by further research in students' workplaces/communities but if confirmed it would very positive because Hockey et al. (2010, p. 532) argues learning generic skills in a setting removed from the workplace such as higher education requires the student to overcome a "far-transfer" challenge (Haskell 2001). Yet wherever they studied the students do not seem to have found this transfer from educational institution to workplace or community difficult.

8 Unanticipated Outcomes from Delivering the FdSc

In order to provide a rounded view of the three years activity around the FdSc, as well as exploring whether the FdSc was meeting its specific objectives in terms of skills, it was also important to explore if there had been any unanticipated outcomes as a result of going through the process of development and delivery of the FdSc.

One additional outcome had been how involvement in the FdSc had, quite directly, impacted on the content and delivery of other more traditional courses within the HEIs. This included instances of refocusing content of existing courses upon sustainability issues as well as transferring knowledge from the development of assessment framework of the FdSc to other programmes. For Milton Keynes and Northampton they had been pleasantly surprised by the wide range of private sector interest in the broad subject area of 'sustainability' and the focus they are now giving to creating working communities. They were engaging with the new Local Enterprise Partnership (LEP) (independent sub-regional bodies established to drive forward local socio-economic interests) to explore potential opportunities and links.

An arguably more significant (in terms of lessons for HEIs generally) unanticipated outcome was the resoundingly positive view of the HEI staff towards the establishment of the FdSc Network. As well as this being supportive in a practical sense, in terms of providing shared learning around validation and resources, it was clear that the Network members received a great deal of intellectual support from their colleagues. It was also expressed by one HEI that the Network could potentially be a trailblazer in delivering new, and established, thinking around sustainability, community development and localism:

> The FdSc network is ahead of the game and could be in the vanguard of making sure that teaching and learning around the 'localism' agenda is taken forward. Saul Alinski model of

'community organising' may well be helpful in presenting FdSc work as a development tool to "map knowledge that was not previously on the map" and to help a range of sectors to create avenues to break down barriers on how to work with communities. But to do this it needs to be more widely based than housing/planning/regeneration groups within HEI's.

The issues with the lack of recruitment to the Foundation Degree Sustainable Communities discussed earlier in the chapter have meant that the Network's main reason for existence dissolved. However, given that there was such a positive view of involvement in the Network when it had a focus on the Foundation Degree it is worth reflecting on why this may have been the case as it may have lessons for future collaboration in HEIs. Members of the Network suggested that one of the key reasons for the success was that the HEIs involved were not directly competing with each other for students. The tension between educational institutions being expected to both compete and collaborate was indentified in the mid 1990s (Bridges and Husband 1996 cited in Connolly et al. 2007) and continues at even greater levels today with the recent increase in student fees in the UK. As Connolly et al. discuss there are various models for successful collaboration with some focusing more on process and structure issues (Connolly et al. 2007) and others such as Weiss (1987 cited in Connolly et al. 2007, p. 161) emphasising the importance of motivation. Yet whichever model is applied it is clear that whilst collaborating organisations need to have a shared common interest 'too much' competition can hinder collaborative working.

It is suggested that in HEIs generally, the likelihood of successful collaboration is likely to be increased if organisations avoid working with rival HEIs. If institutions are to be encouraged to focus on seeking collaborative working with organisations that are not direct rivals this will often mean that they will be working with more geographically disparate organisations. In such circumstances the use of digital environments can potentially help with the collaborative of process.

9 Conclusions

In spite of the many positives highlighted by the research the picture painted by the staff, delivering the FdSc, of the future of the course was bleak. There were a number of perceived challenges facing the FdSc: low student numbers, changing political rhetoric away from 'sustainable communities', reduction in public sector funding and a lack of support within higher education for Foundation Degrees. In addition, there was uncertainty around whether there would be the capacity, within training and staff development budgets of public sector employers, to fund students to undertake the FdSc. It was thought that the FdSc may be seen as a 'risky' option for employers who may instead prefer more established, familiar and 'tried and tested' subjects such as Housing, Planning, Regeneration and Surveying. The end point for the FdSc programme in the HEIs analysed here and for the more specifically was cessation; none of the institutions were able to continue to support this programme as it was originally configured over the long-term. At the end of the research most HEIs were either considering or were actively exploring how the FdSc could be merged with other more established programmes (i.e. social policy, geography, housing) in order to embed the valuable messages around sustainability.[1]

This research has provided an insight as to the journey HEIs have undergone in developing a programme which was perceived as crucial to meeting the needs of a workforce required to deliver a programme closely aligned to a particular political discourse driven by the public sector. On an operational level the FdSc appears to have been successful. To a limited extent it has met many of the aims laid out by the initial earlier report which preceded its establishment:

- It allowed and encouraged progression to further qualifications to allow specialisation
- It extended and enhanced the generic skills of students undertaking the FdSc in a professional context
- It introduced technical skills with an emphasis on cross-disciplinary working
- It was able to adapt to evolving issues
- It provided for flexible learning approaches (although a number of students reportedly prefer more traditional modes of delivery).

The unique features of the FdSc in particular the involvement and support of the HCA and the creation of the FdSc Network can be considered significant successes and have led to a number of positive unanticipated outcomes.

There remains a significant barrier in place in order for the overall aim of the FdSc to be the entry level award necessary in order to meet the skills needs in the field of Sustainable Communities. This barrier is a lack of apparent synergy between those who have a strategic overview and influence of the housing, regeneration and community development sectors, who recognize the need for a workforce who have generic skills and that is literate in cross-sectoral partnership working, and the actual organizations currently working in these sectors. Although there are a number of issues arising from this study that explain the lack of synergy it appears that there are three main reasons:

Firstly, putting to one side the increase in student fees at Universities in the UK, the FdSc was developed and rolled out at an unfortunate time for the target sectors. The recent economic climate has meant that housing development stalled, a new government (with different approaches and priorities) was elected, regeneration programmes were mothballed, public sector budgets have been reduced meaning that there is less job security and potentially fewer staff in post. This has affected the housing, regeneration and community development sectors more than most within the public sector.

[1]It should be noted that at the time of writing Glyndwr University has a Housing and Sustainable Communities Programme.

Secondly, there remains a lack of awareness of the FdSc and its relevance for organisations. There does not appear to have been a successful narrative created around what sustainable communities means. The Foundation degree in Sustainable Communities formed part of a package of ASC/HAC activity including a range of toolkits that has since been criticised by Hockey et al. (2010). They argue the ASC/HAC approach was a top down approach to vocational learning that did not recognise that a more bottom-up approach was required because learning is "… dependent on an intimate knowledge of concrete social detail in the workplace". Hockey et al. also suggest that some workplaces may have questioned the relevance of the more abstract definitions of competences that were described as part of skills agenda (p. 226). For the FdSc this lack of clarity about the key terms exacerbates a lack of understanding about Foundation Degrees generally and specifically what the qualification offers individuals and organisations. Although these issues will arguably require time to resolve the current result is that the FdSc entered into a niche market between more established and 'validated' courses of housing, regeneration, planning and community development.

Thirdly, there was some miss-marketing of the FdSc which has, largely, sought students from the public sector who are already working within housing, regeneration and/or community development. These are areas where there exist reasonably clear pathways for qualifications and professional accreditation set by line and senior managers. The FdSc was, for the most part, not strategically marketed at potential students not yet in these sectors.

The relationship between employers, individuals, providers and government in delivering the skills agenda has had a varied emphasis on the extent to which it is supplier led compared to demand led since 1964. This is discussed in the Leitch Review (2006, p. 48) which also put forward the case for more demand led provision around the time the Egan recommendations were being translated into training and education programmes by the ASC and then HCA. At the time of its development the Foundation Degree in Sustainable Communities appeared to be based on robust evidence of demand (Egan 2004; Homes and Communities Agency 2008). The evidence in this chapter suggests that possibly the methodology for assessing demand had substantial weaknesses or, if the demand was there, it was generated by a desire amongst employers to be seen to be supporting a particular political discourse. As such it has proved vulnerable to the change in government.

The findings reported here have highlighted the danger for HEIs associated with closely aligning a course to the political orthodoxy of the time. The name and notion of 'Sustainable Communities' appears to have been an ongoing barrier for the programme as it was perceived as too ambiguous as well as wedded to a New Labour political discourse. This appears to have been a real barrier to achieving greater numbers of students as well as a potential barrier to achieving high level recognition and future funding. Whilst the politics may have moved on, however, the priorities set out within the concept, such as meeting diverse needs, enhancing local environment, strengthening economic prosperity and promoting social cohesion endure as prominent narratives with which urban professionals are encouraged to engage. There is valuable learning here that should be considered if HEIs are looking at opportunities to develop programmes around the latest political discourse/movement.

References

Academy for Sustainable Communities. (2007). *Mind the skills gap*. http://skills.homesandcommunities.co.uk/whatwedo/mind-the-skills-gap-research. Last accessed 14 July 2011.

Academy for Sustainable Communities. (2008). *Higher education strategy: Delivering sustainable communities skills and knowledge*. October.

Agyeman, J. (2005). *Sustainable communities and the challenge of environmental justice*. New York: NYU Press.

Barton, H. (2000). *Sustainable communities: The potential for eco-neighborhoods*. London: Routledge.

Bell, S., & Morse, S. (2008). *Sustainability indicators. Measuring the immeasurable?* London: Earthscan.

Burke, T. (1995). View from the inside: UK environmental policy seen from a practitioner's perspective. In T. S. Gray (Ed.), *UK environmental policy in the 1990s*. London: MacMillan Press Ltd.

Chartered Institute for Housing and Royal Town Planning Institute Planning for Housing. (2003). *The potential for sustainable communities—A CIH RTPI Policy Paper Summary*. http://www.rtpi.org.uk/download/637/Planning-for-Housing-Summary.pdf. Last accessed 14 July 2011.

Cochrane, A. (2007). *Understanding Urban Policy. A critical approach*. Oxford: Blackwell Publishing Limited.

Connolly, M., Jones, C., & Jones, N. (2007). Managing collaboration across further and higher education: A case in practice. *Journal of Further and Higher Education, 31*(2), 159–169.

Corcoran, P. B., & Wals, A. E. J. (2004). *Higher education and the challenge of sustainability: Problematics, promise, and practice*. London: Kluwer.

Dryzek, J. (1997). *The Politics of the Earth*. Oxford: Oxford University Press.

Egan, J. (2004). *The Egan review: Skills for sustainable communities*. http://www.communities.gov.uk/documents/communities/pdf/152086.pdf. Last accessed 14 July 2011.

Haskell, R. (2001). *Transfer of learning cognition, instruction, and reasoning*. London: Academic Press.

Hockey, A., Jimenez-Bescos, C., Maclean, J., & Spaul, M. (2010). Generic skills for sustainable communities: Design principles for a learning support environment. *Town Planning Review, 81* (5), 523–540.

Homes and Communities Agency. (2008). *Higher education strategy: Delivering sustainable communities skills and knowledge*. Homes and Communities Agency.

Jones, P., & Evans, J. (2008). *Urban regeneration in the UK*. London: Sage.

Leitch, A. (2006). *Leitch review of skills: Prosperity for all in the global economy—World class skills*. Norwich: HMSO.

ODPM. (2003). *Sustainable communities plan (Sustainable communities: Building for the future)*. http://www.communities.gov.uk/publications/communities/sustainablecommunitiesbuilding. Last accessed 12 July 2011.

ODPM (2005) *Sustainable communities: People, places and prosperity—A five year plan from the office of the deputy prime minister*. http://www.communities.gov.uk/archived/publications/corporate/peopleplacesprosperity. Last accessed 12 July 2011.

Raco, M. (2005). Sustainable development, rolled-out neoliberalism and sustainable communities. *Antipode, 37*(2), 324–347.

Rogerson, R., Sadler, S., Wong, C., & Green, A. (2010). Planning sustainable communities—skills and learning to envision future communities: An introduction. *Town Planning Review, 81*(5), 505–521.

Sheffield Hallam University. (2007). *Market research into the development of a foundation degree in Sustainable Communities*. Sheffield Hallam University.

Tallon, A. (2010). *Urban regeneration in the UK*. Abingdon: Routledge.

World Commission on Environment and Development. (1987). *Our common future*. Oxford: Oxford University Press.

MIX
Papier aus verantwortungsvollen Quellen
Paper from responsible sources
FSC® C105338

If you have any concerns about our products,
you can contact us on
ProductSafety@springernature.com

In case Publisher is established outside the EU,
the EU authorized representative is:
Springer Nature Customer Service Center GmbH
Europaplatz 3, 69115 Heidelberg, Germany

Printed by Libri Plureos GmbH
in Hamburg, Germany